HYPERBOLIC BOUNDARY VALUE PROBLEMS

REIKO SAKAMOTO

HYPERBOLIC BOUNDARY VALUE PROBLEMS

Reiko Sakamoto is Professor at the Department of Mathematics
Nara Women's University, Japan

TRANSLATED BY KATSUMI MIYAHARA

CAMBRIDGE UNIVERSITY PRESS
Cambridge
London New York New Rochelle
Melbourne Sydney

CAMBRIDGE UNIVERSITY PRESS
Cambridge, New York, Melbourne, Madrid, Cape Town, Singapore, São Paulo, Delhi

Cambridge University Press
The Edinburgh Building, Cambridge CB2 8RU, UK

Published in the United States of America by Cambridge University Press, New York

www.cambridge.org
Information on this title: www.cambridge.org/9780521107594

Originally published in Japanese by Iwanami Shoten, Publishers, Tokyo, 1978
Translated by K. Miyahara
First published in English by Cambridge University Press 1982

This digitally printed version 2009

A catalogue record for this publication is available from the British Library

Library of Congress Catalogue Card Number: 81-3865

ISBN 978-0-521-23568-6 hardback
ISBN 978-0-521-10759-4 paperback

CONTENTS

[illegible]

A few [illegible] Fourier [illegible] the [illegible] very [illegible]

[illegible]

[illegible] second-order partial differential equations.

Chapter [illegible] boundary value problems [illegible] Laplace transforms.

[illegible] equations [illegible] inequalities.

[illegible] Lebesgue

PREFACE

The book is designed to introduce the reader to the general theory of the existence and uniqueness of solutions of hyperbolic initial value problems in partial differential equations of higher orders. The subject has been my main research theme for more than a decade.

A few years ago when I decided to write this book at the suggestion of Professor Hiroshi Fujita I thought that the project was very tempting in the sense that I could retrace the old course of inquiries along which I had struggled for ten years, and I should be able to put various ideas in perspective.

However, after setting down some ideas, I soon realised that this was an audacious task; I must confess that I often went astray while writing the book.

Even now that the book has eventually taken shape I fear that some readers may blame me for my obsession with logical consistency at the expense of the wider view of the theory which I originally had in mind.

To such a reader, I can only reply by pointing out that my presentation is merely one of many alternatives for expounding the subject matter; drawing a beautiful global map of the present state of the art is beyond my capabilities.

The book is divided into three chapters.

Chapter 1 is a collection of facts later used in the main body of the text, which are basic to the theory of second-order partial differential equations.

Chapter 2 discusses hyperbolic boundary value problems with constant coefficients by the method of Fourier–Laplace transforms.

Chapter 3 deals with hyperbolic boundary value problems with variable coefficients by the method of energy inequalities.

I have presented the above-mentioned subjects while presupposing a minimal background. The reader is assumed to have some prior acquaintance with the elementary theories of a complex variable and Lebesgue integrals, and when extra tools are needed, brief explanations are provided to make the text as self-contained as possible.

I wish to thank my teachers, Professor Shigeru Mizohata and Professor Masaya Yamaguchi for their kind guidance throughout my career as a researcher. My academic indebtedness to them does not end there. Several ideas concerning hyperbolic initial value problems, which have resulted from their vast research, underpin the arguments that I have presented.

Special thanks go to Professor Hiroshi Fujita, who gave me the chance to write this book, and to Mr Hideo Arai, editorial staff of Iwanami Shoten, who showed remarkable patience and gave willing assistance in spite of the slow progress of my writing.

Finally, I must thank my husband, Kazuichi, without whose constant encouragement this book would never have come into being.

February 1978 Reiko Sakamoto

PREFACE TO THE ENGLISH EDITION

I am pleased to see that this English edition has appeared through the enduring labour of Katsumi Miyahara and the skilful arrangement by the editorial staff of Cambridge University Press. As a result, I believe that the book may appeal to a wider audience outside my country.

As the translation progressed, I took the opportunity of incorporating various corrections and revisions in the text. I am indebted to Dr Tatsuo Nishitani and other Japanese readers of the original edition for their valuable suggestions.

All this has led to a significant improvement of the original text. I am sincerely grateful to everyone concerned.

October 1980 Reiko Sakamoto

1

Second-order Hyperbolic Equations

The natural habitat of second-order partial differential equations is in classical mathematical physics. Historically, this type of equation emerged as a description of basic relations in continuous mechanics after Newtonian mechanics was firmly established.

Since then a vast amount of literature about this subject has appeared. In this chapter, we shall confine ourselves to reviewing some well-known results of second-order equations which are closely related to hyperbolic initial boundary value problems, the main theme of this book.

1. Initial value problems

The most general second-order equation in the Euclidean space $\mathbb{E}^{n+1}$ for an unknown function $u(x_0, x_1, \ldots, x_n)$ has the form

$$\sum_{i,j=0}^{n} a_{ij}(x_0, \ldots, x_n)\frac{\partial^2 u}{\partial x_i \partial x_j} + \sum_{i=0}^{n} b_i(x_0, \ldots, x_n)\frac{\partial u}{\partial x_i} + c(x_0, \ldots, x_n)u$$
$$= f(x_0, \ldots, x_n).$$

For brevity, we write Lu for the left hand side of the above equation.† Let us assume that $a_{00}(x_0, \ldots, x_n) \neq 0$ and write

$$x_0 = t, \quad (x_1, \ldots, x_n) = x.$$

Consider the initial value problem with

$$u(0, x) = \varphi(x), \quad \frac{\partial u}{\partial t}(0, x) = \psi(x)$$

at $t = 0$. We seek the solution of $Lu = f$.

We state two principal results in the form of a theorem as follows: Let the coefficients a_{ij}, b_i, c of the above equation be analytic in a neighbourhood of the origin.

† Translator's note: L is a (differential) operator which assigns to each u another function (see §2).

The Cauchy–Kowalevski theorem *(the existence theorem)*
For given data (i.e. a prescribed set of functions) $f(t,x), \varphi(x), \psi(x)$ each of which is analytic in a neighbourhood of the origin, there exists a function which is defined and analytic in a neighbourhood U of the origin and satisfies

$$Lu(t,x) = f(t,x),$$
$$u(0,x) = \varphi(x), \quad \frac{\partial u}{\partial t}(0,x) = \psi(x)$$

[and this solution is unique in the class of analytic functions C^ω].† In particular, if the data are entire functions, then the domain U of the solution u does not depend upon the data.

Holmgren's theorem *(the uniqueness theorem)*
Let U be a neighbourhood of the origin over which a C^2-function $u(t,x)$ is defined. If u satisfies

$$Lu(t,x) = 0,$$
$$u(0,x) = 0, \quad \frac{\partial u}{\partial t}(0,x) = 0$$

then, $u \equiv 0$ in U.‡ ■

In short, these existence and uniqueness theorems show that given the analytic data $\{f, \varphi, \psi\}$ at a neighbourhood of the origin, there is a neighbourhood U of the origin such that an analytic solution u exists and is unique in U. From this observation, within the class of analytic functions, the problem seems to have a satisfactory answer. However, Hadamard pointed out the following example and introduced the notion of 'well-posedness' to initial value problems.

Hadamard's example
For a fixed natural number p, and each $n = 1, 2, 3, \ldots$, the solutions of the equation with the analytic data

$$\frac{\partial^2 u}{\partial t^2} + \frac{\partial^2 u}{\partial x^2} = 0,$$
$$u(0,x) = 0, \quad \frac{\partial u}{\partial t}(0,x) = \frac{1}{n^p}\cos nx \, (\equiv \psi_n(x))$$

† Remarks in square brackets have been added by the translator.
‡ Translator's note: For the proofs of these theorems see R. Courant & D. Hilbert, *Methods of mathematical physics* vol. 2. Interscience, New York (1953, 1962), pp. 39–54; or S. Mizohata, *The theory of partial differential equations*. Cambridge University Press (1973) p. 245.

are given by

$$u(t,x)=\frac{1}{n^{p+1}}\cos nx\cdot\frac{e^{nt}-e^{-nt}}{2}(\equiv u_n(t,x)).$$

Hence, for the initial value,

$$\sup_{\substack{-\infty<x<+\infty\\ q=0,1,\dots,p-1}}\left|\left(\frac{d}{dx}\right)^q\psi_n(x)\right|=\frac{1}{n}$$

holds. However, at $t=t_0>0$ we have

$$u_n(t_0,0)\to+\infty\quad\text{as } n\to+\infty.$$

This example shows that, in general, for given data $\{f,\varphi,\psi\}$ the map which associates the data to the solution u for an initial value problem is not always continuous (for example, consider the topology induced by uniform convergence in the wider sense). This means that, even if the change in the initial data is arbitrarily small, the corresponding change in the solution cannot be guaranteed to be sufficiently small.

Where normal physical phenomena are concerned, the above argument shows that such initial value problems are meaningless because the data are obtained by experimental measurements and a small error in the data does not produce a large error in the solution of the problem. Therefore, in such cases, it is reasonable to expect that the solution of the initial value problem should be continuously dependent upon the initial data.

An initial value problem is said to be *well-posed* if the solution depends continuously on the data of the problem [the existence and uniqueness of the solution are presupposed]. However, in this case we need to make the meaning of 'continuity' more precise. To this end, let us assume that the data $\{f,\varphi,\psi\}$ are entire functions, so that as mentioned before, there exists the domain U of the solution u where the existence and uniqueness of u is guaranteed. Let us also assume that for any compact subset $K\subset U$ and any natural number p, there exist a compact set $K'\subset\mathbb{R}^{n+1}$, a natural number p', and a positive real number C such that

$$\sum_{\alpha_0+\alpha_1+\dots+\alpha_n\leqq p}\sup_{(t,x)\in K}\left|\left(\frac{\partial}{\partial t}\right)^{\alpha_0}\left(\frac{\partial}{\partial x_1}\right)^{\alpha_1}\cdots\left(\frac{\partial}{\partial x_n}\right)^{\alpha_n}u(t,x)\right|$$

$$\leqq C\left\{\sum_{\alpha_0+\dots+\alpha_n\leqq p'}\sup_{(t,x)\in K'}\left|\left(\frac{\partial}{\partial t}\right)^{\alpha_0}\cdots\left(\frac{\partial}{\partial x_n}\right)^{\alpha_n}f(t,x)\right|\right.$$

$$+\sum_{\alpha_1+\dots+\alpha_n\leqq p'}\sup_{(0,x)\in K'}\left|\left(\frac{\partial}{\partial x_1}\right)^{\alpha_1}\cdots\left(\frac{\partial}{\partial x_n}\right)^{\alpha_n}\varphi(x)\right|$$

$$\left.+\sum_{\alpha_1+\dots+\alpha_n\leqq p'}\sup_{(0,x)\in K'}\left|\left(\frac{\partial}{\partial x_1}\right)^{\alpha_1}\cdots\left(\frac{\partial}{\partial x_n}\right)^{\alpha_n}\psi(x)\right|\right\}$$

If these conditions are satisfied, the map $\{f, \varphi, \psi\} \mapsto u$ is said to be *$\mathscr{E}$-continuous.*† Given an initial value problem, if the solution depends $\mathscr{E}$-continuously on the data, the original problem is said to be *$\mathscr{E}$-well-posed.*

Up to now, our argument has been concerned with mapping data of entire functions $\{f, \varphi, \psi\}$ onto an analytic solution u, based on the Cauchy–Kowalevski theorem. However, assuming $\mathscr{E}$-continuity we can establish the following fact:

Given C^∞-data $\{f, \varphi, \psi\}$, there is a unique C^∞-solution $u \in U$.

In fact, for fixed K and p, we can obtain a solution in $\mathscr{B}^p(K)$ (by a polynomial approximation of $\{f, \varphi, \psi\}$ in the topological space $\mathscr{B}^p(K)$).‡ That is, if each member of the set of polynomials $\{f_k, \varphi_k, \psi_k\}$ $(k = 1, 2, \ldots)$ and their derivatives up to order p' (inclusive) converges uniformly to $\{f, \varphi, \psi\}$ in K', etc., then the corresponding sequence of solutions $\{u_k\}$ $(k = 1, 2, \ldots)$ and their derivatives up to order p converges uniformly to u in K, etc., where u is a solution for the given data $\{f, \varphi, \psi\}$. According to Holmgren's theorem, u does not depend upon K and p. Therefore, u is a C^∞-solution in U.

Summing up, we state that if the analytic initial value problem is $\mathscr{E}$-well-posed, then

1° There exists a unique C^∞-solution u in U corresponding to C^∞-data $\{f, \varphi, \psi\}$.

2° The above-mentioned correspondence is $\mathscr{E}$-continuous.

To be more precise, 1° follows from 2°. (To see this fact use Banach's closed graph theorem.*) Therefore, given an initial value problem, if the existence and uniqueness of a C^∞-solution can be established, the solution will be satisfactory with regard to $\mathscr{E}$-continuity.

We now go back to the starting point, and ask what are the characteristics of L satisfying an $\mathscr{E}$-well-posed initial value problem. As we shall see in the following, this problem is, in fact, closely related to the classification of the systems of partial differential equations, and there the relation between $\mathscr{E}$-well-posedness and an algebraic condition 'hyperbolicity' will be clarified.

† $\mathscr{E}(X)$ is a Fréchet space; see Chapter 2 §2.

‡ Translator's note: $\mathscr{B}^p(K)$ is the Banach space of functions having continuous and bounded partial derivatives up to order p. The norm of the space is provided by

$$|f(x)|_p = \sum_{\substack{|\alpha| \leq p \\ x \in K}} \sup |D^\alpha f(x)| \quad \text{for } f \in \mathscr{B}^p(K).$$

* See Y. Yoshida, *Functional analysis* (second edition). Springer, Berlin (1968) pp. 77–9.

In passing, we note that here the order of $L(=2$ in this section) does not affect the essence of our argument. In fact, the same argument applies to any higher-order L as will be seen in Chapter 2 onwards.

2. Types of partial differential equations

In the previous section we considered partial differential equations and their initial conditions. In this section, however, these additional conditions are not considered because here our main concern is a rough classification of equations.

In $\mathbb{R}^n$ we consider a *partial differential operator*

$$L=\sum_{i,j=1}^{n} a_{ij}(x)\frac{\partial^2}{\partial x_i \partial x_j}+\sum_{i=1}^{n} b_i(x)\frac{\partial}{\partial x_i}+c(x) \quad (a_{ij}=a_{ji} \text{ real})$$

and a change of variables such that

$$\begin{aligned} y_1 &= \varphi_1(x_1,\dots,x_n) \\ &\vdots \\ y_n &= \varphi_n(x_1,\dots,x_n) \end{aligned} \qquad \det\begin{bmatrix} \dfrac{\partial\varphi_1}{\partial x_1} & \cdots & \dfrac{\partial\varphi_n}{\partial x_1} \\ \vdots & & \vdots \\ \dfrac{\partial\varphi_1}{\partial x_n} & \cdots & \dfrac{\partial\varphi_n}{\partial x_n} \end{bmatrix} \neq 0$$

As the result, by the transformation

$$\begin{bmatrix} \dfrac{\partial}{\partial x_1} \\ \vdots \\ \dfrac{\partial}{\partial x_n} \end{bmatrix} = \begin{bmatrix} \dfrac{\partial\varphi_1}{\partial x_1} & \cdots & \dfrac{\partial\varphi_n}{\partial x_1} \\ \vdots & & \vdots \\ \dfrac{\partial\varphi_1}{\partial x_n} & \cdots & \dfrac{\partial\varphi_n}{\partial x_n} \end{bmatrix} \begin{bmatrix} \dfrac{\partial}{\partial y_1} \\ \vdots \\ \dfrac{\partial}{\partial y_n} \end{bmatrix}$$

L becomes

$$\tilde{L}=\sum_{i,j=1}^{n} \tilde{a}_{ij}(y)\frac{\partial^2}{\partial y_i \partial y_j}+\sum_{i=1}^{n} \tilde{b}_i(y)\frac{\partial}{\partial y_i}+\tilde{c}(y)$$

where

$$\begin{bmatrix} \tilde{a}_{11} & \cdots & \tilde{a}_{1n} \\ \vdots & & \vdots \\ \tilde{a}_{n1} & \cdots & \tilde{a}_{nn} \end{bmatrix} = \begin{bmatrix} \dfrac{\partial\varphi_1}{\partial x_1} & \cdots & \dfrac{\partial\varphi_1}{\partial x_n} \\ \vdots & & \vdots \\ \dfrac{\partial\varphi_n}{\partial x_1} & \cdots & \dfrac{\partial\varphi_n}{\partial x_n} \end{bmatrix} \begin{bmatrix} a_{11} & \cdots & a_{1n} \\ \vdots & & \vdots \\ a_{n1} & \cdots & a_{nn} \end{bmatrix} \begin{bmatrix} \dfrac{\partial\varphi_1}{\partial x_1} & \cdots & \dfrac{\partial\varphi_n}{\partial x_1} \\ \vdots & & \vdots \\ \dfrac{\partial\varphi_1}{\partial x_n} & \cdots & \dfrac{\partial\varphi_n}{\partial x_n} \end{bmatrix}$$

Consider the correspondence between L and the quadratic form $\sum_{i,j=1}^{n} a_{ij}(x)\xi_i\xi_j$ with signature (p,q). Since (p,q) is invariant with regard to changes of variables, a classification of L can be obtained according to the value of (p,q) as follows: For fixed $x = x_0$

Case 1. $p = 0$ or $q = 0$,
Case 2. $p = 1$ or $q = 1$,
Case 3. $p \geqq 2$ and $q \geqq 2$,

L is said to be, in case 1 *elliptic*, in case 2 *hyperbolic*, in case 3 *hyper-hyperbolic* at $x = x_0$. In particular, if, in addition, $p + q = n$ is true for each case, L is said to be, in case 1 *strongly elliptic*, in case 2 *strongly hyperbolic*, in case 3 *strongly hyper-hyperbolic*. Putting

$$P(x,\xi) = \sum_{i,j=1}^{n} a_{ij}(x)\xi_i\xi_j$$

cases 1 and 2 can be characterised as

Case 1′. $P(x_0, \xi) \neq 0, \xi \in \mathbb{R}^n \backslash \{0\}$,
Case 2′. There exists $\eta_0 \in \mathbb{R}^n \backslash \{0\}$ such that $P(x_0, \xi + i\eta_0) \neq 0, \xi \in \mathbb{R}^n$.

If $a_{ij}(x)$ does not depend on x, we write $a_{ij} = a_{ij}(x)$. Then consider

$$A = \begin{bmatrix} a_{11} \dots a_{1n} \\ \vdots \quad\quad \vdots \\ a_{n1} \dots a_{nn} \end{bmatrix}$$

and a constant matrix

$$T = \begin{bmatrix} t_{11} \dots t_{1n} \\ \vdots \quad\quad \vdots \\ t_{n1} \dots t_{nn} \end{bmatrix}, \quad \det(T) \neq 0$$

such that

$$ {}^tTAT = \begin{bmatrix} \left.\begin{matrix} 1 & & \\ & \ddots & \\ & & 1 \end{matrix}\right\} p & & \\ & \left.\begin{matrix} -1 & & \\ & \ddots & \\ & & -1 \end{matrix}\right\} q & \\ & & \begin{matrix} 0 & & \\ & \ddots & \\ & & 0 \end{matrix} \end{bmatrix},$$

where superscript t denotes the transpose. Then, by the linear transformation

$$\begin{bmatrix} y_1 \\ \vdots \\ y_n \end{bmatrix} = {}^{t}T \begin{bmatrix} x_1 \\ \vdots \\ x_n \end{bmatrix}$$

we have

$$\tilde{L} = \left(\frac{\partial^2}{\partial {y_1}^2} + \ldots + \frac{\partial^2}{\partial {y_p}^2}\right) - \left(\frac{\partial^2}{\partial {y_{p+1}}^2} + \ldots + \frac{\partial^2}{\partial {y_{p+q}}^2}\right) + \sum_{i=1}^{n} \tilde{b}_i(y)\frac{\partial}{\partial y_i} + \tilde{c}(y).$$

On the other hand, if $a_{ij}(x)$ depends upon x, it is impossible, in general, to find a simultaneous change of variables $\varphi : x \mapsto y$ in a domain that will give us a normal form for each variable, except in the case $n = 2$. That is, if L is strongly elliptic or strongly hyperbolic in a certain domain, we can transform L into

$$\tilde{L} = \pm\left(\frac{\partial^2}{\partial {y_1}^2} + \frac{\partial^2}{\partial {y_2}^2}\right) + \tilde{b}_1\frac{\partial}{\partial y_1} + \tilde{b}_2\frac{\partial}{\partial y_2} + \tilde{c}$$

and

$$\tilde{L} = \pm\left(\frac{\partial^2}{\partial {y_1}^2} - \frac{\partial^2}{\partial {y_2}^2}\right) + \tilde{b}_1\frac{\partial}{\partial y_1} + \tilde{b}_2\frac{\partial}{\partial y_2} + \tilde{c},$$

respectively.

Of course, these classifications are too 'coarse' in some cases. For example, if L belongs to a certain category according to the first (weak) classification, the situation is complicated by the behaviour of the lower-order terms of L. Therefore, some finer classification is necessary in such cases. In particular, if the equations have constant coefficients, and we consider the lower-order terms of L as

$$P(\xi) = \sum_{i,j=1}^{n} a_{ij}\xi_i\xi_j + \sum_{i=1}^{n} b_i\xi_i + c$$

then by using a similar method to the one employed for the classification 1′ and 2′, another classification can be obtained.

For example, when we study 'hyperbolic' equations with constant coefficients in Chapter 2, the term 'hyperbolic' will be used in a stronger sense with regard to lower-order terms. In Chapter 3, we shall also study strongly hyperbolic equations with variable coefficients.

These arguments are all closely connected to the notion of the

$\mathscr{E}$-well-posedness of initial value problems (recall the remark in §1). However, in order to understand the significance of the above statement, the reader must wait until $\mathscr{E}$-well-posed initial value problems are fully explored in the subsequent chapters.

3. Vibrating strings: problems and their solutions

In this section, we consider the vibrations of stretched strings, one of the most simple physical phenomena, and see how the most typical hyperbolic equations arise in a natural way. Their initial value problems and solutions are presented in such a way that the reader can visualise the entire argument in concrete terms.

3.1

(a) A finite string with fixed ends (I)

Consider a light homogeneous string with both ends fixed, stretching between 0 and 1 along the x-axis (see Fig. 1). Suppose that the string

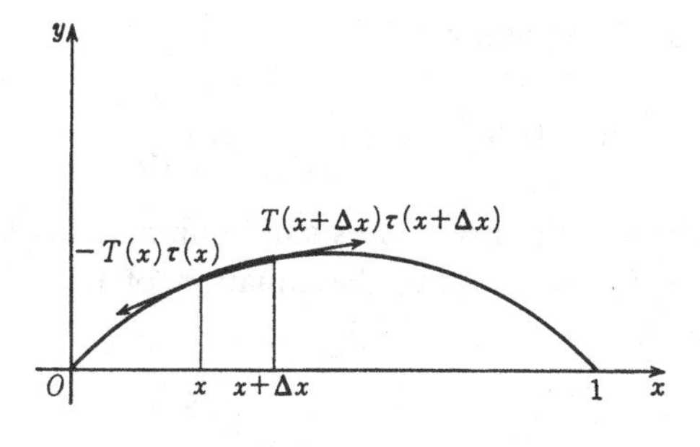

Fig. 1

vibrates in the vertical direction (parallel to the y-axis) and that its location at t is $y = u(t, x)$ and $\partial u/\partial x$ is small. For a fixed t, the unit tangent vector τ at x along $y = u(t, x)$ is given by

$$\tau = \left(\frac{1}{\sqrt{\{1 + (\partial u/\partial x)^2\}}}, \frac{\partial u/\partial x}{\sqrt{\{1 + (\partial u/\partial x)^2}}\right).$$

Let $T = T(t, x)$ be the tension of the string, and consider the portion of the string which has its ends at x and $x + \Delta x$. Since the sum of tensions in this portion is

$$T(x + \Delta x)\tau(x + \Delta x) - T(x)\tau(x) = \frac{\partial}{\partial x}\{T(x)\tau(x)\}\Delta x + O(\Delta x^2)$$

the equation of motion is expressed as

$$\frac{\partial}{\partial x}\left(T\frac{1}{\sqrt{\{1+(\partial u/\partial x)^2\}}}\right)\Delta x + O(\Delta x^2) = 0,$$

$$\Delta x \rho_0 \frac{\partial^2 u}{\partial t^2} = \frac{\partial}{\partial x}\left(T\frac{\partial u/\partial x}{\sqrt{\{1+(\partial u/\partial x)^2\}}}\right)\Delta x + O(\Delta x^2)$$

where ρ_0 denotes the linear density of the string at a static position (in this case it is a constant). Therefore we have

$$\frac{\partial}{\partial x}\left(T\frac{1}{\sqrt{\{1+(\partial u/\partial x)^2\}}}\right) = 0,$$

$$\rho_0 \frac{\partial^2 u}{\partial t^2} = \frac{\partial}{\partial x}\left(T\frac{\partial u/\partial x}{\sqrt{\{1+(\partial u/\partial x)^2\}}}\right).$$

Let us set

$$\frac{T}{\sqrt{\{1+(\partial u/\partial x)^2\}}} = H$$

and rewrite the above equation as

$$H = H(t) \quad \text{and} \quad \rho_0 \frac{\partial^2 u}{\partial t^2} = H(t)\frac{\partial^2 u}{\partial x^2}.$$

Since the vibrations of the string are small in amplitude ($\partial u/\partial x$ is small), by setting $H(t) = T_0$ (=constant), the equation of the vibrating string becomes

$$\frac{1}{c^2}\frac{\partial^2 u}{\partial t^2} = \frac{\partial^2 u}{\partial x^2}, \quad c = \sqrt{\left(\frac{T_0}{\rho_0}\right)}.$$

(b) An infinitely long string with no fixed ends

We now consider an imaginary string stretching infinitely in both directions in order to avoid special considerations at the fixed ends of the string. We observe the vibration of the string for $-\infty < t < +\infty$. Assume

$$\frac{1}{c^2}\frac{\partial^2 u}{\partial t^2} = \frac{\partial^2 u}{\partial x^2}$$

for $-\infty < t < +\infty$, $-\infty < x < +\infty$. By the change of variables

$$\xi = x + ct,$$

$$\eta = x - ct,$$

we have

$$\partial^2 u/\partial \xi \partial \eta = 0$$

for $-\infty < \xi < +\infty$, $-\infty < \eta < +\infty$. By integration we find the solution

$$u = f(\xi) + g(\eta),$$

where f and g are arbitrary functions, and in terms of the original variables this becomes

$$u(t, x) = f(x + ct) + g(x - ct).$$

Since for $x + ct = x_0$ (= constant), $f(x + ct) = f(x_0)$ (= constant), it follows that $f(x + ct)$ and $g(x - ct)$ represent waves travelling to the left and right at a constant speed c while preserving their shapes.

Now let us view the vibration of the string in the framework of an initial value problem. More precisely, at $t = 0$ let us make the location of the string u and its speed $\partial u/\partial t$ satisfy the following conditions

$$u(0, x) = u_0(x), \quad \frac{\partial u}{\partial t}(0, x) = u_1(x)$$

and seek a solution. Suppose

$$u(t, x) = f(x + ct) + g(x - ct)$$

satisfies the initial conditions. Then obviously

$$f(x) + g(x) = u_0(x),$$

$$f'(x) - g'(x) = \frac{1}{c} u_1(x).$$

Differentiate the first equation with respect to x and pair with the second equation to obtain

$$f'(x) = \frac{1}{2}\left\{ u_0'(x) + \frac{1}{c} u_1(x) \right\},$$

$$g'(x) = \frac{1}{2}\left\{ u_0'(x) - \frac{1}{c} u_1(x) \right\},$$

which means that

$$f(x) = k + \frac{1}{2}\left\{ u_0(x) + \frac{1}{c}\int_0^x u_1(x)\,\mathrm{d}x \right\},$$

$$g(x) = -k + \frac{1}{2}\left\{ u_0(x) - \frac{1}{c}\int_0^x u_1(x)\,\mathrm{d}x \right\}.$$

Hence

$$(*) \qquad u(t, x) = \frac{1}{2}\{u_0(x + ct) + u_0(x - ct)\} + \frac{1}{2c}\int_{x-ct}^{x+ct} u_1(x)\,\mathrm{d}x.$$

Conversely, if u_0 is of class C^2 and u_1 is of class C^1, then u defined by the above equation (*) becomes of class C^2; therefore

$$\frac{1}{c^2}\frac{\partial^2 u}{\partial t^2} = \frac{\partial^2 u}{\partial x^2}$$

and at $t = 0$ it satisfies the initial conditions

$$u(0,x) = u_0(x), \quad \frac{\partial u}{\partial t}(0,x) = u_1(x).$$

(c) A semi-infinite string with one end fixed

Next we wish to look at the state of the fixed end of the string when it is stretched along the interval $x > 0$ on the x-axis. Let $u(t,x)$ satisfy the equation of the vibrating string in $\{x > 0, \ -\infty < t < +\infty\}$; then it is represented there as

$$u(t,x) = f(x+ct) + g(x-ct).$$

Notice that the right hand side of the equation is defined also for $x \le 0$, therefore u can be extended to $\tilde{u}$ for $-\infty < x < +\infty$. Assume the following boundary condition

$$u(t,0) = 0, \quad -\infty < t < +\infty.$$

Then, from

$$f(ct) + g(-ct) = 0, \quad -\infty < t < +\infty$$

we have

$$g(t) = -f(-t), \quad -\infty < t < +\infty.$$

Now, as

$$\tilde{u}(t,x) = f(x+ct) - f(-x+ct),$$

$\tilde{u}$ is an odd function with respect to x. Suppose u satisfies the initial conditions

$$u(0,x) = u_0(x), \quad x > 0, \quad \frac{\partial u}{\partial t}(0,x) = u_1(x), \quad x > 0.$$

Let $\tilde{u}_0(x)$, $\tilde{u}_1(x)$ be odd functions which are the extensions of $u_0(x)$, $u_1(x)$ to $-\infty < x < +\infty$; then we can rewrite the initial conditions as

$$\tilde{u}(0,x) = \tilde{u}_0(x), \quad -\infty < x < +\infty,$$

$$\frac{\partial \tilde{u}}{\partial t}(0,x) = \tilde{u}_1(x), \quad -\infty < x < +\infty.$$

Using the formula which was obtained in the previous case (*b*), we obtain

$$(**)\qquad \tilde{u}(t,x) = \frac{1}{2}\{\tilde{u}_0(x+ct) + \tilde{u}_0(x-ct)\} + \frac{1}{2c}\int_{x-ct}^{x+ct} \tilde{u}(x)\mathrm{d}x.$$

If u is of class C^2 for $x \geq 0$, $-\infty < t < +\infty$, then the initial data $u_0(x)$, $u_1(x)$ become C^2, C^1 for $x \geq 0$ respectively. Since u satisfies $u(t,0) = 0$, we have

$$u(0,0) = \frac{\partial u}{\partial t}(0,0) = \frac{\partial^2 u}{\partial t^2}(0,0) = 0.$$

Hence

$$u_0(0) = u(0,0) = 0,$$

$$u_1(0) = \frac{\partial u}{\partial t}(0,0) = 0,$$

$$u_0''(0) = \frac{\partial^2 u}{\partial x^2}(0,0) = \frac{1}{c^2}\frac{\partial^2 u}{\partial t^2}(0,0) = 0,$$

This is called the *compatibility condition* between the initial and boundary data.

Now, conversely, if $u_0(x)$, $u_1(x)$ belong to C^2, C^1 for $x \geqq 0$ with the compatibility condition then the odd extensions of u_0, u_1 still belong to the classes C^2, C^1 respectively. Therefore, $\tilde{u}$ defined by the right hand side of the equation (**) becomes a solution of the initial value problem.

(d) A finite string with fixed ends (II)

We now reconsider the same problem posed in (*a*). If there exists a solution u for the equation of the vibrating string, then u satisfies

$$u(t,x) = f(x+ct) + g(x-ct)$$

for $0 < x < 1$, $\quad -\infty < t < +\infty$. However, as we mentioned before, the right hand side of the above equation shows that we can extend u to $\tilde{u}$ which is a solution for $-\infty < x < +\infty$, $\quad -\infty < t < +\infty$.

Also, from the boundary condition

$$u(t,0) = u(t,1) = 0, \qquad -\infty < t < +\infty$$

we see that

$$f(ct) + g(-ct) = 0, \qquad -\infty < t < +\infty,$$
$$f(1+ct) + g(1-ct) = 0, \qquad -\infty < t < +\infty.$$

From the first equation, we have

$$g(t) = -f(-t)$$

and, by substitution in the second equation we obtain

$$f(t+2) = f(t)$$

which shows that f is a periodic function with period 2. Using f we write

$$\tilde{u}(t,x) = f(x+ct) - f(-x+ct).$$

This shows that $\tilde{u}(t,x)$ is an odd function with respect to x, and, at the same time, a periodic function with period 2.

Let us assume that u satisfies the initial conditions

$$u(0,x) = u_0(x), \quad 0 < x < 1,$$

$$\frac{\partial u}{\partial t}(0,x) = u_1(x), \quad 0 < x < 1.$$

We extend $u_0(x)$, $u_1(x)$ to $-1 < x < +1$ as odd functions and then extend again to $u_0(x)$, $\tilde{u}_1(x)$ to make them periodic functions with period 2 for $-\infty < x < +\infty$. Then we can rewrite the initial conditions as

$$\tilde{u}(0,x) = \tilde{u}_0(x), \quad -\infty < x < +\infty,$$

$$\frac{\partial \tilde{u}}{\partial t}(0,x) = \tilde{u}_1(x), \quad -\infty < x < +\infty.$$

Therefore

$$(*) \qquad \tilde{u}(t,x) = \frac{1}{2}\{\tilde{u}_0(x+ct) + \tilde{u}_0(x-ct)\} + \frac{1}{2c}\int_{x-ct}^{x+ct} \tilde{u}_1(x)\,dx.$$

Conversely, let $u_0(x)$, $u_1(x)$ be of class C^2, C^1 for $0 \leq x \leq 1$ with the compatibility conditions

$$u_0(0) = u_1(0) = u_0''(0) = 0,$$
$$u_0(1) = u_1(1) = u_0''(1) = 0$$

then $\tilde{u}$, which is defined by the right hand sides of the above equations, becomes a C^2-solution.

3.2

Consider the vibration of a stretched string which has its ends fixed to two springs (see Fig. 2). As we assumed in §3.1, the string itself is stretched along $0 < x < 1$ on the x-axis. At the end where $x = 0$, we have the

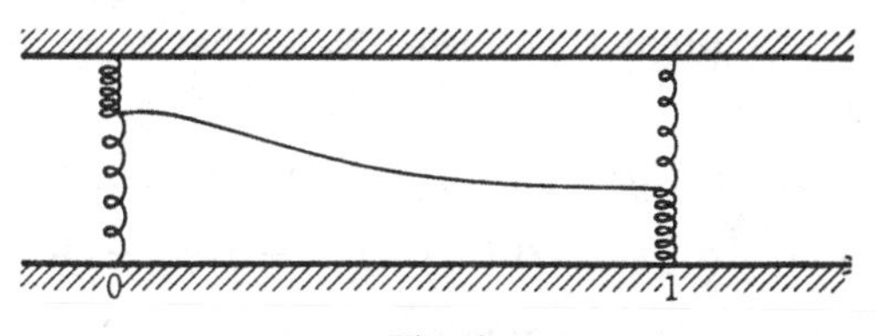

Fig. 2

equilibrium of force,

$$-\kappa_0 u + T\frac{\partial u/\partial x}{\sqrt{\{1+(\partial u/\partial x)^2\}}} = 0,$$

where κ_0 is the elastic modulus of the string at $x = 0$. Since we assume that the vibrations of the string are small in amplitude we have

$$\partial u/\partial x - \sigma_0 u = 0, \quad \sigma_0 = \kappa_0/T_0.$$

Similarly, at $x = 1$, we have

$$\partial u/\partial x + \sigma_1 u = 0, \quad \sigma_1 = \kappa_1/T_0$$

where κ_1 is the elastic modulus of the string at $x = 1$.

Let the equation of the vibrating string be

$$\frac{1}{c^2}\frac{\partial^2 u}{\partial t^2} = \frac{\partial^2 u}{\partial x^2}, \quad 0 < x < 1, \quad -\infty < t < +\infty.$$

We impose the boundary conditions

$$\frac{\partial u}{\partial x}(t, 0) - \sigma_0 u(t, 0) = 0, \quad -\infty < t < +\infty,$$

$$\frac{\partial u}{\partial x}(t, 1) + \sigma_1 u(t, 1) = 0, \quad -\infty < t < +\infty$$

and the initial conditions

$$u(0, x) = u_0(x), \quad 0 < x < 1,$$

$$\frac{\partial u}{\partial t}(0, x) = u_1(x), \quad 0 < x < 1.$$

We shall solve the problem by using the same argument as we employed in §3.1. Recall that if there is a solution u, from the equation of the vibrating string for $0 < x < 1$, $-\infty < t < +\infty$, the following equality holds:

$$u(t, x) = f(x + ct) + g(x - ct).$$

Then we obtain the extension $\tilde{u}$ of u to $-\infty < x < +\infty$ because the right hand side of the above equation is defined for $-\infty < x < +\infty$.

We return to the present problem. Note that from the boundary conditions we have

$$f'(ct) - \sigma_0 f(ct) + g'(-ct) - \sigma_0 g(-ct) = 0,$$
$$f'(1+ct) + \sigma_1 f(1+ct) + g'(1-ct) + \sigma_1 g(1-ct) = 0$$

and from the first equation we obtain

$$\begin{aligned}\frac{\partial \tilde{u}}{\partial x} - \sigma_0 \tilde{u} &= f'(x+ct) - \sigma_0 f(x+ct) \\ &\quad + g'(x-ct) - \sigma_0 g(x-ct) \\ &= \{f'(x+ct) - \sigma_0 f(x+ct)\} \\ &\quad - \{f'(-x+ct) - \sigma_0 f(-x+ct)\}.\end{aligned}$$

This shows that $(\partial\tilde{u}/\partial x - \sigma_0\tilde{u})(t,x)$ is an odd function with respect to x. On the other hand, the second equation shows that $(\partial\tilde{u}/\partial x + \sigma_1\tilde{u})(t,x)$ is an odd function with respect to $x-1$. In general, given a C^1-function $u(x)$ for $0 \leqq x \leqq 1$ satisfying the condition

$$u'(0) - \sigma_0 u(0) = u'(1) + \sigma_1 u(1) = 0$$

we define an extension $\tilde{u}$ of u to $-\infty < x < +\infty$ as follows: To start with, for $\{x \mid 0 \leqq x \leqq 1\}$ we set

$$\left(\frac{\mathrm{d}}{\mathrm{d}x} - \sigma_0\right)u(x) = v(x), \quad \left(\frac{\mathrm{d}}{\mathrm{d}x} + \sigma_1\right)u(x) = w(x).$$

Then we divide our argument into three steps:
Step 1. Let $v^{(1)}(x)$ be an odd function and an extension of $v(x)$ to $-1 \leqq x \leqq +1$, i.e. set

$$v^{(1)}(x) = \begin{cases} v(x), & 0 \leqq x \leqq 1, \\ -v(-x), & -1 \leqq x \leqq 0. \end{cases}$$

Consider a C^1-extension $u^{(1)}(x)$ of $u(x)$ such that it satisfies

$$\left(\frac{\mathrm{d}}{\mathrm{d}x} - \sigma_0\right)u^{(1)}(x) = v^{(1)}(x), \quad -1 \leqq x \leqq 1.$$

We set

$$\left(\frac{\mathrm{d}}{\mathrm{d}x} + \sigma_0\right)u^{(1)}(x) = w^{(1)}(x), \quad -1 \leqq x \leqq 1.$$

Thus $w^{(1)}(x)$ is an extension of $w(x)$.
Step 2. Let $w^{(2)}(x)$ be an odd function of $x-1$ and an extension of $w^{(1)}(x)$ to $-1 \leqq x \leqq +3$. Then we choose $u^{(2)}(x)$, a C^1-function, and an extension

of $u^{(1)}(x)$ such that

$$\left(\frac{d}{dx}+\sigma_1\right)u^{(2)}(x) = w^{(2)}(x), \quad -1 \leqq x \leqq 3.$$

Set

$$\left(\frac{d}{dx}-\sigma_0\right)u^{(2)}(x) = v^{(2)}(x).$$

We see that $v^{(2)}(x)$ is an extension of $v^{(1)}(x)$.
Step 3. We now proceed as in step 1. Let $v^{(3)}(x)$ be an odd function and an extension of $v^{(2)}(x)$ to $-3 \leqq x \leqq +3$. Setting

$$\left(\frac{d}{dx}-\sigma_0\right)u^{(3)}(x) = v^{(3)}(x), \quad -3 \leqq x \leqq 3,$$

$$\left(\frac{d}{dx}+\sigma_1\right)u^{(3)}(x) = w^{(3)}(x), \quad -3 \leqq x \leqq 3$$

we obtain $u^{(3)}, w^{(3)}$ which are the extensions of $u^{(2)}, w^{(2)}$. We repeat this process until we reach $\tilde{u}$, $\tilde{v}$, $\tilde{w}$ $(-\infty < x < +\infty)$ from u, v, $w(0 \leqq x \leqq 1)$, where $\tilde{v}$ is an odd function of x, and $\tilde{w}$ is an odd function of $x-1$.

Let us look at the initial conditions. By the extensions defined above we now have $\tilde{u}_0, \tilde{u}_1$ instead of u_0, u_1, so that we can rewrite

$$u(0,x) = u_0(x), \quad \frac{\partial u}{\partial t}(0,x) = u_1(x) \qquad \text{for } 0 < x < 1$$

as

$$\tilde{u}(0,x) = \tilde{u}_0(x), \quad \frac{\partial \tilde{u}}{\partial t}(0,x) = \tilde{u}_1(x) \quad \text{for } -\infty < x < +\infty.$$

From this we conclude that

$$(*) \qquad \tilde{u}(t,x) = \frac{1}{2}\{\tilde{u}_0(x+ct) + \tilde{u}_0(x-ct)\} + \frac{1}{2c}\int_{x-ct}^{x+ct} \tilde{u}_1(x)\,dx.$$

We now consider the reverse process to see whether $\tilde{u}$ which satisfies the last equality $(*)$, is in fact a solution. To see this let u_0, u_1 be of class C^2, C^1 for $0 \leqq x \leqq 1$ satisfying the compatibility conditions.

$$\left(\frac{d}{dx}-\sigma_0\right)u_i(0) = 0, \quad \left(\frac{d}{dx}+\sigma_1\right)u_i(1) = 0 \quad (i = 0, 1).$$

Then $\tilde{u}_0, \tilde{u}_1$ become C^2, C^1 for $-\infty < x < +\infty$, $\tilde{u}$ can be obtained from $\tilde{u}_0, \tilde{u}_1$ and becomes the solution to this problem.

3.3

Up to now we have considered initial boundary value problems based on the fact that the solutions of the equations of the vibrating strings should be in the form

$$u(t,x) = f(x+ct) + g(x-ct).$$

In this subsection we return to the problem of the vibrating finite string with both ends fixed.

The difference between the present approach and that of §3.1(*a*) is that this time a Fourier series is used and a 'stable' solution is sought for the same problem. By a *stable solution* we mean a solution

$$u(t,x) = T(t)X(x)$$

satisfying

$$\frac{1}{c^2}\frac{\partial^2 u}{\partial t^2} = \frac{\partial^2 u}{\partial x^2}, \quad 0 < x < 1, \quad -\infty < t < +\infty,$$

$$u(t,0) = u(t,1) = 0, \quad -\infty < t < +\infty$$

(without the initial conditions).

If a solution exists at all, then it must satisfy the equation of the vibrating string

$$\frac{1}{c^2}\frac{T''(t)}{T(t)} = \frac{X''(x)}{X(x)}.$$

Notice that the left side of the equation is a function of t only, and the right hand side a function of x only, therefore both sides must be equal to a constant $-\lambda$ which does not depend upon t or x, i.e.

$$\frac{1}{c^2}\frac{T''(t)}{T(t)} = -\lambda, \quad \frac{X''(x)}{X(x)} = -\lambda.$$

Hence, the boundary condition is

$$X(0) = X(1) = 0.$$

From this, firstly, with respect to X we see that

$$X''(x) + \lambda X(x) = 0, \quad 0 < x < 1,$$
$$X(0) = X(1) = 0.$$

That is, the problem is reduced to an eigenvalue problem in the theory of ordinary differential equations. Therefore, for the values $\lambda = n^2\pi^2 (n = 1, 2, \ldots)$ only, we have the solutions

$$X(x) = C \sin n\pi x \quad (n = 1, 2, \ldots)$$

where C is an arbitrary constant.

Secondly, with respect to T we have

$$T''(t) + \lambda c^2 T(t) = 0.$$

From this we see that

$$T(t) = a\cos\sqrt{\lambda}ct + b\sin\sqrt{\lambda}ct,$$

where a, b are arbitrary constants. Therefore, for $n = 1, 2, \ldots$ we have

$$u(t, x) = (a_n \cos n\pi ct + b_n \sin n\pi ct)\sin n\pi x$$

where a_n, b_n are arbitrary constants. These constitute an infinite number of stable solutions.

But it is remarkable that any solution of the vibrating string with fixed ends can be represented as the sum of an infinite number of these stable solutions. To see this, let us assume that for arbitrary $\{a_n\}$, $\{b_n\}$, the conditions

$$\sum_{n=1}^{\infty} |a_n| n^2 < +\infty, \quad \sum_{n=1}^{\infty} |b_n| n^2 < +\infty$$

are satisfied. Then, $u(t, x)$ which is defined as

$$u(t, x) = \sum_{n=1}^{\infty} (a_n \cos n\pi ct + b_n \sin n\pi ct) \sin n\pi x$$

is obviously a C^2-function satisfying the equation of the vibrating string and the boundary conditions at $x = 0, 1$ and moreover

$$u(0, x) = \sum_{n=1}^{\infty} a_n \sin n\pi x,$$

$$\frac{\partial u}{\partial t}(0, x) = \sum_{n=1}^{\infty} b_n n\pi c \sin n\pi x.$$

Now for arbitrary u_0, u_1, we wish to know if there are some $\{a_n\}$, $\{b_n\}$ such that

$$u_0(x) = \sum_{n=1}^{\infty} a_n \sin n\pi x,$$

$$u_1(x) = \sum_{n=1}^{\infty} b_n n\pi c \sin n\pi x.$$

In order to see this, we use a Fourier series to obtain

$$a_n = 2\int_0^1 u_0(x) \sin n\pi x \, dx,$$

$$b_n n\pi c = 2\int_0^1 u_1(x) \sin n\pi x \, dx.$$

In particular, if u_0, u_1 are of class C^4, C^3 satisfying the conditions

$$u_0(0) = u_0''(0) = u_0(1) = u_0''(1) = 0,$$
$$u_1(0) = u_1''(0) = u_1(1) = u_1''(1) = 0,$$

then, by integration by parts of the defining equations of a_n, b_n, the inequality

$$|a_n|, |b_n| \leqq C/n^4$$

can be obtained. Therefore,

$$\sum_{n=1}^{\infty} |a_n| n^2, \quad \sum_{n=1}^{\infty} |b_n| n^2 < +\infty.$$

Hence,

$$u_0(x) = \sum_{n=1}^{\infty} a_n \sin n\pi x, \quad u_1(x) = \sum_{n=1}^{\infty} b_n n\pi c \sin n\pi x,$$

are uniformly convergent. Since we know that the hypothesis imposed on $\{a_n, b_n\}$ is satisfied, we see that the above-mentioned u in the form of a convergent series is, in fact, a solution for the initial value problem.

3.4

In this subsection, we consider the problem of a small vibration of a light homogeneous membrane which is stretched over a certain bounded domain Ω in the (x, y)-plane, and is fixed on the boundary $\partial\Omega$ of Ω. We assume that the membrane vibrates vertically with respect to the (x, y)-plane. We then use a similar argument to that of §3.1 to show that the displacement $u(t, x, y)$, which is perpendicular to the (x, y)-plane, must satisfy the relation

$$\frac{1}{c^2}\frac{\partial^2 u}{\partial t^2} = \frac{\partial^2 u}{\partial x^2} + \frac{\partial^2 u}{\partial y^2}, \qquad c = \sqrt{\left(\frac{T_0}{\rho_0}\right)},$$

where ρ_0, T_0 are the surface density and the tension of the membrane at the position $u = 0$. Compared with the equation of the vibrating string, the equation of the vibrating membrane has an extra dimension which adds to the complexity of the problem. In the former case we can obtain a 'general solution' by the integration of the equation, i.e.

$$u(t, x) = f(x + ct) + g(x - ct)$$

but in the latter case we cannot use the integration process and the meaning of 'general solution' itself becomes very much more complicated. This is a classic example of the essential difficulty in mathematics caused by increasing the number of the dimensions under consideration by just one (in this case from one to two dimensions).

We approach the problem by Fourier's method as we did in §3.3 and seek a stable solution

$$u(t, x, y) = T(t)U(x, y)$$

satisfying

$$\frac{1}{c^2}\frac{\partial^2 u}{\partial t^2} = \frac{\partial^2 u}{\partial x^2} + \frac{\partial^2 u}{\partial y^2}, \quad (x, y) \in \Omega,$$

$$u = 0, \quad (x, y) \in \partial\Omega.$$

From the equation of the vibrating membrane we have

$$\frac{1}{c^2}\frac{T''}{T} = \frac{\partial^2 U/\partial x^2 + \partial^2 U/\partial y^2}{U} = -\lambda$$

therefore, for U the problem becomes an elliptic (two-dimensional) eigenvalue problem such that

$$\frac{\partial^2 U}{\partial x^2} + \frac{\partial^2 U}{\partial y^2} + \lambda U = 0, \quad (x, y) \in \Omega,$$

$$U = 0, \quad (x, y) \in \Omega.$$

It is not easy to solve this problem, but we know that there exists a series

$$0 < \lambda_1 \leqq \lambda_2 \leqq \lambda_3 \leqq \cdots \to +\infty$$

and for each n there exists $U_n(x, y)$ such that

$$\frac{\partial^2 U_n}{\partial x^2} + \frac{\partial^2 U_n}{\partial y^2} + \lambda_n U_n = 0, \quad (x, y) \in \Omega$$

$$U_n = 0, \quad (x, y) \in \partial\Omega$$

and $\{U_n(x, y)\}$ is a complete orthonormal system of functions of $L^2(\mathbb{R}^2)$-space. Using this fact, under the initial conditions

$$u(0, x, y) = u_0(x, y), \quad (x, y) \in \Omega,$$

$$\frac{\partial u}{\partial t}(0, x, y) = u_1(x, y), \quad (x, y) \in \Omega,$$

by choosing $\{a_n\}, \{b_n\}$ as

$$u_0(x, y) = \sum_{n=1}^{\infty} a_n U_n(x, y),$$

$$u_1(x, y) = \sum_{n=1}^{\infty} b_n \sqrt{\lambda_n} c U_n(x, y),$$

we obtain a solution whose general form is

$$u(t, x, y) = \sum_{n=1}^{\infty} (a_n \cos\sqrt{\lambda_n} ct + b_n \sin\sqrt{\lambda_n} ct) U_n(x, y).$$

In this way, it seems to have succeeded in solving the problem of the vibrating membrane in the same way as that of the vibrating string. However, the real difficulties remain in the form of the elliptic eigenvalue problem, although in some cases when Ω is in an 'ordinary shape', solving a problem of this kind can be simple and straightforward. For example, let Ω be the rectangular domain $(0, a) \times (0, b)$. We seek eigenfunctions of the form $U(x, y) = X(x) \cdot Y(y)$. From the original equation the factors X, Y must satisfy

$$X''/X + Y''/Y + \lambda = 0.$$

That is

$$X''/X + \nu = 0, \quad Y''/Y + \mu = 0, \quad \nu + \mu = \lambda.$$

Under the boundary condition we obtain

$$X = \sin\frac{n\pi}{a}x, \quad Y = \sin\frac{m\pi}{b}y, \quad \frac{n^2\pi^2}{a^2} + \frac{m^2\pi^2}{b^2} = \lambda.$$

Therefore, for eigenvalues

$$\lambda_{nm} = \frac{n^2\pi^2}{a^2} + \frac{m^2\pi^2}{b^2},$$

we have eigenfunctions

$$U_{nm}(x, y) = \sin\frac{n\pi}{a}x \sin\frac{m\pi}{b}y,$$

which are defined in Ω.

In many cases, by means of the Fourier method, we arrive at an *elliptic eigenvalue problem.* Then, if the shape of Ω is ordinary as in the foregoing example, all the solutions can be obtained in the form of infinite linear combinations of 'simple' eigenfunctions.†

The case of the rectangular domain is important in the sense that in the subsequent chapters we develop the same method by replacing a Fourier series with a Fourier transform. That is, if a given domain is a half-space, a Fourier transform can be applied in the tangential direction leaving intact the normal direction with respect to the domain. Admittedly, the application of Fourier transforms seems to restrict severely the shape of the domain, but, for a domain which is surrounded by a smooth surface, if a local transformation of variables can successfully map the domain into a half-space, the above method will yield some fruitful results.

† Translator's note: the above U_{nm} are such eigenfunctions, for example.

4. The problem of the vibrating membrane and energy inequalities

In this section we shall not seek the solution for the problem of the vibrating membrane. Instead, we shall focus our attention on two classical problems, namely Dirichlet's and Neumann's problems concerning 'fixed' and 'free' membranes (the precise definitions of these concepts are given below), and demonstrate that our argument eventually leads to two different inequalities. We begin by clarifying the difference between these problems and at the end give their precise formulations in equational terms.

Let us consider the vertical vibration of a membrane stretched over a domain Ω which is surrounded by a curve S in the (x, y)-plane. Assume that an external force with a certain magnitude is acting in the vertical direction over the (x, y)-plane. We have a kinetic equation

$$\rho_0 \frac{\partial^2 u}{\partial t^2} - T_0\left(\frac{\partial^2 u}{\partial x^2} + \frac{\partial^2 u}{\partial y^2}\right) = f(t, x, y), \quad (x, y) \in \Omega,$$

where u denotes the vertical displacement of the membrane, f the external force acting upon the unit area of Ω, ρ_0 the surface density, and T_0 the tension of the membrane. Note that ρ_0 and T_0 are constants.

We impose the initial conditions

$$u|_{t=0} = u_0(x, y), \quad (x, y) \in \Omega,$$
$$(\partial u/\partial t)|_{t=0} = u_1(x, y); \quad (x, y) \in \Omega,$$

and the boundary condition

$$u|_S = g_0(t, s), \quad s \in S$$

(where g_0 is a given displacement on S) in the case of *fixed membrane* or, in the case of *free membrane*,

$$T_0(\partial u/\partial n)|_S = g_1(t, s), \quad s \in S,$$

where $\partial/\partial n$ represents the directional differential along the interior normal unit vector (n_x, n_y), and g_1 the vertical component (in relation to the (x, y)-plane) of the external force acting upon the unit length of S.

In other words, by imposing these different boundary conditions, we obtain two different kinds of initial value problem for vibrating strings. These problems are called Dirichlet's problem in the fixed membrane case and Neumann's problem in the free membrane case.

In the subsequent argument, to simplify things we impose the normalisation conditions by setting $\rho_0 = T_0 = 1$, and, for the same reason, we set

$$\partial u/\partial t = u_t, \quad \partial^2 u/\partial t^2 = u_{tt}, \quad \ldots$$

and set

$$u_{tt} - u_{xx} - u_{yy} = \Box u.$$

We can now write these problems in more compact forms as follows:

Dirichlet's problem

$$\begin{aligned} \Box u &= f(t,x,y), \quad (x,y)\in\Omega, \quad t>0,\\ u|_S &= g_0(t,s), \quad s\in S, \quad t>0,\\ u|_{t=0} &= u_0(x,y), \quad u_t|_{t=0} = u_1(x,y), \quad (x,y)\in\Omega. \end{aligned}$$

Neumann's problem

$$\begin{aligned} \Box u &= f(t,x,y), \quad (x,y)\in\Omega, \quad t>0,\\ (\partial u/\partial n)|_S &= g_1(t,s), \quad s\in S, \quad t>0,\\ u|_{t=0} &= u_0(x,y), \quad u_t|_{t=0} = u_1(x,y), \quad (x,y)\in\Omega. \end{aligned}$$

In what follows we deal only with real-valued functions.

4.1

In order to deal with the above problems we need Green's formula. In this subsection, therefore, we shall briefly discuss this important formula.

Let S be the positively oriented boundary of Ω which is parametrically represented by a length parameter s. This means that

$$S = \{(x,y);\, x = x(s),\, y = y(s)\},$$

where s increases in the positive direction and the functions x, y are of class C^2. We introduce a local coordinate (s,n) in a neighbourhood of S by setting

$$\begin{aligned} x &= x(s) + n_x(s)n\\ y &= y(s) + n_y(s)n. \end{aligned}$$

In fact

$$\begin{bmatrix} x'(s)\\ y'(s)\end{bmatrix} = \begin{bmatrix} n_y(s)\\ -n_x(s)\end{bmatrix}, \quad \begin{bmatrix} n_x'(s)\\ n_y'(s)\end{bmatrix} = -\rho(s)\begin{bmatrix} n_y(s)\\ -n_x(s)\end{bmatrix}$$

(where $\rho(s)$ is the curvature) so that

$$\begin{bmatrix} \partial x/\partial s & \partial y/\partial s\\ \partial x/\partial n & \partial y/\partial n\end{bmatrix} = \begin{bmatrix} (1-\rho n)n_y & -(1-\rho n)n_x\\ n_x & n_y\end{bmatrix}.$$

Hence, for small n we have

$$\partial(x,y)/\partial(s,n) = 1 - \rho n \neq 0.$$

This shows that (s,n) is a local coordinate as we claimed. By this coordinate transformation, we obtain a new differential operator

$$\begin{bmatrix} \partial/\partial s \\ \partial/\partial n \end{bmatrix} = \begin{bmatrix} (1-\rho n)n_y & -(1-\rho n)n_x \\ n_x & n_y \end{bmatrix} \begin{bmatrix} \partial/\partial x \\ \partial/\partial y \end{bmatrix}.$$

Therefore,

$$\begin{bmatrix} \dfrac{1}{1-\rho n}\dfrac{\partial}{\partial s} \\ \dfrac{\partial}{\partial n} \end{bmatrix} = \begin{bmatrix} n_y & -n_x \\ n_x & n_y \end{bmatrix} \begin{bmatrix} \dfrac{\partial}{\partial x} \\ \dfrac{\partial}{\partial y} \end{bmatrix}, \quad \begin{bmatrix} \dfrac{\partial}{\partial x} \\ \dfrac{\partial}{\partial y} \end{bmatrix} = \begin{bmatrix} n_y & n_x \\ -n_x & n_y \end{bmatrix} \begin{bmatrix} \dfrac{1}{1-\rho n}\dfrac{\partial}{\partial s} \\ \dfrac{\partial}{\partial n} \end{bmatrix}.$$

Finally, we obtain

$$\Delta u = \frac{\partial^2 u}{\partial x^2} + \frac{\partial^2 u}{\partial y^2} = \frac{1}{1-\rho n}\frac{\partial}{\partial s}\left(\frac{1}{1-\rho n}\frac{\partial u}{\partial s}\right)$$
$$+ \frac{1}{1-\rho n}\frac{\partial}{\partial n}\left\{(1-\rho n)\frac{\partial u}{\partial n}\right\}.$$

To make things simpler, we write

$$\frac{1}{1-\rho n}\frac{\partial}{\partial s} = \partial_{\tilde{s}}, \quad \frac{\rho}{1-\rho n} = \tilde{\rho},$$

and then we see that

$$\partial_n \partial_{\tilde{s}} - \partial_{\tilde{s}} \partial_n = \tilde{\rho}\partial_{\tilde{s}}, \quad \Delta = \partial_{\tilde{s}}{}^2 + \partial_n{}^2 - \tilde{\rho}\partial_n,$$

We write the inner products of the spaces $L^2(\Omega)$, $L^2(S)$ as $(\,,)$, $\langle\,,\rangle$. Assume that the support of u is contained in a neighbourhood of S. By replacing integration in (x, y)-space with integration in (s, n)-space, and by integration by parts we obtain

$$(\partial_{\tilde{s}}u, v) + (u, \partial_{\tilde{s}}v) = 0,$$
$$(\partial_n u, v) + (u, \partial_n v) = -\langle u, v\rangle + (\tilde{\rho}u, v),$$

or, in general,

$$(\varphi\partial_{\tilde{s}}u, v) + (u, \varphi\partial_{\tilde{s}}v) = -((\partial_{\tilde{s}}\varphi)u, v),$$
$$(\varphi\partial_n u, v) + (u, \varphi\partial_n v) = -\langle\varphi u, v\rangle + (\tilde{\rho}\varphi u, v) - ((\partial_n\varphi)u, v).$$

Using this fact we now use integration by parts twice and obtain the desired formula as follows: To begin with we calculate

$$\int_0^t (\Box u, v)\,dt$$
$$= \int_0^t \{(u_{tt}, v) - (u_{\tilde{s}\tilde{s}}, v) - (u_{nn}, v) + (\tilde{\rho}u_n, v)\}\,dt$$

$$= (u_t(t), v(t)) - (u_t(0), v(0)) + \int_0^t \langle u_n, v \rangle \, \mathrm{d}t$$

$$+ \int_0^t \{-(u_t, v_t) + (u_{\tilde{s}}, v_{\tilde{s}}) + (u_n, v_n)\} \, \mathrm{d}t.$$

We consider the following three cases:

$$(\alpha)\ v = u_n, \quad (\beta)\ v = u_{\tilde{s}}, \quad (\gamma)\ v = \dot{u}_t.$$

By using integration by parts again we find

Case α

$$\int_0^t (\Box u, u_n) \, \mathrm{d}t = (u_t(t), u_n(t)) - (u_t(0), u_n(0)) + \int_0^t \langle u_n \rangle^2 \, \mathrm{d}t$$

$$+ \int_0^t \{-(u_t, \partial_n u_t) + (u_{\tilde{s}}, \partial_n u_{\tilde{s}}) - (u_{\tilde{s}}, \tilde{\rho} u_{\tilde{s}}) + (u_n, \partial_n u_n)\} \, \mathrm{d}t$$

$$= (u_t(t), u_n(t)) - (u_t(0), u_n(0)) + \int_0^t \langle u_n \rangle^2 \mathrm{d}t$$

$$+ \int_0^t \Big[-\tfrac{1}{2}\{-\langle u_t \rangle^2 + (u_t, \tilde{\rho} u_t)\}$$

$$+ \tfrac{1}{2}\{-\langle u_{\tilde{s}} \rangle^2 + (u_{\tilde{s}}, \tilde{\rho} u_{\tilde{s}})\} - (u_{\tilde{s}}, \tilde{\rho} u_{\tilde{s}})$$

$$+ \tfrac{1}{2}\{-\langle u_n \rangle^2 + (u_n, \tilde{\rho} u_n)\} \Big] \mathrm{d}t$$

$$= (u_t(t), u_n(t)) - (u_t(0), u_n(0))$$

$$+ \tfrac{1}{2} \int_0^t \{\langle u_t \rangle^2 - \langle u_s \rangle^2 + \langle u_n \rangle^2\} \, \mathrm{d}t$$

$$+ \tfrac{1}{2} \int_0^t \{-(\tilde{\rho} u_t, u_t) - (\tilde{\rho} u_{\tilde{s}}, u_{\tilde{s}}) + (\tilde{\rho} u_n, u_n)\} \, \mathrm{d}t,$$

Case β

$$\int_0^t (\Box u, u_s) \, \mathrm{d}t = (u_t(t), u_{\tilde{s}}(t)) - (u_t(0), u_{\tilde{s}}(0)) + \int_0^t \langle u_n, u_s \rangle \, \mathrm{d}t$$

$$+ \int_0^t \{-(u_t, \partial_{\tilde{s}} u_t) + (u_{\tilde{s}}, \partial_{\tilde{s}} u_{\tilde{s}}) + (u_n, \partial_{\tilde{s}} u_n) + (u_n, \tilde{\rho} u_{\tilde{s}})\} \, \mathrm{d}t$$

$$= (u_t(t), u_{\tilde{s}}(t)) - (u_t(0), u_{\tilde{s}}(0)) + \int_0^t \langle u_n, u_s \rangle \, \mathrm{d}t + \int_0^t (\tilde{\rho} u_n, u_{\tilde{s}}) \, \mathrm{d}t,$$

Case γ

$$\int_0^t (\Box u, u_t)\,dt = \|u_t(t)\|^2 - \|u_t(0)\|^2 + \int_0^t \langle u_n, u_t\rangle\,dt$$

$$+ \int_0^t \{-(u_t, \partial_t u_t) + (u_{\tilde{s}}, \partial_t u_{\tilde{s}}) + (u_n, \partial_t u_n)\}\,dt$$

$$= \|u_t(t)\|^2 - \|u_t(0)\|^2\} + \int_0^t \langle u_n, u_t\rangle\,dt$$

$$-\tfrac{1}{2}\{\|u_t(t)\|^2 - \|u_t(0)\|^2\}$$

$$+\tfrac{1}{2}\{\|u_{\tilde{s}}(t)\|^2 - \|u_{\tilde{s}}(0)\|^2\}$$

$$+\tfrac{1}{2}\{\|u_n(t)\|^2 - \|u_n(0)\|^2\}$$

$$= \tfrac{1}{2}\|\nabla u(t)\|^2 - \tfrac{1}{2}\|\nabla u(0)\|^2 + \int_0^t \langle u_n, u_t\rangle\,dt.$$

where

$$\|\nabla u\|^2 = \|u_t\|^2 + \|u_x\|^2 + \|u_y\|^2 = \|u_t\|^2 + \|u_{\tilde{s}}\|^2 + \|u_n\|^2.$$

By setting

$$(\tilde{\rho} u, v) = (u, v)_{,\tilde{\rho}}$$

We obtain the following formula:

Lemma 4.1 *(Green's formula : the original form)*
If the support of u lies in a neighbourhood of S (except in the case γ), the following equalities hold:

(α)
$$\int_0^t (\Box u, u_n)\,dt = (u_t(t), u_n(t)) - (u_t(0), u_n(0))$$

$$+\tfrac{1}{2}\int_0^t \{\langle u_t\rangle^2 - \langle u_s\rangle^2 + \langle u_n\rangle^2\}\,dt$$

$$+\tfrac{1}{2}\int_0^t \{-(u_t, u_t)_{\tilde{\rho}} - (u_{\tilde{s}}, u_{\tilde{s}})_{\tilde{\rho}} + (u_n, u_n)_{\tilde{\rho}}\}\,dt,$$

(β)
$$\int_0^t (\Box u, u_{\tilde{s}})\,dt = (u_t(t), u_{\tilde{s}}(t)) - (u_t(0), u_{\tilde{s}}(0))$$

$$+ \int_0^t \langle u_n, u_s\rangle\,dt + \int_0^t (u_n, u_{\tilde{s}})_{\tilde{\rho}}\,dt,$$

$$(\gamma) \qquad \int_0^t (\Box u, u_t)\mathrm{d}t = \tfrac{1}{2}(\|\nabla u(t)\|^2 - \|\nabla u(0)\|^2) + \int_0^t \langle u_n, u_t \rangle \mathrm{d}t.$$

Note. (γ) is valid without any restrictions on supp $[u]$.

4.2

First, as an application of Green's formula, we establish the energy inequality for Dirichlet's problem. The problem we are interested in can be stated as follows: Suppose there exists a solution u for Dirichlet's problem such that

$$\Box u = f, \quad (x, y)\in\Omega, \quad t>0,$$
$$u|_S = g, \quad s\in S, \quad t>0,$$
$$u|_{t=0} = u_0, \quad u_t|_{t=0} = u_1, \quad (x, y)\in\Omega.$$

We wish to know how to estimate u by the data $\{f, g, u_0, u_1\}$ in L^2 sense. In order to apply Green's formula, we decompose u into two parts, v and w, such that v has its support in a neighbourhood of S, while w has not. In other words, if φ, ψ are smooth functions such that φ has its support in a small neighbourhood N of S and ψ has not, and they satisfy the condition $\varphi + \psi = 1$, then by setting

$$\varphi u = v, \quad \psi u = w,$$

the following equalities hold

$$\Box v = \varphi f - [\varphi, \Box]u \equiv \Phi,$$
$$v|_S = g;$$
$$\Box w = \psi f - [\psi, \Box]u \equiv \Psi,$$
$$w|_N = 0.$$

Using this fact and Green's formula (γ) for w, we have

$$\int_0^t (\Psi, w_t)\,\mathrm{d}t = \tfrac{1}{2}(\|\nabla w(t)\|^2 - \|\nabla w(0)\|^2).$$

Therefore

$$\|\nabla w(t)\|^2 \leqq \|\nabla w(0)\|^2 + 2\int_0^t \|\Psi\|\cdot\|\nabla w\|\,\mathrm{d}t.$$

For v, from Green's formulas (γ) and (α), considering $(\gamma) + (\alpha)\times\varepsilon$ for

$0<\varepsilon<1$ we have

$$\int_0^t (\Phi, v_t+\varepsilon v_n)\,dt = \int_0^t \langle v_n, g_t\rangle + \frac{\varepsilon}{2}\{\langle g_t\rangle^2 - \langle g_s\rangle^2 + \langle v_n\rangle^2\}\,dt$$
$$+\{\tfrac{1}{2}\|\nabla v(t)\|^2 + \varepsilon(v_t(t), v_n(t))\}$$
$$-\{\tfrac{1}{2}\|\nabla v(0)\|^2 + \varepsilon(v_t(0), v_n(0))\}$$
$$+\frac{\varepsilon}{2}\int_0^t \{-(v_t, v_t)_{\hat\rho} + (v_n, v_n)_{\tilde\rho} - (v_{\tilde s}, v_{\tilde s})_{\tilde\rho}\}\,dt,$$

Hence

$$\int_0^t \langle v_n\rangle^2\,dt + \|\nabla v(t)\|^2 \leqq C\left\{\int_0^t (\langle g_t\rangle^2 + \langle g_s\rangle^2)\,dt + \|\nabla v(0)\|^2 + \int_0^t (\|\nabla v\|^2 + \|\Phi\|\cdot\|\nabla v\|)\,dt\right\}.$$

Combining the two inequalities for v and w we obtain

$$\|\nabla u\| = \|\nabla v + \nabla w\| \leqq \|\nabla v\| + \|\nabla w\|,$$
$$\|\nabla v\| + \|\nabla w\| = \|(\nabla\varphi)u + \varphi(\nabla u)\| + \|(\nabla\psi)u + \psi(\nabla u)\|$$
$$\leqq C(\|u\| + \|\nabla u\|),$$
$$\|\Phi\| + \|\Psi\| \leqq C(\|f\| + \|u\| + \|\nabla u\|).$$

So that

$$\int_0^t \langle u_n\rangle^2\,dt + \|\nabla u(t)\|^2 \leqq C\left[\int_0^t (\langle g_t\rangle^2 + \langle g_s\rangle^2)\,dt + \|u(0)\|^2 + \|\nabla u(0)\|^2 + \int_0^t \{(\|u\|^2 + \|\nabla u\|^2) + \|f\|(\|u\| + \|\nabla u\|)\}\,dt\right].$$

On the other hand we have an estimation of the norm of u in the space L^2 as

$$\|u(t)\|^2 \leqq \|u(0)\|^2 + 2\int_0^t (u_t, u)\,dt$$
$$\leqq \|u(0)\|^2 + 2\int_0^t \|u_t\|\cdot\|u\|\,dt.$$

Writing

$$\|u\|_1{}^2 = \|u\|^2 + \|u_x\|^2 + \|u_y\|^2,$$
$$|||u|||_1{}^2 = \|u\|^2 + \|\nabla u\|^2,$$
$$\langle u\rangle_1{}^2 = \langle u\rangle^2 + \langle u_s\rangle^2,$$
$$\langle\!\langle u\rangle\!\rangle_1{}^2 = \langle u\rangle^2 + \langle \nabla u\rangle^2,$$

from the previous result we have

Proposition 4.2 *(the fundamental inequality for Dirichlet's problem)*
For $t > 0$,

$$\int_0^t \langle\!\langle u \rangle\!\rangle_1{}^2\,\mathrm{d}t + |||u(t)|||_1{}^2 \leqq C\left\{\int_0^t (\langle g \rangle_1{}^2 + \langle g_t \rangle^2)\,\mathrm{d}t + |||u(0)|||_1{}^2 + \int_0^t (\|f\|\cdot|||u|||_1 + |||u|||_1{}^2)\,\mathrm{d}t\right\}$$

where $\Box u = f$, $u|_S = g$.

We now wonder if we can make a better estimation by using certain equivalent norms instead of $|||\cdot|||$ and $\langle\!\langle \cdot \rangle\!\rangle_1$. This is in fact the case as we see below. Adding a relatively large parameter $\gamma > 0$ to the terms of lower orders we define new norms $|||\cdot|||_{1,\gamma}$ and $\langle\!\langle \cdot \rangle\!\rangle_{1,\gamma}$ such that

$$|||u|||_{1,\gamma}{}^2 = \gamma^2\|u\|^2 + \|\nabla u\|^2,$$
$$\langle\!\langle u \rangle\!\rangle_{1,\gamma}{}^2 = \gamma^2\langle u \rangle^2 + (\nabla u \rangle^2.$$

Then it is easy to see that these old and new norms are equivalent. Furthermore, they are tied to each other as follows:

Lemma 4.3
For $t > 0$, $\gamma > 0$,

$$\int_0^t e^{-2\gamma t}\langle\!\langle u(t) \rangle\!\rangle_{1,\gamma}{}^2\mathrm{d}t + e^{-2\gamma t}|||u(t)|||_{1,\gamma}{}^2 + \gamma\int_0^t e^{-2\gamma t}|||u(t)|||_{1,\gamma}{}^2\mathrm{d}t$$

$$\leqq C\left\{\int_0^t e^{-2\gamma t}\langle\!\langle u(t) \rangle\!\rangle_1{}^2\mathrm{d}t + e^{-2\gamma t}|||u(t)|||_1{}^2 + \gamma\int_0^t e^{-2\gamma t}|||u(t)|||_1{}^2\mathrm{d}t + |||u(0)|||_{1,\gamma}{}^2\right\}.$$

Proof
First we see that

$$\begin{aligned} e^{-2\gamma t}\|u(t)\|^2 - \|u(0)\|^2 &= \int_0^t \frac{\mathrm{d}}{\mathrm{d}t}\{e^{-2\gamma t}\|u(t)\|^2\}\,\mathrm{d}t \\ &\leqq -2\gamma\int_0^t e^{-2\gamma t}\|u(t)\|^2\,\mathrm{d}t \\ &\quad + 2\int_0^t e^{-2\gamma t}\|u(t)\|\cdot\|u_t(t)\|\,\mathrm{d}t \\ &\leqq -\gamma\int_0^t e^{-2\gamma t}\|u(t)\|^2\,\mathrm{d}t \\ &\quad + \frac{1}{\gamma}\int_0^t e^{-2\gamma t}\|u_t(t)\|^2\,\mathrm{d}t. \end{aligned}$$

That is,

$$e^{-2\gamma t}\|u(t)\|^2+\gamma\int_0^t e^{-2\gamma t}\|u(t)\|^2\,dt$$
$$\leqq \|u(0)\|^2+\frac{1}{\gamma}\int_0^t e^{-2\gamma t}|||u(t)|||_1{}^2\,dt.$$

Therefore

$$e^{-2\gamma t}\gamma^2\|u(t)\|^2+\gamma\int_0^t e^{-2\gamma t}\gamma^2\|u(t)\|^2\,dt$$
$$\leqq |||u(0)|||_{1,\gamma}{}^2+\gamma\int_0^t e^{-2\gamma t}|||u(t)|||_1{}^2\,dt.$$

Similarly

$$e^{-2\gamma t}\langle u(t)\rangle^2+\gamma\int_0^t e^{-2\gamma t}\langle u(t)\rangle^2\,dt$$
$$\leqq \langle u(0)\rangle^2+\frac{1}{\gamma}\int_0^t e^{-2\gamma t}\langle\langle u(t)\rangle\rangle_1{}^2\,dt.$$

For the first term on the right hand side of this inequality we have

$$\langle u(0)\rangle^2=\langle(\varphi u)(0)\rangle^2=-\iint_\Omega\frac{\partial}{\partial n}(\varphi u)^2(0,x,y)\frac{1}{1-\rho n}\,dx\,dy$$
$$\leqq C\|\nabla u(0)\|\cdot\|u(0)\|$$
$$\leqq C\left(\frac{1}{\gamma}\|\nabla u(0)\|^2+\gamma\|u(0)\|^2\right)$$
$$=C\frac{1}{\gamma}|||u(0)|||_{1,\gamma}{}^2.$$

Therefore we obtain

$$\int_0^t e^{-2\gamma t}\gamma^2\langle u(t)\rangle^2\,dt\leqq C\left\{|||u(0)|||_{1,\gamma}{}^2+\int_0^t e^{-2\gamma t}\langle\langle u(t)\rangle\rangle_1{}^2\,dt\right\}.$$

On the other hand obviously

$$\int_0^t e^{-2\gamma t}\langle\langle\nabla u\rangle\rangle^2\,dt+e^{-2\gamma t}\|\nabla u(t)\|^2+\gamma\int_0^t e^{-2\gamma t}\|\nabla u\|^2\,dt$$
$$\leqq\int_0^t e^{-2\gamma t}\langle\langle u\rangle\rangle_1{}^2\,dt+e^{-2\gamma t}|||u(t)|||_1{}^2+\gamma\int_0^t e^{-2\gamma t}|||u|||_1{}^2\,dt.$$

Term-by-term addition of the last inequality and the inequalities which have just been established yields the desired result. ■

Keeping this fact in mind we now proceed as follows: We return to the fundamental inequality for Dirichlet's problem (Proposition 4.2). Multiplying the terms on both sides by $e^{-2\gamma t}(\gamma > 0)$ and integrating them with respect to t we have

$$\int_0^{t_0} e^{-2\gamma t}\left(\int_0^t f(\tau)\,d\tau\right)dt = \int_0^{t_0} \frac{e^{-2\gamma\tau} - e^{-2\gamma t_0}}{2\gamma} f(\tau)\,d\tau$$
$$\leqq \frac{1}{2\gamma}\int_0^{t_0} e^{-2\gamma\tau} f(\tau)\,d\tau$$

for an arbitrary $f(\tau) \geq 0$. Therefore we obtain

(A)
$$\int_0^{t_0} \frac{e^{-2\gamma t} - e^{-2\gamma t_0}}{2\gamma} \langle\langle u(t)\rangle\rangle_1{}^2\,dt + \int_0^{t_0} e^{-2\gamma t} |||u(t)|||_1{}^2\,dt$$
$$\leqq \frac{C}{2\gamma}\left\{\int_0^{t_0} e^{-2\gamma t}(\langle g(t)\rangle_1{}^2 + \langle g_t(t)\rangle^2)dt + |||u(0)|||_1{}^2\right.$$
$$\left. + \int_0^{t_0} e^{-2\gamma t}(\|f(t)\|\cdot|||u(t)|||_1 + |||u(t)|||_1{}^2)dt\right\}.$$

Also, multiplying the terms on both sides in the same inequality by $e^{-2\gamma t_0}$ yields

(B)
$$e^{-2\gamma t_0}\int_0^{t_0} \langle\langle u(t)\rangle\rangle_1{}^2\,dt + e^{-2\gamma t_0}|||u(t_0)|||_1{}^2$$
$$\leqq C\left\{\int_0^{t_0} e^{-2\gamma t}(\langle g(t)\rangle_1{}^2 + \langle g_t(t)\rangle^2)\,dt + |||u(0)|||_1{}^2\right.$$
$$\left. + \int_0^{t_0} e^{-2\gamma t}(\|f(t)\|\cdot|||u(t)|||_1 + |||u(t)|||_1{}^2)\,dt\right\}.$$

By term-by-term addition 2γ(A) + (B) we see that

$$\int_0^{t_0} e^{-2\gamma t}\langle\langle u(t)\rangle\rangle_1{}^2\,dt + e^{-2\gamma t_0}|||u(t_0)|||_1{}^2 + 2\gamma\int_0^{t_0} e^{-2\gamma t}|||u(t)|||_1{}^2\,dt$$
$$\leqq 2C\left\{\int_0^{t_0} e^{-2\gamma t}(\langle g(t)\rangle_1{}^2 + \langle g_t(t)\rangle^2)dt + |||u(0)|||_1{}^2\right.$$
$$\left. + \int_0^{t_0} e^{-2\gamma t}(\|f(t)\|\cdot|||u(t)|||_1 + |||u(t)|||_1{}^2)\,dt\right\}.$$

For the right hand term containing f we have

$$\int_0^{t_0} e^{-2\gamma t}\|f(t)\|\cdot|||u(t)|||_1\,dt \leqq \frac{2C}{\gamma}\int_0^{t_0} e^{-2\gamma t}\|f(t)\|^2\,dt + \frac{\gamma}{2C}\int_0^{t_0} e^{-2\gamma t}|||u(t)|||_1{}^2\,dt.$$

From this we find

$$\begin{aligned}\int_0^{t_0} & e^{-2\gamma t}\langle\langle u(t)\rangle\rangle_1{}^2\,dt + e^{-2\gamma t_0}|||u(t_0)|||_1{}^2 + \gamma\int_0^{t_0} e^{-2\gamma t}|||u(t)|||_1{}^2\,dt \\ &\leqq C'\left\{\int_0^{t_0} e^{-2\gamma t}(\langle g(t)\rangle_1{}^2 + \langle g_t(t)\rangle^2)\,dt + |||u(0)|||_1{}^2 \right.\\ &\left.\quad + \frac{1}{\gamma}\int_0^{t_0} e^{-2\gamma t}\|f(t)\|^2\,dt + \int_0^{t_0} e^{-2\gamma t}|||u(t)|||_1{}^2\,dt\right\}.\end{aligned}$$

We now choose a sufficiently large $\gamma_0 > 0$. Then, for any $\gamma \geqq \gamma_0$, the last term on the left hand side absorbs the counterpart on the right hand side.

Applying Lemma 4.3, we establish

Proposition 4.4 *(the inequality (H) for Dirichlet's problem)*
For $\gamma \geqq \gamma_0$ with a sufficiently large $\gamma_0 > 0$,

$$\text{(H)}\qquad \begin{aligned}\int_0^{t_0} & e^{-2\gamma t}\langle\langle u(t)\rangle\rangle_{1,\gamma}{}^2\,dt + e^{-2\gamma t_0}|||u(t_0)|||_{1,\gamma}{}^2 \\ &+ \gamma\int_0^{t_0} e^{-2\gamma t}|||u(t)|||_{1,\gamma}{}^2\,dt \\ &\leqq C\left\{\int_0^{t_0} e^{-2\gamma t}(\langle g(t)\rangle_1{}^2 + \langle g_t(t)\rangle^2)\,dt + |||u(0)|||_{1,\gamma}{}^2\right.\\ &\left.\quad + \frac{1}{\gamma}\int_0^{t_0} e^{-2\gamma t}\|f(t)\|^2\,dt\right\}\end{aligned}$$

where $\Box u = f$, $u|_S = g$, and C is a positive constant that does not depend on u, $t_0(>0)$, or $\gamma(\geqq \gamma_0)$.

In particular

$$\int_0^{t_0}\langle\langle u(t)\rangle\rangle_1{}^2\,dt + |||u(t_0)|||_1{}^2 \leqq C' e^{-2\gamma_0 t_0}\left\{\int_0^{t_0}(\langle g(t)\rangle_1{}^2 + \langle g_t(t)\rangle^2)\,dt + |||u(0)|||_1{}^2 + \int_0^{t_0}\|f(t)\|^2\,dt\right\}.$$

(To see this set $\gamma = \gamma_0$)

4.3
We shall now derive the energy inequality for Neumann's problem. Let u satisfy

$$\Box u = f, \quad (\partial u/\partial n)|_S = g.$$

Applying Green's formula (γ) we find

$$\int_0^t (f, u_t)\,\mathrm{d}t = \int_0^t \langle g, u_t\rangle\,\mathrm{d}t + \tfrac{1}{2}(\|\nabla u(t)\|^2 - \|\nabla u(0)\|^2).$$

In order to estimate $\langle g, u_t\rangle$ in the equality in an efficient way, we must use the notion of a differential operator Λ with a fractional power α. This enables us to write

$$\Lambda^\alpha = (-(\partial^2/\partial s^2) + 1)^{\alpha/2}, \quad \langle u\rangle_\alpha = \langle \Lambda^\alpha u\rangle.$$

We establish

Lemma 4.5
For $\Box u = f$ and $t > 0$.

$$\int_0^t \langle \nabla u\rangle_{-1/2}{}^2\,\mathrm{d}t \leqq C\Big\{ |||u(t)|||_1{}^2 + |||u(0)|||_1{}^2 + \int_0^t (\|f\|\cdot|||u|||_1 + |||u|||_1{}^2)\,\mathrm{d}t\Big\}.$$

Proof

Let φ be the same as in §4.2, i.e. φ is smooth, takes the value 0 near S and has its support in a neighbourhood of S. we set

$$v = \varphi\Lambda^{-1/2}\varphi u,$$

$$\Box v = \varphi\Lambda^{-1/2}\varphi f - [\varphi\Lambda^{-1/2}\varphi, \Box]u \equiv \Phi$$

and apply Green's formula (α) to obtain

$$\int_0^t (\Phi, v_n)\,\mathrm{d}t = \tfrac{1}{2}\int_0^t (\langle v_t\rangle^2 + \langle v_n\rangle^2 - \langle v_s\rangle^2)\,\mathrm{d}t + (v_t(t), v_n(t)) - (v_t(0), v_n(0)) + \tfrac{1}{2}\int_0^t \{-(v_t, v_t)^0 + (v_n, v_n)_{\tilde{\rho}} - (v_{\tilde{s}}, v_{\tilde{s}})_{\tilde{\rho}}\}\,\mathrm{d}t.$$

Therefore

$$\int_0^t \langle v_t\rangle^2 + \langle v_n\rangle^2\,\mathrm{d}t \leqq \int_0^t \langle v_s\rangle^2\,\mathrm{d}t + C\Big\{ |||v(t)|||_1{}^2 + |||v(0)|||_1{}^2 + \int_0^t (\|\Phi\|\cdot|||v|||_1 + |||v|||_1{}^2)\,\mathrm{d}t\Big\}.$$

Also, we have

$$\langle v_t \rangle = \langle u_t \rangle_{-1/2}, \quad \langle v_n \rangle = \langle u_n \rangle_{-1/2},$$
$$\langle v_s \rangle = \langle u_s \rangle_{-1/2} \leqq \langle u \rangle_{1/2} \leqq \|u\|_1$$

and

$$|||v|||_1 \leqq C|||u|||_1,$$
$$\|\Phi\| \leqq C(\|f\| + \|u\|_1).$$

From these facts we obtain the desired result. ■

Now,

$$\int_0^t \langle g, u_t \rangle \,\mathrm{d}t \leqq \int_0^t \langle g \rangle_{1/2} \langle u_t \rangle_{-1/2} \,\mathrm{d}t$$
$$\leqq \frac{1}{\varepsilon} \int_0^t \langle g \rangle_{1/2}{}^2 \,\mathrm{d}t + \varepsilon \int_0^t \langle u_t \rangle_{-1/2} \,\mathrm{d}t.$$

Applying Lemma 4.5 to the second term on the right hand side and considering the inequality.

$$\|u(t)\|^2 \leqq \|u(0)\|^2 + \int_0^t |||u|||_1{}^2 \,\mathrm{d}t$$

which appeared in the case of Dirichlet's problem, from the previous inequality (the result obtained after the application of Green's formula (γ); see the second equation of §4.3), we have

Proposition 4.6 *(the fundamental inequality for Neumann's problem)*

For $t > 0$,

$$|||u(t)|||_1{}^2 \leqq C\left\{\int_0^t \langle g \rangle_{1/2}{}^2 \,\mathrm{d}t + |||u(0)|||_1{}^2 + \int_0^t (\|f\| \cdot |||u|||_1 + |||u|||_1{}^2) \,\mathrm{d}t\right\}$$

where $\square u = f$, $(\partial u/\partial n)|_S = g$.

With the additional observation similar to the case of Dirichlet's problem we have

Corollary 1 of Proposition 4.6 *(the energy inequality (H) for Neumann's problem)*

There exist $\gamma_0 > 0$ and $C > 0$, for $\gamma \geq \gamma_0$, $t > 0$, such that

$$\text{(H)} \qquad e^{-2\gamma t}|||u(t)|||_{1,\gamma}{}^2 + \gamma \int_0^t e^{-2\gamma\tau}|||u(\tau)|||_{1,\gamma}{}^2 \,\mathrm{d}\tau$$

$$\leqq C\left\{\int_0^t e^{-2\gamma\tau}\langle g(\tau)\rangle_{1/2}{}^2\,d\tau + |||u(0)|||_{1,\gamma}{}^2 + \frac{1}{\gamma}\int_0^t e^{-2\gamma\tau}\|f(\tau)\|^2\,d\tau\right\}$$

where $\square u = f$, $(\partial u/\partial n)|_S = g$.

Corollary 2 of Proposition 4.6
For $t > 0$

$$|||u(t)|||_1{}^2 \leqq Ce^{2\gamma_0 t}\left\{\int_0^t \langle g(\tau)\rangle_{1/2}{}^2\,d\tau + |||u(0)|||_1{}^2 + \int_0^t \|f(\tau)\|^2\,d\tau\right\}$$

where $\square u = f$, $(\partial u/\partial n)|_S = g$.

4.4

So far we have completed the task of establishing the energy inequalities for both Dirichlet's and Neumann's problems as we originally planned. In this subsection, however, we briefly explore the problem concerning boundary conditions in their general form. It is hoped that the reader will gain some insight into energy inequalities. We wish to establish an energy inequality for the general initial value problem

$$\square u = f,$$
$$Bu|_S = g,$$
$$u|_{t=0} = u_0,\ (\partial u/\partial t)|_{t=0} = u_1$$

where

$$B = a\frac{\partial}{\partial n} + b\frac{\partial}{\partial s} + c\frac{\partial}{\partial t}$$

and a, b, c are real-valued functions on S such that

$$a^2 + b^2 = 1, \quad a > 0.$$

First, instead of Green's formulas (α), (β) (see §4.1), we consider the following alternative formulas $(\alpha)', (\beta)'$ assuming that the support of u is contained in a neighbourhood of S:

$$(\alpha)' \quad \int_0^t (\square u, \alpha u_n)\,dt - \int_0^t (-\alpha_t u_t + \alpha_s u_s + \alpha_n u_n,\ u_n)dt$$
$$= (u_t(t), \alpha(t)u_n(t)) - (u_t(0), \alpha(0)u_n(0))$$

$$+\tfrac{1}{2}\int_0^t \{\langle u_t, \alpha u_t\rangle - \langle u_s, \alpha u_s\rangle + \langle u_n, \alpha u_n\rangle\}\,dt$$

$$+\tfrac{1}{2}\int_0^t \{-u_t, \alpha u_t)_{\bar\rho} - (u_{\bar s}, \alpha u_{\bar s})_{\bar\rho} + (u_n, \alpha u_n)_{\bar\rho}\}\,dt$$

$$+\tfrac{1}{2}\int_0^t \{(u_t, \alpha_n u_t) - (u_{\bar s}, \alpha_n u_{\bar s}) - (u_n, \alpha_n u_n)\}\,dt,$$

$$(\beta)' \quad \int_0^t (\Box u, \beta u_{\bar s})\,dt - \int_0^t (-\beta_t u_t + \beta_{\bar s} u_{\bar s} + \beta_n u_n, u_{\bar s})\,dt$$

$$= (u_t(t), \beta(t) u_{\bar s}(t)) - (u_t(0), \beta(0) u_{\bar s}(0))$$

$$+ \int_0^t \langle u_n, \beta u_{\bar s}\rangle\,dt + \int_0^t (u_n, \beta u_{\bar s})_{\bar\rho}\,dt$$

$$+\tfrac{1}{2}\int_0^t \{(u_t, \beta_{\bar s} u_t) - (u_{\bar s}, \beta_{\bar s} u_{\bar s}) - (u_n, \beta_{\bar s} u_n)\}\,dt,$$

i.e. $(\alpha)'$, $(\beta)'$ are the weighted versions of $(\alpha), (\beta)$ leaving some degree of freedom. Consider $(\gamma) - (\alpha)' - (\beta)'$. We have

$$\int_0^t (\Box u, u_t - \alpha u_n - \beta u_{\bar s})\,dt$$

$$= \{\tfrac{1}{2}\|\nabla u(t)\|^2 - (u_t(t), \alpha(t) u_n(t)) - (u_t(t), \beta(t) u_{\bar s}(t))\}$$

$$- \{\tfrac{1}{2}\|\nabla u(0)\|^2 - (u_t(0), \alpha(0) u_n(0)) - (u_t(0), \beta(0) u_{\bar s}(0))\}$$

$$+ \int_0^t \Big\{\langle u_n, u_t\rangle - \langle u_n, \beta u_s\rangle - \tfrac{1}{2}(\langle u_t, \alpha u_t\rangle - \langle u_s, \alpha u_s\rangle + \langle u_n, \alpha u_n\rangle)\Big\}\,dt$$

$$+ \int_0^t [-\tfrac{1}{2}\{-(u_t, \alpha u_t)_{\bar\rho} - (u_{\bar s}, \alpha u_{\bar s})_{\bar\rho} + (u_n, \alpha u_n)_{\bar\rho}\}$$

$$-(-\alpha_t u_t + \alpha_{\bar s} u_{\bar s} + \alpha_n u_n, u_n) - (u_n, \beta u_{\bar s})_\rho$$

$$-\tfrac{1}{2}\{(u_t, \alpha_n u_t) - (u_{\bar s}, \alpha_n u_{\bar s}) - (u_n, \alpha_n u_n)\}$$

$$-(-\beta_t u_t + \beta_{\bar s} u_{\bar s} + \beta_n u_n, u_{\bar s})$$

$$-\tfrac{1}{2}\{(u_t, \beta_s u_t) - (u_{\bar s}, \beta_{\bar s} u_{\bar s}) - (u_n, \beta_{\bar s} u_n)\}]\,dt.$$

Letting

$$E = \tfrac{1}{2}\{\|u_n - \alpha u_t\|^2 + \|u_{\bar s} - \beta u_t\|^2 + ((1 - \alpha^2 - \beta^2) u_t, u_t)\},$$

$$F = \frac{1}{2}\left\langle \begin{bmatrix} -\alpha & -\beta & 1 \\ -\beta & \alpha & 0 \\ 1 & 0 & -\alpha \end{bmatrix} \begin{bmatrix} u_n \\ u_s \\ u_t \end{bmatrix}, \begin{bmatrix} u_n \\ u_s \\ u_t \end{bmatrix} \right\rangle,$$

$$G = \frac{1}{2}\left(\begin{bmatrix} -\alpha_n + \beta_{\bar{s}} - \alpha\tilde{\rho} & -\beta_n - \alpha_s - \beta\tilde{\rho} & \alpha_t \\ -\beta_n - \alpha_{\bar{s}} - \beta\tilde{\rho} & \alpha_n - \beta_{\bar{s}} + \alpha\tilde{\rho} & \beta_t \\ \alpha_t & \beta_t & -\alpha_n - \beta_{\bar{s}} + \alpha\tilde{\rho} \end{bmatrix}\begin{bmatrix} u_n \\ u_s \\ u_t \end{bmatrix}, \begin{bmatrix} u_n \\ u_{\bar{s}} \\ u_t \end{bmatrix}\right)$$

We see that the right hand side of the equation is

$$E(t) - E(0) + \int_0^t F(t)\,dt + \int_0^t G(t)\,dt.$$

On the other hand

$$Bu = au_n + bu_s + cu_t.$$

Hence

$$\begin{bmatrix} u_n \\ u_s \\ u_t \end{bmatrix} = \frac{1}{a}\begin{bmatrix} 1 & -b & -c \\ 0 & a & 0 \\ 0 & 0 & a \end{bmatrix}\begin{bmatrix} Bu \\ u_s \\ u_t \end{bmatrix}.$$

From this, we obtain

$$F = \frac{1}{2}\left\langle \frac{1}{a^2}\begin{bmatrix} -\alpha & b\alpha - a\beta & c\alpha + a \\ b\alpha - a\beta & (a^2 - b^2)\alpha + 2ab\beta & -bc\alpha + ca\beta - ab \\ c\alpha + a & -bc\alpha + ca\beta - ab & -(c^2 + a^2)\alpha - 2ac \end{bmatrix}\begin{bmatrix} Bu \\ u_s \\ u_t \end{bmatrix}, \begin{bmatrix} Bu \\ u_s \\ u_t \end{bmatrix}\right\rangle$$

To make this simpler we use the orthogonal transformation $(\alpha, \beta) \to (\alpha', \beta')$ such that

$$\begin{bmatrix} \alpha' \\ \beta' \end{bmatrix} = \begin{bmatrix} a & b \\ -b & a \end{bmatrix}\begin{bmatrix} \alpha \\ \beta \end{bmatrix}$$

to obtain

$$F = \frac{1}{2}\left\langle \frac{1}{a^2}\begin{bmatrix} -(a\alpha' - b\beta') & -\beta' & c(a\alpha' - b\beta') + a \\ -\beta' & a\alpha' + b\beta' & c\beta' - ab \\ c(a\alpha' - b\beta') + a & c\beta' - ab & -(a^2 + c^2)(a\alpha' - b\beta') - 2ac \end{bmatrix}\cdot\begin{bmatrix} Bu \\ u_s \\ u_t \end{bmatrix}\cdot\begin{bmatrix} Bu \\ u_s \\ u_t \end{bmatrix}\right\rangle$$

We now look for the domain of (α', β') such that

$$\beta = \begin{bmatrix} a\alpha' + b\beta' & c\beta' - ab \\ c\beta' - ab & -(a^2 + c^2)(a\alpha' - b\beta') - 2ac \end{bmatrix} \geqq 0.$$

There are three types of domain as follows:

(i) $$(1/a^2)\det(\beta) = (1-\alpha'^2-\beta'^2)(c^2-b^2)-(\alpha'+c)^2$$

$$= -(a^2+c^2)\left(\alpha'+\frac{c}{a^2+c^2}\right)^2$$

$$-(c^2-b^2)\beta'^2+\frac{a^2(c^2-b^2)}{a^2+c^2} \geqq 0,$$

(ii) $a\alpha'+b\beta' \geqq 0,$

(iii) $a\alpha'-b\beta' \leqq -\dfrac{2ac}{a^2+c^2}.$

In particular, setting

$$\alpha' = -\frac{c}{a^2+c^2}, \quad \beta'=0,$$

we see that for $\beta \geqq 0$,

$$c^2-b^2 \geqq 0, \quad c \leqq 0,$$

but for $\beta > 0$,

$$c^2-b^2>0, \quad c<0.$$

We now assume that the boundary S is compact, and consider the following two cases in S:

Case 1. $c^2-b^2>0, \quad c<0,$

Case 2. $c^2-b^2 \geqq 0, \quad c \leqq 0.$

We have, in case 1

$$F \geqq c\{\langle u_s\rangle^2+\langle u_t\rangle^2\}-C\langle g\rangle^2,$$

and in case 2

$$F \geqq -C\{\langle g\rangle^2+\langle g\rangle_{1/2}\langle u_t\rangle_{-1/2}\}.$$

If $c^2 \geqq b^2$, then

$$1-\alpha^2-\beta^2 = 1-\alpha'^2-\beta'^2 = 1-\left(\frac{c}{a^2+c^2}\right)^2$$

$$= \frac{a^2(a^2+c^2)+c^2(c^2-b^2)}{(a^2+c^2)^2}$$

$$\geqq \frac{a^2}{a^2+c^2} > 0,$$

therefore

$$c_1\|\nabla u\|^2 \leqq E \leqq c_2\|\nabla u\|^2$$

and, of course,

$$|G| \leqq C\|\nabla u\|^2.$$

So that in cases 1 and 2 we obtain

$$\|\nabla u(t)\|^2 + \int_0^t \langle u_t\rangle^2 + \langle u_y\rangle^2 \,\mathrm{d}t \leqq C\left\{\|\nabla u(0)\|^2 + \int_0^t \langle g\rangle^2 \,\mathrm{d}t + \int_0^t (\|f\|\cdot\|\nabla u\| + \|\nabla u\|^2)\,\mathrm{d}t\right\},$$

and

$$\|\nabla u(t)\|^2 \leqq C\left\{\|\nabla u(0)\|^2 + \int_0^t (\langle g\rangle_{1/2}\langle u_t\rangle_{-1/2} + \langle g\rangle^2 \,\mathrm{d}t + \int_0^t (\|f\|\cdot\|\nabla \mathrm{u}\| + \|\nabla \mathrm{u}\|^2)\,\mathrm{d}t\right\}$$

respectively.

If the support of u has no restrictions, we proceed as in the §4.2, i.e. we decompose u into two parts; one possessing its support in a neighbourhood of S and the other not, and apply the last inequalities to the former and Green's formula (γ) to the latter.

The result is

Proposition 4.7
In the above case 1 for $t > 0$

$$\int_0^t \langle\!\langle u\rangle\!\rangle_1{}^2 \,\mathrm{d}t + |||u(t)|||_1{}^2 \leqq C\left\{\int_0^t \langle g\rangle^2 \,\mathrm{d}t + |||u(0)|||_1{}^2 + \int_0^t (\|f\|\cdot|||u|||_1 + |||u|||_1{}^2)\,\mathrm{d}t\right\},$$

and in case 2, for $t > 0$

$$|||u(t)|||_1{}^2 \leqq C\left\{\int_0^t \langle g\rangle_{1/2}{}^2 \,\mathrm{d}t + |||u(0)|||_1{}^2 + \int_0^t (\|f\|\cdot|||u|||_1 + |||u|||_1{}^2)\,\mathrm{d}t\right\}$$

where

$$\Box u = f, \quad Bu|_S = g.$$

Notice that in the above cases 1 and 2 we obtained the fundamental inequalities of Dirichlet and Neumann types, respectively. Furthermore, we can easily see that in case 1

$$\int_0^t \langle\!\langle u(t)\rangle\!\rangle_1{}^2\,\mathrm{d}t + |||u(t)|||_1{}^2 \leqq C\mathrm{e}^{ct}\left\{\int_0^t \langle g(t)\rangle^2\,\mathrm{d}t + |||u(0)|||_1{}^2 + \int_0^t \|f(t)\|^2\,\mathrm{d}t\right\}$$

and in case 2.

$$|||u(t)|||_1{}^2 \leqq C\mathrm{e}^{ct}\left\{\int_0^t \langle g(t)\rangle_{1/2}{}^2\,\mathrm{d}t + |||u(0)|||_1{}^2 + \int_0^t \|f(t)\|^2\,\mathrm{d}t\right\}.$$

2

Hyperbolic Boundary Value Problems With Constant Coefficients

In chapter 1 we discussed a typical hyperbolic initial boundary value problem by a method which was, in a sense, technically complicated. We might naturally ask if there is any other way of dealing with the same type of problem that gives better perspectives. In this chapter therefore we take a different stand; we set the problem in a mathematically 'nicer' way allowing the use of a more general method in order to gain better insight.

To begin with, we assume that all equations are defined in a half-space with constant coefficients and, at the same time, all boundary conditions are given by linear partial differential operators with, again, constant coefficients. In such cases we know that, generally speaking, the problem can be reduced to the one of an ordinary differential equation with constant coefficients which is defined over a half-line, by using a Fourier–Laplace transform. Then, in the main, we concentrate our attention on the behaviour of the parameters which are introduced when a Fourier–Laplace transform is used. In other words, the behaviour of these parameters is a decisive factor in obtaining a successful transform.

1. Fourier–Laplace transforms

In this section, we present a minimal 'tool-kit' of Fourier–Laplace transforms which we intend to use.

1.1

Before giving the definition of a Fourier–Laplace transform, we shall briefly sketch the notions of 'functional space' and 'Fourier transform'.

The definitions of $\mathscr{D}$, $\mathscr{D}'$

In the following, we use 'Schwartz distributions' as a general framework for our work on families of functions. We therefore start with a definition of this useful concept. Consider the family of complex-valued functions φ on $\mathbb{R}^n$ which are infinitely differentiable and possess compact supports. Then it is clear that the entire family of such functions forms a vector space, which we denote by $\mathscr{D}$. For $j = 1, 2, \ldots$, by

$$\varphi_j \to 0 \ (\mathscr{D}) \quad \text{as } j \to \infty$$

we mean that the union of the supports of φ_j is bounded, and, at the same time, for an arbitrary $\nu = (\nu_1, \ldots, \nu_n)$,

$$\sup_{x \in \mathbb{R}^n} |\mathrm{D}^\nu \varphi_j(x)| \to 0 \quad \text{as } j \to \infty.$$

Notice that the notation D^ν is used according to the familiar convention:

$$\mathrm{D}^\nu = \mathrm{D}_{x_1}{}^{\nu_1} \ldots \mathrm{D}_{x_n}{}^{\nu_n}, \quad \mathrm{D}_{x_j} = \frac{1}{\mathrm{i}} \frac{\partial}{\partial x_j} \quad (\mathrm{i} = \sqrt{(-1)}).$$

Henceforth we shall follow this convention for denoting a differential operator.

Then, the dual of $\mathscr{D}$, written $\mathscr{D}'$, is defined as the vector space of all continuous linear functionals defined over $\mathscr{D}$, which means that for each $f \in \mathscr{D}'$ and $\varphi \in \mathscr{D}$ there is a corresponding complex number $\langle f, \varphi \rangle$, which satisfies the following conditions:

$$\text{Continuity } \varphi_j \to 0 (\mathscr{D}) \Rightarrow \langle f, \varphi_j \rangle \to 0,$$

$$\text{Linearity } \langle f, c_1\varphi_1 + c_2\varphi_2 \rangle = c_1 \langle f, \varphi_1 \rangle + c_2 \langle f, \varphi_2 \rangle$$

(c_1, c_2 are complex numbers).

On the other hand, by

$$f_j \to 0 \ (\mathscr{D}') \quad \text{as} \quad j \to \infty$$

we mean that for an arbitrary $\varphi \in \mathscr{D}$,

$$\langle f, \varphi \rangle \to 0 \qquad \text{as } j \to \infty.$$

We call each element of $\mathscr{D}'$ a *distribution.*

In particular, if f is a locally differentiable function of $\mathbb{R}^n$, then a distribution f can be defined by

$$\mathscr{D} \ni \varphi \mapsto \langle f, \varphi \rangle = \int f(x) \varphi(x) \, \mathrm{d}x.$$

Note that, in this sense, obviously $\mathscr{D} \subset \mathscr{D}'$.

We can define the support of a distribution as follows: Consider a distribution $f \in \mathscr{D}'$ such that $\langle f, \varphi \rangle = 0$ is true for a function $\varphi \in \mathscr{D}$ whose support is contained in an open set U of $\mathbb{R}^n$. Then, such f is said to be *zero on the open set U*. It is easy to see that if f is zero on each U_α, then f is also zero on $\bigcup_\alpha U_\alpha$. Therefore, there exists a maximal open set with this property. We call the complement of this set the *support of the distribution* f, written supp $[f]$.

The definitions of $\mathscr{S}$, $\mathscr{S}'$

Following L. Schwartz we introduce function spaces $\mathscr{S}$, $\mathscr{S}'$ which play important rôles in Fourier transforms. Let us write $\mathscr{S}$ for the vector space of all infinitely differentiable complex-valued functions defined on $\mathbb{R}^n$ whose derivatives remain bounded on $\mathbb{R}^n$ even if they are multiplied by any polynomials. We can introduce a semi-norm in $\mathscr{S}$

$$p_m(\varphi) = \sup_{x \in \mathbb{R}^n} (1 + |x|)^m \sum_{|\nu| \leq m} \left| \left(\frac{\partial}{\partial x} \right)^\nu \varphi(x) \right| \quad (m = 0, 1, 2, \ldots).$$

Now, by

$$\varphi_j \to 0 \, (\mathscr{S}) \quad \text{as } j \to \infty$$

we mean that for arbitrary m,

$$p_m(\varphi_j) \to 0 \quad \text{as } j \to \infty.$$

The dual $\mathscr{S}'$ of $\mathscr{S}$ is similarly defined as the vector space of all continuous linear functionals. That is, for each $f \in \mathscr{S}'$ and $\varphi \in \mathscr{S}$ there is a corresponding complex number $\langle f, \varphi \rangle$, which satisfies the following conditions:

$$\text{Continuity } \varphi_j \to 0 \, (\mathscr{S}) \Rightarrow \langle f, \varphi_j \rangle \to 0,$$

$$\text{Linearity } \langle f, c_1\varphi_1 + c_2\varphi_2 \rangle = c_1 \langle f, \varphi_1 \rangle + c_2 \langle f, \varphi_2 \rangle$$

$$(c_1, c_2 \text{ are complex numbers}).$$

Also, by

$$f_j \to 0 \, (\mathscr{S}') \quad \text{as } j \to \infty$$

we mean for any $\varphi \in \mathscr{S}$,

$$\langle f_j, \varphi \rangle \to 0 \quad \text{as } j \to \infty.$$

Obviously $\mathscr{S}' \subset \mathscr{D}'$. We call each element of $\mathscr{S}'$ a *tempered distribution.*

Note that when we defined the continuity of $f \in \mathscr{S}'$ we meant it in the weak sense, but we can demonstrate that this implies continuity in the stronger sense of the topology defined by the semi-norm in $\mathscr{S}$. To see this we prove

Lemma 1.1

For $f \in \mathscr{S}'$ there exist C and m such that

$$|\langle f, \varphi \rangle| \leq C p_m(\varphi), \quad \varphi \in \mathscr{S}.$$

Proof

Assume that this is not true. Then, for $j = 1, 2, \ldots$, there exist $\varphi_j \in \mathscr{S}$ such that

$$|\langle f, \varphi_j \rangle| \geqq j p_j(\varphi_j).$$

This implies

$$\left| \left\langle f, \frac{1}{j} \frac{\varphi_j}{p_j(\varphi_j)} \right\rangle \right| \geqq 1, \quad j = 1, 2, \ldots.$$

But for arbitrary m, we have

$$p_m\left(\frac{1}{j}\frac{\varphi_j}{p_j(\varphi_j)}\right) = \frac{1}{j}\frac{p_m(\varphi_j)}{p_j(\varphi_j)} \leqq \frac{1}{j}, \quad j \geqq m.$$

Hence

$$\frac{1}{j}\frac{\varphi_j}{p_j(\varphi_j)} \to 0\ (\mathscr{S}) \quad \text{as } j \to \infty.$$

Therefore, from the continuity of f in the weak sense, we have

$$\left\langle f, \frac{1}{j}\frac{\varphi_j}{p_j(\varphi_j)} \right\rangle \to 0 \quad \text{as } j \to \infty.$$

a contradiction. ∎

We now introduce the Fourier transforms of distributions into $\mathscr{S}$. First, for $\varphi \in \mathscr{S}$ we define a bounded continuous function $\mathscr{F}\varphi$ of $\mathbb{R}^n$ † such that

$$(\mathscr{F}\varphi)(\xi) = \int_{\mathbb{R}^n} e^{-ix\cdot\xi}\varphi(x)\,dx, \quad x\cdot\xi = \sum_{j=1}^{n} x_j\xi_j, \quad \xi \in \mathbb{R}^n.$$

Then, for an arbitrary ν we have

$$\text{(1)} \qquad \begin{cases} \mathscr{F}[D_x{}^\nu\varphi](\xi) = \xi^\nu \mathscr{F}[\varphi](\xi), \\ \mathscr{F}[x^\nu\varphi](\xi) = (-D_\xi)^\nu \mathscr{F}[\varphi](\xi). \end{cases}$$

This shows that $\mathscr{F}\varphi \in \mathscr{S}_\xi$, and $\mathscr{F}$ is a continuous linear map from $\mathscr{S}_x$ into $\mathscr{S}_\xi$ such that $\mathscr{S}_x \ni \varphi \to \mathscr{F}\varphi \in \mathscr{S}_\xi$, where $\mathscr{S}_{x(\text{or } \xi)} (= \mathscr{S})$ indicates that the independent variables of the functions ($\in \mathscr{S}$) are x(or ξ). We call $\mathscr{F}$ a *Fourier transform over* $\mathscr{S}$. Furthermore, By Fourier's inversion formula‡ $\mathscr{F}$ gives a bijection from $\mathscr{S}_x$ onto $\mathscr{S}_\xi$. The inverse map $\bar{\mathscr{F}}$ of $\mathscr{F}$ is given

† Translator's note: $\mathscr{F}\varphi$ is a function of the dual $(\mathbb{R}^n)'$ of $\mathbb{R}^n$, but, by identification, $\mathscr{F}\varphi$ becomes a function of $\mathbb{R}^n$.

‡ Translator's note: See L. Schwartz, *Mathematics for the physical sciences*. Addison–Wesley, Reading, Mass. (1966); or Y. Yoshida, *Functional analysis* (second edition). Springer, Berlin (1968).

by

$$(\bar{\mathscr{F}}\psi)(x) = \frac{1}{(2\pi)^n}\int_{\mathbb{R}^n} e^{ix\cdot\xi}\psi(\xi)\,d\xi, \quad \psi\in\mathscr{S}_\xi.$$

For $a\in\mathbb{R}^n$, letting

$$(\tau_a\varphi)(x) = \varphi(x-a)$$

we have

(2) $$\begin{cases} \mathscr{F}[\tau_a\varphi](\xi) = e^{-ia\cdot\xi}\mathscr{F}[\varphi](\xi), \\ \mathscr{F}[e^{ia\cdot x}\varphi](\xi) = (\tau_a\mathscr{F}[\varphi])(\xi). \end{cases}$$

We also see that an important property in L^2-theory

$$\|f\|_{L_x^2} = (2\pi)^{-n/2}\|\mathscr{F}f\|_{L_\xi^2}$$

is valid.

So far we have seen that $\mathscr{F}$ is a continuous linear map defined over $\mathscr{S}$, and so are

$$\left(\frac{\partial}{\partial x}\right)^\nu : \mathscr{S}\ni\varphi(x)\to\left(\frac{\partial}{\partial x}\right)^\nu\varphi(x)\in\mathscr{S},$$

$$\alpha(x) : \mathscr{S}\ni\varphi(x)\to\alpha(x)\varphi(x)\in\mathscr{S},$$

$$\tau_a : \mathscr{S}\ni\varphi(x)\to\varphi(x-a)\in\mathscr{S}$$

(here the independent variable x is made explicit) where α is an infinitely differentiable function such that for every ν the inequality

$$|(\partial/\partial x)^\nu\alpha(x)| \leqq C_\nu(1+|x|)^{k_\nu}$$

holds. In general, if we write such a continuous linear map defined over $\mathscr{S}$ as T, then T induces a continuous linear map T' over $\mathscr{S}'$ defined by

$$\langle T'f,\varphi\rangle = \langle f, T\varphi\rangle, \quad f\in\mathscr{S}', \quad \varphi\in\mathscr{S}.$$

We call T' the *transpose map* of T.

Using this terminology, we can say that for a Fourier transform $\mathscr{F}$ over $\mathscr{S}$, the transpose map $\mathscr{F}'$ over $\mathscr{S}'$ is defined as

$$\langle\mathscr{F}'f,\varphi\rangle = \langle f,\mathscr{F}\varphi\rangle, \quad f\in\mathscr{S}', \quad \varphi\in\mathscr{S}.$$

Since $\mathscr{S}\subset\mathscr{S}'$, for $f\in\mathscr{S}$,

$$\langle\mathscr{F}'f,\varphi\rangle = \langle f,\mathscr{F}\varphi\rangle = \int f(x)(\mathscr{F}\varphi)(x)\,dx = \int f(x)\left(\int e^{-ix\cdot\xi}\varphi(\xi)\,d\xi\right)dx$$

$$= \int\varphi(\xi)\left(\int e^{-ix\cdot\xi}f(x)\,dx\right)d\xi = \langle\mathscr{F}f,\varphi\rangle, \quad \varphi\in\mathscr{S},$$

therefore $\mathscr{F}'f = \mathscr{F}f$. This shows that the transpose map $\mathscr{F}'$ over $\mathscr{S}'$

induced by $\mathscr{F}$ is an extension of the Fourier transform $\mathscr{F}$ over $\mathscr{S}$ to $\mathscr{S}'$. Let us write $\mathscr{F}$ instead of $\mathscr{F}'$, and call this $\mathscr{F}$ a *Fourier transform over* $\mathscr{S}'$. Similarly, we write the transposed maps of $\mathscr{F}$, α, $(\partial/\partial x)^\nu$, τ_a as $\mathscr{F}$ (unchanged), α (unchanged), $(-\partial/\partial x)^\nu$, τ_{-a} respectively. Thus, we obtain the extensions of these continuous linear maps over $\mathscr{S}$ to $\mathscr{S}'$. In this way, we can see that the formulas (1), (2) for a Fourier transform over $\mathscr{S}$ are also valid for the corresponding Fourier transform over $\mathscr{S}'$. In particular, if an element of $\mathscr{S}'$ is representable in the following sense, then the image of its Fourier transform can be represented by an infinitely differentiable function, i.e.

Lemma 1.2

If $f = \alpha(x) f_1$, $\alpha \in \mathscr{S}_x, f_1 \in \mathscr{S}'$, then

$$\mathscr{F} f = \langle f, e^{-ix\cdot\xi} \rangle,$$

where

$$\langle f, e^{-ix\cdot\xi} \rangle = \langle \alpha(x) f_1, e^{-ix\cdot\xi} \rangle = \langle f_1, \alpha(x) e^{-ix\cdot\xi} \rangle.$$

In this case, $\mathscr{F} f$ is an infinitely differentiable function such that for any ν,

$$|(\partial/\partial\xi)^\nu (\mathscr{F} f)(\xi)| \leqq C_\nu (1 + |\xi|)^m.$$

Proof

Note that $f_1 \in \mathscr{S}_x'$, $\alpha e^{-ix\cdot\xi} \in \mathscr{S}_x$.

By setting

$$g(\xi) = \langle f_1, \alpha(x) e^{-ix\cdot\xi} \rangle_x, \quad \xi \in \mathbb{R}^n,$$

from Lemma 1.1, we have

$$|g(\xi)| \leqq C p_m(\alpha(x) e^{-ix\cdot\xi}) \leqq C(1 + |\xi|)^m.$$

Notice that we can perform differentiation of $\langle f_1, \alpha e^{-ix\cdot\xi} \rangle$ with respect to ξ within $\langle , \rangle$. In fact, by letting a be a real number and $e^{(l)} = (0, \ldots, 0, 1, 0, \ldots, 0)$, where the 1 is in the lth position, we see that

$$\alpha(x) \left(\frac{e^{-ix\cdot(\xi + ae^{(l)})} - e^{-ix\cdot\xi}}{a} + ix_l e^{-ix\cdot\xi} \right)$$
$$= \alpha(x) e^{-ix\cdot\xi} (-ix_l)^2 a \int_0^1 (1 - s) e^{-ix_l as} \, ds.$$

Therefore,

$$p_m\left(\alpha(x) e^{-ix\cdot\xi} (-ix_l)^2 a \int_0^1 (1 - s) e^{-ix_l as} \, ds \right) \to 0 \quad \text{as } a \to 0.$$

Hence,

$$\frac{\partial}{\partial \xi_l} \langle f_1, \alpha(x) e^{-ix\cdot\xi} \rangle = \langle f, \alpha(x)(-ix_l) e^{-ix\cdot\xi} \rangle.$$

In general

$$\left|\left(\frac{\partial}{\partial\xi}\right)^{\nu} g(\xi)\right| = |\langle f_1, \alpha(x)(-\mathrm{i}x)^{\nu}\mathrm{e}^{-\mathrm{i}x\cdot\xi}\rangle| \leqq Cp_m(\alpha(x)(-\mathrm{i}x)^{\nu}\mathrm{e}^{-\mathrm{i}x\cdot\xi})$$
$$\leqq C_{\nu}(1+|\xi|)^m.$$

On the other hand, for $\varphi \in \mathscr{S}_{\xi}$ we have

$$\langle g(\xi), \varphi(\xi)\rangle_{\xi} = \int g(\xi)\varphi(\xi)\,\mathrm{d}\xi = \int \langle f_1, \alpha(x)\mathrm{e}^{-\mathrm{i}x\cdot\xi}\rangle_x \varphi(\xi)\,\mathrm{d}\xi$$
$$= \left\langle f_1, \alpha(x)\int \mathrm{e}^{-\mathrm{i}x\cdot\xi}\varphi(\xi)\,\mathrm{d}\xi\right\rangle_x$$

from a similar observation in the case of differentiation. Therefore

$$\langle g(\xi), \varphi(\xi)\rangle_{\xi} = \langle f_1, \alpha(x)(\mathscr{F}\varphi)(x)\rangle_x = \langle f, (\mathscr{F}\varphi)(x)\rangle_x = \langle \mathscr{F}f, \varphi(x)\rangle_x.$$

That is,

$$\mathscr{F}f = g(\xi). \quad \blacksquare$$

1.2

Let us introduce the Fourier–Laplace transform as the extension of the Fourier transform. Recall that in §1.1 a Fourier transform was regarded as a map from a function or a distribution of $\mathbb{R}^n_x$ to a function or a distribution of $\mathbb{R}^n_{\xi}$, where $\mathbb{R}^n_{x(\text{or }\xi)}(=\mathbb{R}^n)$ means that each element of $\mathbb{R}^n$ is represented by the variable x (or ξ). For this reason, let us consider $\mathbb{R}^n$ as embedded in the complex n-dimensional space $\mathbb{C}^n$.

For example, if $\varphi \in \mathscr{D}_x$, then

$$(\mathscr{F}\varphi)(\xi) = \int \mathrm{e}^{-\mathrm{i}x\cdot\xi}\varphi(x)\,\mathrm{d}x, \quad \xi \in \mathbb{R}^n$$

whose analytic continuation to the entire space $\mathbb{C}^n$ exists in the form of an entire function. We also have

$$(\mathscr{F}\varphi)(\xi+\mathrm{i}\eta) = \int \mathrm{e}^{-\mathrm{i}x\cdot(\xi+\mathrm{i}\eta)}\varphi(x)\mathrm{d}x = \int \mathrm{e}^{-\mathrm{i}x\cdot\xi}\cdot\mathrm{e}^{x\cdot\eta}\varphi(x)\mathrm{d}x$$
$$= \mathscr{F}[\mathrm{e}^{x\cdot\eta}\varphi(x)](\xi),$$

so that, for a fixed imaginary part η, the Fourier transform $\mathscr{F}$ can be regarded as a map which associates $\mathrm{e}^{x\cdot\eta}\varphi(x)$ with a function defined over $\mathbb{R}^n_{\xi} + \mathrm{i}\eta$, by a parallel translation of the real part of $\mathbb{C}^n$.

Let φ be infinitely differentiable. By a *Fourier–Laplace transform* of φ, written $\mathscr{L}[\varphi]$, or $\mathscr{L}\varphi$, we mean a transform such that

$$\mathscr{L}[\varphi](\xi+\mathrm{i}\eta) = \mathscr{F}[\mathrm{e}^{x\cdot\eta}\varphi(x)] = \int_{\mathbb{R}^n} \mathrm{e}^{-\mathrm{i}x\cdot(\xi+\mathrm{i}\eta)}\varphi(x)\,\mathrm{d}x,$$

provided that there exists $\eta \in \mathbb{R}^n$ satisfying $e^{x \cdot \eta}\varphi \in \mathscr{S}_x$. That is, if $\Delta_\varphi = \{\eta \in \mathbb{R}^n | e^{x \cdot \eta}\varphi \in \mathscr{S}_x\}$, then $\mathscr{L}[f](\xi + i\eta)$ is a function defined over $\mathbb{R}^n + i\Delta_\varphi$ such that, for each fixed imaginary part η, $\mathscr{L}[f]$ belongs to $\mathscr{S}_\xi$ for some $\xi \in \mathbb{R}^n$. By extending this argument to the case of distributions, we obtain a definition of the Fourier–Laplace transform for distributions. To see this we set $\Delta_f = \{\eta \in \mathbb{R}^n | e^{x \cdot \eta} f \in \mathscr{S}_x', f \in \mathscr{D}'\}$, and define

$$(\mathscr{L}f)(\xi + i\eta) = \mathscr{F}[e^{x \cdot \eta} f], \quad \xi + i\eta \in \mathbb{R}^n + i\Delta_f.$$

We see that Δ_f is a convex set because if

$$\frac{1}{p} + \frac{1}{q} = 1, \quad p > 0, \quad q > 0,$$

then we have

$$e^{x(\eta/p + \eta'/q)} = (e^{x \cdot \eta})^{1/p}(e^{x \cdot \eta'})^{1/q} \leqq \frac{e^{x \cdot \eta}}{p} + \frac{e^{x \cdot \eta'}}{q}.$$

In the following we see that if Δ_f possesses interior points, some distinct relations exist between a Fourier–Laplace transform and its inverse. We start by proving

Lemma 1.3

Let $\eta \in \mathring{\Delta}_f$ (the interior of Δ_f). For an arbitrary ε such that

$$0 < \varepsilon < \operatorname{dis}(\eta, \partial\Delta_f), \quad e^{\varepsilon\sqrt{\{1 + |x|^2\}} + x \cdot \eta} f \in \mathscr{S}'.$$

Proof

We put $\operatorname{dis}(\eta^0, \partial\Delta_f) = d$, and let $0 < \varepsilon < \varepsilon_1 < d$. Consider $\varphi_j(x)$ which is a homogeneous function of degree zero and is infinitely differentiable for $x \neq 0$. Assume that the support of $\varphi_j(x)$ is contained in a sufficiently small angular domain, such that the angle θ between any pair of points in the support of $\varphi_j(x)$ satisfies the inequality $\cos\theta > \varepsilon/\varepsilon_1$ and, at the same time,

$$\sum_{j=1}^{N} \varphi_j(x) = 1.$$

In a neighbourhood of $x = 0$, we take a function $\varphi_0 \in \mathscr{D}$ such that $\varphi_0(x) = 1$ near $x = 0$, and write

$$\tilde{\varphi}_j(x) = \varphi_j(x)(1 - \varphi_0(x)).$$

(See Fig. 3.) Then we see that $\tilde{\varphi}_j \in \mathscr{B}$, where $\mathscr{B}$ represents the family of functions whose derivatives are bounded.

Choose a point $c^{(j)} \in \operatorname{supp}[\varphi_j(x)]$ such that $|c^{(j)}| = 1$. Then, letting $\eta^{(j)} = \eta^0 + \varepsilon_1 c^{(j)}$ we see that $\eta^{(j)} \in \Delta_f$, and therefore $e^{x \cdot \eta^{(j)}} f \in \mathscr{S}'$.

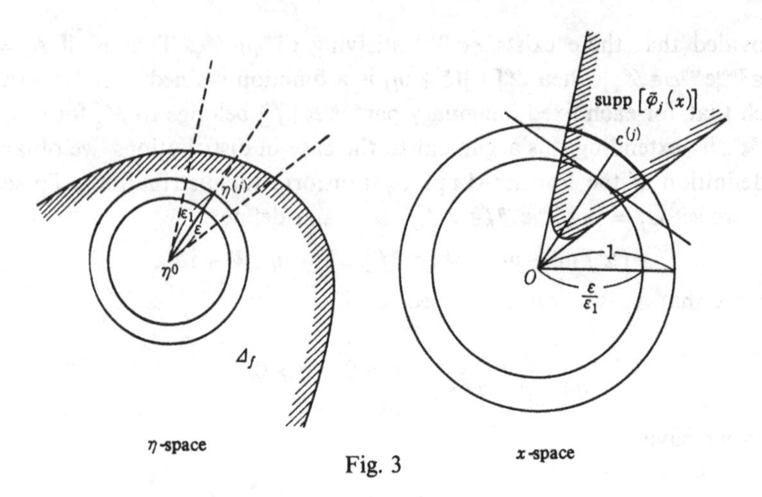

Fig. 3

On the other hand, over the support of $\varphi_j(x)$ we have

$$\varepsilon_1 c^{(j)}\cdot x > \varepsilon|x|.$$

From this we see that

$$e^{\varepsilon\sqrt{\{1+|x|^2\}}-\varepsilon_1 c^{(j)}\cdot x}\tilde{\varphi}_j \in \mathscr{B}.$$

Hence

$$e^{\varepsilon\sqrt{\{1+|x|^2\}}+x\cdot\eta^0}\tilde{\varphi}_j f = (e^{\varepsilon\sqrt{\{1+|x|^2\}}-\varepsilon_1 c^{(j)}\cdot x}\tilde{\varphi}_j)(e^{x\cdot\eta^{(j)}}f) \in \mathscr{S}'.$$

We conclude that

$$e^{\varepsilon\sqrt{\{1+|x|^2\}}+x.\eta^0}f = \sum_{j=1}^{N} e^{\varepsilon\sqrt{\{1+|x|^2\}}+x\cdot\eta^0}\tilde{\varphi}_j f$$

$$+ e^{\varepsilon\sqrt{\{1+|x|^2\}}+x\cdot\eta^0}\varphi_0 f \in \mathscr{S}'. \quad \blacksquare$$

By Lemma 1.3 (which we have just obtained), if $\eta \in \mathring{\Delta}_f, 0 < \varepsilon < \operatorname{dis}(\eta, \partial\Delta_f)$, then, letting

$$e^{\varepsilon\sqrt{\{1+|x|^2\}}}f = f_1, \quad e^{-\varepsilon\sqrt{\{1+|x|^2\}}} = \alpha(x),$$

we have the following equality

$$e^{x\cdot\eta}f = \alpha(x)\cdot e^{x\cdot\eta}f_1, \quad \alpha(x) \in \mathscr{S}, \quad e^{x\cdot\eta}f_1 \in \mathscr{S}'.$$

Therefore, from Lemma 1.2 we see that

$$(\mathscr{L}f)(\xi + i\eta) = \mathscr{F}[e^{x\cdot\eta}f](\xi) = \langle e^{x\cdot\eta}f_1, \alpha(x)e^{-ix\cdot\xi}\rangle = \langle f, e^{-ix\cdot(\xi+i\eta)}\rangle.$$

This is a basic representation of $\mathscr{L}f$ in the case where Δ_f possesses interior points. We shall see the importance of such 'direct' representations later when we prove the three fundamental theorems about Fourier–Laplace transforms. But, first of all, let us explore the link between Fourier–Laplace transforms and their inverses in general.

Theorem 1(a)

Let $f\in\mathscr{D}'$ and $\mathring{\Delta}\neq\phi$. $\mathring{\Delta}_f=\Omega, g(\zeta)=(\mathscr{L}f)(\zeta)$, then the following statements are true:

(i) $g(\zeta)$ is regular in $\mathbb{R}^n+\mathrm{i}\Omega$,

(ii) for any compact set $K\subset\Omega$,

$$|g(\zeta)|\leqq C_k(1+|\xi|)^{m_k},\quad \zeta=\xi+\mathrm{i}\eta\in\mathbb{R}^n+\mathrm{i}K.$$

Conversely, if there exist a domain Ω and a function $g(\zeta)$ satisfying the above conditions (i) and (ii), then the inverse transform $\mathscr{L}^{-1}g$ defined by

$$(\mathscr{L}^{-1}g)(x)=\mathrm{e}^{-x\cdot\eta}\bar{\mathscr{F}}_{\xi\to x}[g(\xi+\mathrm{i}\eta)],\quad \eta\in\Omega$$

is a constant element f of $\mathscr{D}_x'$ for all $\eta\in\Omega$; and $\mathscr{L}f=g$ in $\mathbb{R}^n+\mathrm{i}\Omega$.

Proof

(i) We shall prove the differentiability of $(\mathscr{L}f)(\zeta)$ at $\zeta^0\in\mathbb{R}^n+\mathrm{i}\Omega$. To this end, we let $0<\varepsilon<\operatorname{dis}(\operatorname{Im}\zeta^0,\partial\Omega)$. Then, for $|\zeta-\zeta^0|<\varepsilon$, we have

$$(\mathscr{L}f)(\zeta)=\langle \mathrm{e}^{\varepsilon\sqrt{\{1+|x|^2\}}-\mathrm{i}x\cdot\zeta^0}f,\ \mathrm{e}^{-\varepsilon\sqrt{\{1+|x|^2\}}-\mathrm{i}x\cdot(\zeta-\zeta^0)}\rangle,$$

$$\mathrm{e}^{\varepsilon\sqrt{\{1+|x|^2\}}-\mathrm{i}x\cdot\zeta^0}f\in\mathscr{S}',\quad \mathrm{e}^{-\varepsilon\sqrt{\{1+|x|^2\}}-\mathrm{i}x\cdot(\zeta-\zeta^0)}\in\mathscr{S}.$$

Let $\mathrm{e}^{(l)}=(0,\ldots 0,1,0,\ldots,0)$, where the 1 is in the lth position. For $z\in\mathbb{C}^1, |z|<\varepsilon$, let

$$I_z=\frac{1}{z}\{(\mathscr{L}f)(\zeta^0+z\mathrm{e}^{(l)})-(\mathscr{L}f)(\zeta^0)\}-\mathscr{L}[-\mathrm{i}x_lf](\zeta^0)$$

$$=\left\langle \mathrm{e}^{\varepsilon\sqrt{\{1+|x|^2\}}-\mathrm{i}x\cdot\zeta^0}f,\ \mathrm{e}^{-\varepsilon\sqrt{\{1+|x|^2\}}}\left\{\frac{\mathrm{e}^{-\mathrm{i}x_lz}-1}{z}+\mathrm{i}x_l\right\}\right\rangle$$

$$=\left\langle \mathrm{e}^{\varepsilon\sqrt{\{1+|x|^2\}}-\mathrm{i}x\cdot\zeta^0}f,\ \mathrm{e}^{-\varepsilon\sqrt{\{1+|x|^2\}}}(-\mathrm{i}x_l)^2z\int_0^1(1-s)^{-\mathrm{i}x_lzs}\,\mathrm{d}s\right\rangle.$$

Then we see that

$$|I_z|\leqq Cp_m\left(\mathrm{e}^{-\varepsilon\sqrt{\{1+|x|^2\}}}(-\mathrm{i}x_l)^2z\int_0^1(1-s)\mathrm{e}^{-\mathrm{i}x_lzs}\,\mathrm{d}s\right)\to 0\quad\text{as } z\to 0.$$

This proves (i).

(ii) For $|\eta-\eta^0|<\varepsilon/2$, we have

$$|(\mathscr{L}f)(\zeta)|\leqq Cp_m(\mathrm{e}^{-\varepsilon\sqrt{\{1+|x|^2\}}-\mathrm{i}x\cdot(\zeta-\zeta^0)})\leqq C'(1+|\xi|)^m.$$

Therefore, since the compact set $K\subset\Omega$ can be covered with a finite number of such balls of $|\eta-\eta^0|<\varepsilon/2$, (ii) is true.

For the converse, let

$$\mathrm{e}^{-x\cdot\eta}\bar{\mathscr{F}}_{\xi\to x}[g(\xi+\mathrm{i}\eta)]=f_\eta,\quad \eta\in\Omega.$$

Since $e^{x\cdot\eta}f_\eta \in \mathscr{S}_x'$, we have $f_\eta \in \mathscr{D}'$. For $\varphi \in \mathscr{D}$, we have

$$\langle f_\eta, \varphi(x)\rangle = \langle \mathscr{F}_{\xi\to x}[g(\xi+i\eta)], e^{-x\cdot\eta}\varphi(x)\rangle_x$$
$$= \langle g(\xi+i\eta), \bar{\mathscr{F}}_{x\to\xi}[e^{-x\cdot\eta}\varphi(x)]\rangle_\xi.$$

Letting

$$\bar{\mathscr{F}}_{x\to\xi}[e^{-x\cdot\eta}\varphi(x)] = \frac{1}{(2\pi)^n}\int e^{ix\cdot\xi}e^{-x\cdot\eta}\varphi(x)\,dx$$
$$= \frac{1}{(2\pi)^n}\int e^{ix\cdot(\xi+i\eta)}\varphi(x)\,dx = \psi(\xi+i\eta),$$

we see that $\psi(\zeta)$ is an entire function of ζ, and $\psi(\xi+i\eta)\in\mathscr{S}_\xi$ for every η. Therefore we have

$$\langle g(\xi+i\eta), \psi(\xi+i\eta)\rangle_\xi = \int_{\mathbb{R}^n} g(\xi+i\eta)\psi(\xi+i\eta)\,d\xi.$$

From Cauchy's integral formula, we see that the right hand side of the equation does not depend on $\eta\in\Omega$. Hence f_η does not depend on $\eta\in\Omega$. Setting $f_\eta = f$, obviously we have $\mathscr{L}f = g$. This completes the proof. ■

In particular, if we replace $\mathscr{S}'$ in Theorem 1(a) with $\mathscr{S}$, the following fact can be easily established.

Theorem 1(b)
Let $\Delta_f \neq \varnothing$ and $e^{\eta\cdot x}f(x)\in\mathscr{S}$ for an arbitrary $\eta\in\mathring{\Delta}_f$. If $\mathring{\Delta}_f = \Omega$, $g(\zeta) = (\mathscr{L}f)(\zeta)$, then the following statements are true:

(i) $g(\zeta)$ is regular in $\mathbb{R}^n + i\Omega$,
(ii) for any compact set $K\subset\Omega$, and any $N>0$,

$$|g(\zeta)| \leqq C_{K,N}(1+|\xi|)^{-N}, \quad \zeta = \xi + i\eta \in \mathbb{R}^n + iK.$$

Conversely, if there exist a domain Ω and a function $g(\zeta)$ such that they satisfy the above conditions (i) and (ii), then, for $\eta\in\Omega$, $e^{\eta\cdot x}\mathscr{L}^{-1}g\in\mathscr{S}$.

1.3
We now study the relation between the support of f and that of $\mathscr{L}f$. For this purpose it is convenient to 'measure' the support of f by a 'cone'. We need some definitions.

We say that $\Gamma(\subset\mathbb{R}^n)$ is a *cone* if for any $t>0$, $t\Gamma\subset\Gamma$. Given a domain $\Omega(\subset\mathbb{R}^n)$ and a cone Γ, if $\Omega+\Gamma\subset\Omega$, then we say that Ω *stretches in the direction of* Γ. For any Ω $(\subset\mathbb{R}^n)$, if Γ is convex, $\Omega+\Gamma$ stretches in the direction of Γ, and

$$\Gamma' = \{x \in \mathbb{R}^n | x \cdot \eta \geqq 0, \forall \eta \in \Gamma\}$$

is a closed convex cone, called the *adjoint cone* (or dual convex cone)† of Γ.

Lemma 1.4

Let Γ be a cone, and B_a be a ball of radius a. If the support of f is contained in the set $B_a - \Gamma'$ (B_a stretched in the direction of $-\Gamma'$), then Δ_f stretches in the direction of Γ. (See Fig. 4.)

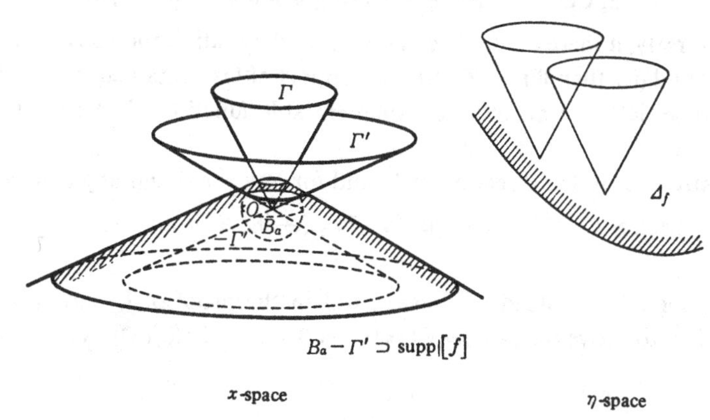

Fig. 4

Proof

It is sufficient to prove that $\eta^0 + \Gamma \subset \Delta_f$ for $\eta^0 \in \Delta_f$. And we can simplify the proof further by taking $\eta^0 = 0$. This is because if we set

$$\tilde{f}(x) = e^{\eta_0 x} f(x)$$

then we see that the support of $\tilde{f}$ is invariant,

$$(\mathscr{L}\tilde{f})(\zeta) = (\mathscr{L}f)(\zeta + i\eta^0),$$

and $\Delta_{\tilde{f}} = \Delta_f - \eta_0$. For this reason, we assume $0 \in \Delta_f$; in other words $f \in \mathscr{S}'$.

Now we show that $\Gamma \subset \Delta_f$. Let α be an infinitely differentiable function such that $\alpha(x) = 1$ for $x \in B_a - \Gamma'$ and the support of α is contained in the δ-neighbourhood $B_{a+\delta} - \Gamma'$. Then we can write

$$e^{x \cdot \eta} f = (\alpha(x) e^{x \cdot \eta}) f, \quad f \in \mathscr{S}'.$$

Also, for $\eta \in \Gamma$ we have

$$\sup_x \left| \left(\frac{\partial}{\partial x} \right)^\nu (\alpha(x) e^{x \cdot \eta}) \right| \leqq C_\nu (1 + |\eta|)^{|\nu|} e^{(a+\delta)|\eta|}.$$

and from this $\alpha(x) e^{x \cdot \eta} \in \mathscr{B}_x$. Therefore, for $\eta \in \Gamma$, $e^{x \cdot \eta} f \in \mathscr{S}'$. Hence $\Gamma \subset \Delta_f$. ∎

† Translator's note: or adjoint convex cone.

Theorem 1 (c)
Let $\mathring{\Delta}_f \neq \varnothing$ for $f \in \mathscr{D}'$, and assume that the support of f is contained in $B_a - \Gamma'$, where B_a, Γ' are as before. If $\mathring{\Delta}_f = \Omega$ and $\mathscr{L}f = g$, then the following statements hold:

(i) $g(\zeta)$ is regular in $\mathbb{R}^n + \mathrm{i}\Omega$,

(ii) Ω stretches in the direction of Γ, and for any $\eta^0 \in \Omega$ and any $\varepsilon > 0$, there exist C, m such that

$$|g(\zeta)| \leqq C\mathrm{e}^{(a+\varepsilon)|\eta|}(1+|\xi|)^m, \quad \zeta = \xi + \mathrm{i}\eta \in \mathbb{R}^n + \mathrm{i}(\eta^0 + \Gamma).$$

Conversely, if there are Ω, Γ, $g(\zeta)$ such that they satisfy the above conditions (i) and (ii), then if $f = \mathscr{L}^{-1}g$, the support of f is contained in $B_a - \Gamma'$.
Note. If we let f be an infinitely differentiable function, then condition (ii) becomes
(ii)′ Ω stretches in the direction of Γ, and for any $\eta^0 \in \Omega$ and any ε, $N > 0$,

$$|g(\zeta)| \leqq C\mathrm{e}^{(a+\varepsilon)|\eta|}(1+|\xi|)^{-N}, \quad \zeta = \xi + \mathrm{i}\eta \in \mathbb{R}^n + \mathrm{i}(\eta^0 + \Gamma).$$

Proof
Consider $\mathrm{e}^{x \cdot \eta^0} f$ instead of f as we did in the proof of Lemma 1.4. It is sufficient to prove the case for $\eta^0 = 0$. Let $0 < \varepsilon' < \operatorname{dis}(0, \partial\Omega)$. We see that

$$g(\zeta) = \langle \mathrm{e}^{\varepsilon'\sqrt{(1+|x|^2)}} f, \mathrm{e}^{-\varepsilon'\sqrt{(1+|x|^2)}} \alpha(x) \mathrm{e}^{-\mathrm{i}x\cdot\zeta} \rangle,$$
$$\mathrm{e}^{\varepsilon'\sqrt{(1+|x|^2)}} f \in \mathscr{S}', \quad \mathrm{e}^{-\varepsilon'\sqrt{(1+|x|^2)}} \alpha(x) \mathrm{e}^{-\mathrm{i}x\cdot\zeta} \in \mathscr{S},$$

where $\alpha(x)$ is the same as in the proof of Lemma 1.4. Therefore, for $\eta \in \Gamma$, we find

$$\begin{aligned}|g(\zeta)| &= Cp_m(\mathrm{e}^{-\varepsilon'\sqrt{(1+|x|^2)}} \alpha(x) \mathrm{e}^{-\mathrm{i}x\cdot(\xi+\mathrm{i}\eta)}) \\ &\leqq C'(1+|\xi|+|\eta|)^m \mathrm{e}^{(a+\delta)|\eta|} \\ &\leqq C''(1+|\xi|)^m \mathrm{e}^{(a+2\delta)|\eta|}.\end{aligned}$$

Conversely, since

$$B_a - \Gamma' = \left\{ x \in \mathbb{R}^n \,\middle|\, \sup_{\eta \in \Gamma} x \cdot \frac{\eta}{|\eta|} \leqq a \right\},$$

we have

$$(B_a - \Gamma')^C = \bigcup_{\eta \in \Gamma} \left\{ x \in \mathbb{R}^n \,\middle|\, x \cdot \frac{\eta}{|\eta|} > a \right\}.$$

Let us choose an arbitrary $\eta^1 \in \Gamma$. Now if we fix η^1, we wish to show that $f = 0$ in the half-space

$$H = \left\{ x \in \mathbb{R}^n \,\middle|\, x \cdot \frac{\eta^1}{|\eta^1|} > a \right\}.$$

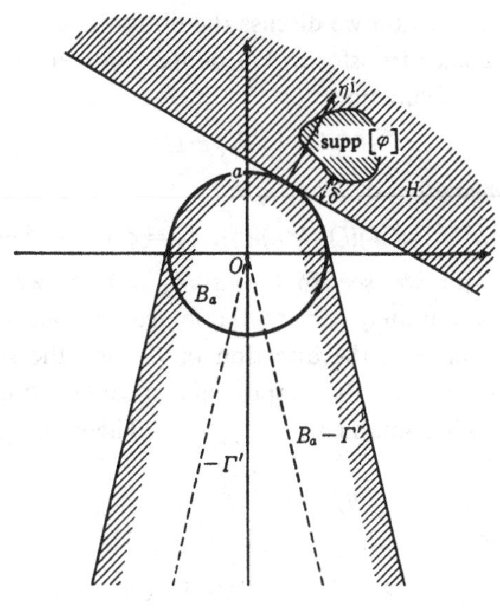

Fig. 5

To this end we choose $\varphi \in \mathscr{D}$ with supp $[\varphi]$ contained in H. (See Fig. 5.) For $\eta \in \Omega$ we obtain

$$\begin{aligned}\langle f, \varphi \rangle &= \langle e^{-x\cdot\eta} \bar{\mathscr{F}}[g(\xi + i\eta)], \varphi(x) \rangle_x = \langle g(\xi + i\eta), \bar{\mathscr{F}}[e^{-x\cdot\eta}\varphi(x)] \rangle_\xi \\ &= \langle g(\xi + i\eta), \psi(\xi + i\eta) \rangle_\xi\end{aligned}$$

where

$$\psi(\zeta) = \frac{1}{(2\pi)^n} \int e^{ix\cdot\zeta} \varphi(x)\, dx.$$

Since the support of φ is contained in H, there exists $\delta > 0$ which is contained in

$$\left\{ x \,\middle|\, x \cdot \frac{\eta^1}{|\eta^1|} \geqq a + \delta \right\}.$$

Therefore, in the particular case when $\eta = \lambda\eta^1 (\lambda > 0)$, we see that

$$|\psi(\xi + i\eta)| \leqq C_N (1 + |\xi|)^{-N} e^{-(a+\delta)|\eta|}$$

where N is any natural number. Hence, by letting $\varepsilon = \delta/2$ (the same ε as in condition (ii)), we have

$$\{\lambda\eta^1 \,|\, \lambda > 0\} \subset \Gamma.$$

That is,

$$|\langle f, \varphi \rangle| \leqq C e^{-(\delta/2)\lambda|\eta^1|}.$$

If we let $\lambda \to +\infty$, then it is easy to see that $\langle f, \varphi \rangle = 0$. ∎

In the rest of this section we discuss the notion of convolution defined by means of a Laplace transform. Let us fix a domain $\Omega(\subset \mathbb{R}^n)$, For Ω, consider the family of functions

$$\mathscr{S}_{\Omega}' = \{f | \Delta_f \supset \Omega\}.$$

We define the *convolution* of f and g as

$$f*g = \mathscr{L}^{-1}[(\mathscr{L}f)(\zeta)\cdot(\mathscr{L}g)(\zeta)], \quad \zeta = \xi + i\eta \in \mathbb{R}^n + i\Omega.$$

From Theorem 1(a) we see that $f*g \in \mathscr{S}_{\Omega}'$. Also, we know that if either f or g is infinitely differentiable, so is $f*g$ (Theorem 1(b)), and that if Ω stretches in the direction of Γ, then the support of f is contained in $B_a - \Gamma'$; and if the support of g is contained in $B_b - \Gamma'$, then the support of $f*g$ is contained in $B_{a+b} - \Gamma'$ (Theorem 1(c)).
By

$$g_j(x) \to g(x)(\mathscr{S}_{\Omega}')$$

we mean that, for any $\eta \in \Omega$,

$$e^{x\cdot\eta} g_j(x) \to e^{x\cdot\eta} g(x)(\mathscr{S}_x'),$$

i.e.

$$(\mathscr{L}g_j)(\xi + i\eta) \to (\mathscr{L}g)(\xi + i\eta) \quad (\mathscr{S}_{\xi}').$$

Similarly, by

$$g_j(x) \to g(x) \quad (\mathscr{S}_{\Omega}),$$

we mean, for an arbitrary $\eta \in \Omega$,

$$e^{x\cdot\eta} g_j(x) \to e^{x\cdot\eta} g(x) \quad (\mathscr{S}_x),$$

i.e.

$$(\mathscr{L}g_j)(\xi + i\eta) \to (\mathscr{L}g)(\xi + i\eta) \quad (\mathscr{S}_{\xi}).$$

We shall prove the following

Proposition 1.5

(i) For a fixed $f \in \mathscr{S}_{\Omega}'$,

$$g_j \to g \ (\mathscr{S}_{\Omega}') \Rightarrow f*g_j \to f*g \quad (\mathscr{S}_{\Omega}').$$
$$g_j \to g \ (\mathscr{S}_{\Omega}) \Rightarrow f*g_j \to f*g \quad (\mathscr{S}_{\Omega}).$$

(ii) In particular, if Ω stretches in the direction of Γ then

$$\operatorname{supp}[f*g] \subset \operatorname{supp}[f] + \operatorname{supp}[g] - \Gamma'.$$

Proof

(i) For a fixed $\eta \in \Omega$

$$\mathscr{L}[f*g_j](\xi + i\eta) = (\mathscr{L}f)(\xi + i\eta)\cdot(\mathscr{L}g_j)(\xi + i\eta).$$

From Theorem 1(a) we have

$$|(\partial/\partial\xi)^{\nu}(\mathscr{L}f)(\xi + i\eta)| \leqq C_{\nu}(1 + |\xi|)^m.$$

Therefore, if

$$(\mathscr{L}g_j)(\xi+i\eta)\to(\mathscr{L}g)(\xi+i\eta) \quad (\mathscr{S}_\xi'),$$

then

$$(\mathscr{L}f)(\xi+i\eta)(\mathscr{L}g_j)(\xi+i\eta)\to(\mathscr{L}f)(\xi+i\eta)(\mathscr{L}g)(\xi+i\eta) \quad (\mathscr{S}_\xi'),$$

and if

$$(\mathscr{L}g_j)(\xi+i\eta)\to(\mathscr{L}g)(\xi+i\eta) \quad (\mathscr{S}_\xi),$$

then

$$(\mathscr{L}f)(\xi+i\eta)(\mathscr{L}g_j)(\xi+i\eta)\to(\mathscr{L}f)(\xi+i\eta)(\mathscr{L}g)(\xi+i\eta) \quad (\mathscr{S}_\xi).$$

(ii) Let

$$\operatorname{supp}[f]\subset x^0+B_a-\Gamma', \quad \operatorname{supp}[g]\subset x^1+B_a-\Gamma'.$$

Then, putting

$$f_1(x)=f(x-x^0), \quad g_1(x)=g(x-x^1),$$

we see that

$$\operatorname{supp}[f_1], \quad \operatorname{supp}[g_1]\subset B_a-\Gamma'$$

therefore

$$\operatorname{supp}[f_1*g_1]\subset B_{2a}-\Gamma'.$$

On the other hand

$$(f_1*g_1)(x)=(f*g)(x-(x^0+x^1)).$$

From this we have

$$\operatorname{supp}[f*g]\subset x^0+x^1+B_{2a}-\Gamma'.$$

For the general case we use the partition of unity in x-space,

$$\sum_{j=1}^{\infty}\alpha_j(x)=1$$

in order to decompose the supports of f, g, and then apply the results of the above process to obtain the desired inclusion property. To see this, notice that the diameter of the support of $\alpha_j\in\mathscr{D}$ is smaller than a, and that if

$$\sum_{j=1}^{N}\alpha_j(x)=1, \quad |x|<R_N \quad (R_N\to+\infty \text{ as } N\to+\infty),$$

then, for $f\in\mathscr{S}'_\Omega$,

$$f_N(x)=\sum_{j=1}^{N}\alpha_j(x)f(x)\to f(x) \quad (\mathscr{S}_\Omega') \quad \text{as } N\to+\infty,$$

$$g_N(x)=\sum_{j=1}^{N}\alpha_j(x)g(x)\to g(x) \quad (\mathscr{S}_\Omega') \quad \text{as } N\to+\infty.$$

Hence

$$f_M * g_N \to f * g_N \quad (\mathscr{S}_\Omega') \quad \text{as } M \to +\infty,$$
$$\operatorname{supp}[f_M * g_N] \subset \operatorname{supp}[f] + \operatorname{supp}[g] + B_{2a} - \Gamma'.$$

Therefore we find

$$\operatorname{supp}[f * g_N] \subset \operatorname{supp}[f] + \operatorname{supp}[g] + B_{2a} - \Gamma'.$$

We also know that

$$f * g_N \to f * g \quad (\mathscr{S}_\Omega') \quad \text{as } N \to +\infty,$$

so that

$$\operatorname{supp}[f * g] \subset \operatorname{supp}[f] + \operatorname{supp}[g] + B_{2a} - \Gamma'.$$

Letting a be as small as possible, we see that

$$\operatorname{supp}[f * g] \subset \operatorname{supp}[f] + \operatorname{supp}[g] - \Gamma'. \qquad \blacksquare$$

2 $\mathscr{E}$-well-posed boundary value problems, and a necessary condition for $\mathscr{E}$-well-posedness

Here we begin by defining an $\mathscr{E}$-well-posed boundary value problem which is the main concern throughout the rest of this chapter. Later, one of the two necessary conditions for $\mathscr{E}$-well-posedness will be presented in the form of a theorem concerning the zero points of Lopatinski's determinant.

2.1

Before explaining the notion of $\mathscr{E}$-well-posedness for initial boundary value problems, let us consider the half-space Ω in $\mathbb{R}^n$ given by

$$\Omega = \mathbb{R}_+{}^n = \{(x, y) \mid x > 0,\ y = (y_1, \ldots, y_{n-1}) \in \mathbb{R}^{n-1}\}$$

and look at problems concerning linear partial differential equations with constant coefficients in Ω and with a time variable t. Let $\{A = A(\mathrm{D}_t, \mathrm{D}_x, \mathrm{D}_y),$ $B_j = B_j(\mathrm{D}_t, \mathrm{D}_x, \mathrm{D}_y)(j = 1, \ldots, \mu)\}$ be a linear partial differential operator of order $\{m, r_j\}$ with constant coefficients. Given the data

$$\{f, g_j (j = 1, \ldots, \mu), h_j (j = 0, \ldots, m-1)\},$$

we seek a solution for

$$(\mathrm{P}) \begin{cases} Au = f, & t > 0,\ (x, y) \in \Omega \quad \text{(equation of motion)}, \\ B_j u = g_j \quad (j = 1, \ldots, \mu), & t > 0,\ (x, y) \in \partial\Omega \quad \text{(boundary condition)}, \\ \mathrm{D}_t{}^j u = h_j \quad (j = 0, \ldots, m-1), & t = 0,\ (x, y) \in \Omega \quad \text{(initial condition)}, \end{cases}$$

where (P) denotes the *type* of the problem. We wish to implement the notion of $\mathscr{E}$-well-posedness (see Chapter 1, §1) for (P).

Let $\mathscr{E}(X)$ be a set of infinitely differentiable functions, f, defined over X. Then, $\mathscr{E}(X)$ becomes a *Fréchet space*† with the semi-norm

$$|f|_{l,K} = \sum_{|\nu| \leqq l} \sup_{x < K} \left| \left(\frac{\partial}{\partial x} \right)^{\nu} f(x) \right|,$$

where l is an arbitrary natural number, and K is an arbitrary compact set such that $K \subset X$. The initial boundary value problem (P) is said to be $\mathscr{E}$-*well-posed* if the following statements 1° and 2° are true.

1° For arbitrary data $\{f \in \mathscr{E}([0, \infty) \times \bar{\Omega}), g_j \in \mathscr{E}([0, \infty) \times \partial\Omega)(j = 1, \ldots, \mu), h_j \in \mathscr{E}(\bar{\Omega})(j = 0, \ldots, m-1)\}$ there exists a unique solution

$$u \in \mathscr{E}([0, \infty) \times \bar{\Omega}).$$

The above data are subject to a certain compatibility condition which is explained in §2.2.

2° The map from the data to a solution is continuous in the sense of $\mathscr{E}$ (see the definition given in Chapter 1 §1).

Now we explain a dependence existing between these conditions. The expression '*to be continuous in the sense of* $\mathscr{E}$' should be interpreted as follows: There exist a natural number l', a compact set $K' \subset [0, \infty) \times \bar{\Omega}$, and a positive number C such that

$$|u|_{l,K} \leqq C \left\{ |f|_{l',K'} + \sum_{j=1}^{m} |g_j|_{l',K_1'} + \sum_{j=0}^{m-1} |h_j|_{l',K_2'} \right\}$$

where

$$K' \cap \{x = 0\} = K_1', \quad K' \cap \{t = 0\} = K_2'.$$

Since $\mathscr{E}(X)$ is a Fréchet space, we can apply Banach's closed graph theorem (as in chapter 1) to see how condition 2° follows from condition 1°.

In this chapter we wish to determine the type of $\{A, B_j\}$ for which (P) becomes $\mathscr{E}$-well-posed. To begin with, we deal with the case of simple initial problems without boundary conditions. We understand that the initial problem

$$\begin{aligned} &A(D_t, D_x, D_y)u = f, && 0 < t < +\infty, \quad (x, y) \in \mathbb{R}^n, \\ &D_t^j u = h_j \quad (j = 0, \ldots, m-1), && t = 0, \qquad\quad (x, y) \in \mathbb{R}^n \end{aligned}$$

is said to be $\mathscr{E}$-*well-posed*, if for arbitrary data $\{f \in \mathscr{E}([0, \infty) \times \mathbb{R}^n), h_j \in \mathscr{E}(\mathbb{R}^n)$ $(j = 0, \ldots, m-1)\}$ there exists a unique solution $u \in \mathscr{E}([0, \infty) \times \mathbb{R}^n)$ and the correspondence is continuous in the sense of $\mathscr{E}$.

There is a key theorem concerning this.

† Translator's note: i.e. a complete metrisable locally convex space.

Gårding's theorem†
Let the initial problem be the same as before. The necessary and sufficient condition for the problem being $\mathscr{E}$-well-posed is that there exists a positive number γ_0 for which

$$A(\tau,\xi,\eta)\neq 0, \quad \operatorname{Im}\tau < -\gamma_0, \quad (\xi,\eta)\in\mathbb{R}^n,$$
$$A_0(1,0,0)\neq 0,$$

holds, where $A_0(\tau,\xi,\eta)$ is the part of $A(\tau,\xi,\eta)$ consisting of the terms that are homogeneous of degree m (called the *principal part* of $A(\tau,\xi,\eta)$). This is equivalent to saying that $A(\tau,\xi,\eta)$ is hyperbolic in the direction of the t-axis.

Using this fact, from now on we impose

Hypothesis (A)
(i) $A(\tau,\xi,\eta)$ is hyperbolic in the direction of the t-axis, i.e.

$$A(\tau,\xi,\eta)\neq 0, \quad \operatorname{Im}\tau < -\gamma_0, \quad (\xi,\eta)\in\mathbb{R}^n,$$
$$A_0(1,0,0)\neq 0.$$

This is called *Gårding's hyperbolicity condition.*

(ii) The boundary $\{x=0,(t,y)\in\mathbb{R}_+{}^n\}$ is *non-characteristic* with respect to $A_0(\tau,\xi,\eta)$. This means $A_0(0,1,0)\neq 0$.

(iii) $\{B_j(\tau,\xi,\eta), j=1,\ldots,\mu\}$ is *normal.* That is,

$$0\leqq r_j\leqq m-1, \quad r_i\neq r_j \quad (i\neq j),$$
$$B_{j0}(0,1,0), \neq 0,$$

where B_{j0} is the principal part of B_j, i.e. the part consisting of the terms that are homogeneous of degree r_j.

2.2
When we stated the conditions for $\mathscr{E}$-well-posedness in the previous subsection (see condition 1° in §2.1) we said that these data are subject to a compatibility condition. Here we give the details of this condition, and then we see that, in fact, our initial boundary value problem can be reduced to a simple type of boundary problem. Roughly speaking, by a *compatibility condition between data* we mean the infinite number of relations which appear in the intersecting domain $\{t=0, x=0, y\in\mathbb{R}^{n-1}\}$

† For the proof see L. Gårding, Linear hyperbolic partial differential equations with constant coefficients, *Acta Math.* **85** (1950).
(Translator's note: See also F. John, *Partial differential equations* (third edition). Springer, Berlin (1978) p. 125.)

of two boundary surfaces $\{x=0,(t,y)\in\mathbb{R}_+{}^n\}$ and $\{t=0,(x,y)\in\mathbb{R}_+{}^n\}$ over which the data $\{f,g_j,h_j\}$ are defined, given that a solution $u\in\mathscr{E}([0,\infty)\times\mathbb{R}_+{}^n)$ exists and satisfies the problem (P).

To find these relations, let us observe the relation of $\{u,f,g_j,h_j\}$ (given by (P)) over the surface $\{t=0,(x,y)\in\mathbb{R}_+{}^n\}$. Since $Au=f$ is true for $\{t>0,(x,y)\in\mathbb{R}_+{}^n\}$, we have

$$\mathrm{D}_t{}^k Au = \mathrm{D}_t{}^k f (k=0,1,\ldots),\quad \{t=0,(x,y)\in\mathbb{R}_+{}^n\}.$$

And from the fact that $B_j u = g_j$ is true for $\{t>0, x=0, y\in\mathbb{R}^{n-1}\}$, we have another set of equalities

$$D_t{}^k B_j u = \mathrm{D}_t{}^k g_j \quad (k=0,1,\ldots),\quad \{t=0,x=0,y\in\mathbb{R}^{n-1}\}.$$

Putting the above two equalities together, we have

Relation over $\{t=0,(x,y)\in\mathbb{R}_+{}^n\}$:

$$\mathrm{D}_t{}^j u = h_j (j=0,\ldots,m-1),\quad \mathrm{D}_t{}^k Au = \mathrm{D}_t{}^k f (k=0,1,\ldots),$$

Relation over $\{t=0,x=0,y\in\mathbb{R}^{n-1}\}$:

$$\mathrm{D}_t{}^k B_j u = \mathrm{D}_t{}^k g_j (j=1,\ldots,\mu, k=0,1,\ldots).$$

Writing

$$A(\tau,\xi,\tau) = c_0\tau^m + c_1(\xi,\eta)\tau^{m-1} + \ldots + c_m(\xi,\eta),$$

we find

$$c_0 = A_0(1,0,0) \neq 0.$$

Therefore, we see that, starting from the given data $\{f,h_0,\ldots,h_{m-1}\}$, we can subsequently define $\{h_m,h_{m+1},\ldots\}$ by way of the relations

$$c_0 h_m(x,y) + c_1(\mathrm{D}_x,\mathrm{D}_y)h_{m-1}(x,y) + \ldots + c_m(\mathrm{D}_x,\mathrm{D}_y)h_0(x,y) = f(0,x,y),$$

$$c_0 h_{m+1}(x,y) + c_1(\mathrm{D}_x,\mathrm{D}_y)h_m(x,y) + \ldots + c_m(\mathrm{D}_x,\mathrm{D}_y)h_1(x,y) = (\mathrm{D}_t f)(0,x,y)$$

From this, the relation over $\{t=0,(x,y)\in\mathbb{R}_+{}^n\}$ (which we mentioned above) can be expressed in a 'compact' form

$$\mathrm{D}_t{}^j u = h_j \quad (j=0,1,2,\ldots),\quad \{t=0,(x,y)\in\mathbb{R}_+{}^n\}.$$

If we write

$$B_j(\tau,\xi,\eta) = c_{j0}\tau^{r_j} + c_{j1}(\xi,\eta)\tau^{r_j-1} + \ldots + c_{jr_j}(\xi,\eta),$$

then, eliminating u, we have the relation over $\{t=0,x=0,y\in\mathbb{R}^{n-1}\}$:

$$(*)\quad \begin{cases} c_{j0}h_{r_j}(0,y) + (c_{j1}(\mathrm{D}_x,\mathrm{D}_y)h_{r_j-1})(0,y) + \ldots \\ \qquad + (c_{jr_j}(\mathrm{D}_x,\mathrm{D}_y)h_0)(0,y) = g_j(0,y), \\ c_{j0}h_{r_j+1}(0,y) + (c_{j1}(\mathrm{D}_x,\mathrm{D}_y)h_{r_j})(0,y) + \ldots \\ \qquad + (c_{jr_j}(\mathrm{D}_x,\mathrm{D}_y)h_1)(0,y) = (\mathrm{D}_t g_j)(0,y), \ldots \end{cases}$$

Obviously, these constitute the infinite number of relations existing among the data $\{f, g_j\ (j=1,\ldots\mu), h_j\ (j=0,\ldots m-1)\}$.

The infinite set of these relations is called the *compatibility condition of infinite order* for the given data.

Under this we now see how the original initial boundary value conditions (P) can be reduced to a set of simple boundary conditions. To this end let the data $\{f, g_j (j=1,\ldots,\mu), h_j\ (j=0,\ldots,m-1)\}$ satisfy a compatibility condition of infinite order. Then if we define $\{h_j\ (j=m,m+1,\ldots)\}$ as we did above, we can obtain $u_0(t,x,y) \in \mathscr{E}([0,\infty)\times\overline{\mathbb{R}_+{}^n})$ which satisfies the relations

$$D_t{}^j u_0|_{t=0} = h_j(x,y) \quad (j=0,1,2,\ldots),$$
$$t \geqq 1 \Rightarrow u_0 \equiv 0.$$

We write

$$f'(t,x,y) = f(t,x,y) - A(D_t, D_x, D_y)u_0(t,x,y), \quad t>0, \quad (x,y)\in \mathbb{R}_+{}^n,$$
$$g_j'(t,x,y) = g_j(t,y) - B_j(D_t, D_x, D_y)u_0(t,0,y), \quad t>0, \quad y\in\mathbb{R}^{n-1}.$$

Because we assumed the compatibility condition, we find that these functions contact 0 at $t=0$ to the infinite order. This means that if we extend $\{f', g_j'\}$ (defined at $t>0$) beyond 0 to $t<0$, and write these extensions with the same symbols $\{f', g_j'\}$, we see that

$$f' \in \mathscr{E}((-\infty,\infty)\times\overline{\mathbb{R}_+{}^n}), \quad g_j' \in \mathscr{E}((-\infty,\infty)\times\mathbb{R}^{n-1}).$$

Writing $u' = u - u_0$, we conclude that the problem (P) can be reduced to the problem of seeking a solution, $u' \in \mathscr{E}([0,\infty)\times\overline{\mathbb{R}_+{}^n})$ for the problem

$$(\mathrm{P}') \quad \begin{cases} Au' = f', & t>0, \quad (x,y)\in\mathbb{R}_+{}^n, \\ B_j u'|_{x=0} = g_j' \quad (j=1,\ldots,\mu), & t>0, \quad y\in\mathbb{R}^{n-1}, \\ D_t{}^j u'|_{t=0} = 0 \quad (j=0,\ldots,m-1), & (x,y)\in\mathbb{R}_+{}^n. \end{cases}$$

Since $f'=0$ for $t<0$, the solution u' for (P′) contacts 0 to the infinite order. Therefore, writing the same u' for the extension of u' beyond 0 to $t<0$, we see that $u' \in \mathscr{E}((-\infty,+\infty)\times\overline{\mathbb{R}_+{}^n})$, and satisfies the problem

$$(\mathrm{P}'') \quad \begin{cases} Au' = f', & -\infty<t<+\infty, \quad (x,y)\in\mathbb{R}_+{}^n, \\ B_j u'|_{x=0} = g_j' \quad (j=1,\ldots,\mu), & -\infty<t<+\infty, \quad y\in\mathbb{R}^{n-1}. \end{cases}$$

This implies that the original initial value problem (P) can be reduced to the boundary problem (P″) for finding $u' \in \mathscr{E}(\mathbb{R}^1\times\overline{\mathbb{R}_+{}^n})$ for the data:

$$f' \in \mathscr{E}(\mathbb{R}^1\times\overline{\mathbb{R}_+{}^n}), \quad g_j \in \mathscr{E}(\mathbb{R}^1\times\mathbb{R}^{n-1})$$

Conversely, we see that if (P″) satisfies an auxiliary condition on the supports: 'if the supports of f', g_j' are contained in $\{t\geqq 0, (x,y)\in\mathbb{R}_+{}^n\}$, then the support of u' is contained in $\{t\geqq 0, (x,y)\in\mathbb{R}_+{}^n\}$', then, needless to say, the solvability of (P) follows from that of (P″).

Up to now we have considered infinitely differentiable functions as possible solutions of initial boundary value problems. In this sense, we remain in the classical territory explored in the pre-distribution period. We can, of course, expand our scope to seek a *distribution as a solution* for the same problem. To demonstrate this, let us consider a map

$$0 \leqq x \longmapsto u(t, x, y) \in \mathscr{D}'(\mathbb{R}_{t,y}{}^n)$$

associating $x \in [0, \infty)$ to a distribution $u(t, x, y)$ in the (t, y)-space by regarding x as a parameter. We assume that this map is infinitely differentiable. By this we mean that for $\varphi(t, y) \in \mathscr{D}(\mathbb{R}_{t,y}{}^n)$,

$$U(x) = \langle u(t, x, y), \varphi(t, y) \rangle_{t,y}$$

is infinitely differentiable for $x \geqq 0$. We write the entire set of such u as $C^\infty([0, \infty); \mathscr{D}'(\mathbb{R}_{t,y}{}^n))$. Then, when $u \in C^\infty([0, \infty); \mathscr{D}'(\mathbb{R}_{t,y}{}^n))$ if we write the term on the left hand side as

$$\langle D_t{}^i D_x{}^j D_y{}^\nu u, \varphi(t, y) \rangle_{t,y} = (-1)^{i+|\nu|} D_x{}^j \langle u, D_t{}^i D_y{}^\nu \varphi(t, y) \rangle_{t,y}$$

we see that

$$D_t{}^i D_x{}^j D_y{}^\nu u \in C^\infty([0, \infty); \mathscr{D}'(\mathbb{R}_{t,y}{}^n))$$

so that we interpret the problem (P″) in this sense. Consequently, the problem (P″) for seeking a distribution as a solution becomes as follows: For the arbitrary data

$$f \in C^\infty([0, \infty); \mathscr{D}'(\mathbb{R}_{t,y}{}^n)), \quad g_j \in \mathscr{D}'(\mathbb{R}_{t,y}{}^n) \quad (j = 1, \ldots, \mu)$$

we obtain the solution $u \in C^\infty([0, \infty); \mathscr{D}'(\mathbb{R}_{t,y}{}^n))$ for

$$(\mathrm{P}'') \qquad \begin{cases} A(D_t, D_x, D_y)u = f, & x > 0, \\ B_j(D_t, D_x, D_y)u|_{x=0} = g_j & (j = 1, \ldots, \mu). \end{cases}$$

Suppose that there exists a distribution (solution) for (P″) such that with respect to (t, y) a Fourier–Laplace transform is applicable. This implies that there exists a domain $\Omega \subset \mathbb{R}^n$ such that

$$u \in C^\infty([0, \infty); \mathscr{S}_\Omega'(\mathbb{R}_{t,y}{}^n)).$$

Then for $x \geqq 0, (\tau, \eta) \in \mathbb{R}^n + i\Omega$ we can define

$$\hat{u}(x; \tau, \eta) = \mathscr{L}_{(t,y) \to (\tau,\eta)}[u(t, x, y)] = \langle u(t, x, y), e^{-i(t\tau + \eta y)} \rangle_{t,y}$$

which is infinitely differentiable for $x \geqq 0$. On the other hand, since

$$Au = f \in C^\infty([0, \infty); \mathscr{S}_\Omega'(\mathbb{R}^n)), \quad x > 0,$$

$$B_j u|_{x=0} = g_j \in \mathscr{S}_\Omega'(\mathbb{R}^n) \quad (j = 1, \ldots, \mu)$$

for $(\tau, \eta) \in \mathbb{R}^n + i\Omega$ if we put

$$\hat{f}(x; \tau, \eta) = \mathscr{L}[f(t, x, y)], \quad \hat{g}_j(\tau, \eta) = \mathscr{L}[g_j(t, y)]$$

then

$$(\hat{\mathrm{P}}) \qquad \begin{cases} A(\tau, \mathrm{D}_x, \eta)\hat{u}(x;\tau,\eta) = \hat{f}(x;\tau,\eta), & x > 0, \\ B_j(\tau, \mathrm{D}_x, \eta)\hat{u}(0;\tau,\eta) = \hat{g}_j(\tau,\eta) & (j = 1, \ldots, \mu) \end{cases}$$

is satisfied. Regarding $(\tau, \eta) \in \mathbb{R}^n + \mathrm{i}\Omega$ as a parameter, we see that the problem $(\hat{\mathrm{P}})$ is, in fact, a boundary problem for an ordinary differential equation at $x > 0$.

Conversely, if the inverse Fourier–Laplace transform is applicable to a solution for $(\hat{\mathrm{P}})$ with regard to (τ, η), then it qualifies as a solution for (P''). Thus we can regard $(\hat{\mathrm{P}})$ as the most simplified problem format.

2.3

Let us now discuss the solvability of the problem $(\hat{\mathrm{P}})$. If we write

$$A(\tau, \xi, \eta) = a_0\xi^m + a_1(\tau,\eta)\xi^{m-1} + \ldots + a_m(\tau,\eta)$$

then, by Hypothesis (A), we see that $a_0 = A_0(0,1,0) \neq 0$. For normalisation we set $a_0 = 1$. According to (i) of Hypothesis (A), if $\operatorname{Im}\tau < -\gamma_0, \eta \in \mathbb{R}^{n-1}$ then none of the m roots of the equation $A(\tau, \xi, \eta) = 0$ with respect to ξ are reals. Therefore it is possible to classify these roots into two different kinds, namely those with positive imaginary parts and those with negative imaginary parts, written

$$\xi_1{}^+(\tau,\eta), \ldots, \xi_{m_+}{}^+(\tau,\eta),$$
$$\xi_1{}^-(\tau,\eta), \ldots, \xi_{m_-}{}^-(\tau,\eta),$$

respectively. Since these roots are continuously dependent on (τ, η), we see that $m_\pm$ is a constant which does not depend on (τ, η). Let us write

$$A_\pm(\tau, \xi, \eta) = \prod_{j=1}^{m_\pm} (\xi - \xi_j{}^\pm(\tau,\eta)).$$

We fix the parameter (τ, η) in $\{\operatorname{Im}\tau < -\gamma_0, \eta \in \mathbb{R}^{n-1}\}$ and seek a solution $\hat{u}$ of the boundary value problem of an ordinary differential equation of the type

$$(\hat{\mathrm{P}}) \qquad \begin{cases} A(\tau, \mathrm{D}_x, \eta)\hat{u} = \hat{f}(x), & x > 0, \\ B_j(\tau, \mathrm{D}_x, \eta)\hat{u}|_{x=0} = \hat{g}_j & (j = 1, \ldots, \mu), \end{cases}$$

such that $\hat{u}(x) \to 0$ as $x \to +\infty$. We begin by seeking a solution for $(\hat{\mathrm{P}})$ in the case of $\hat{f}(x) \equiv 0$. Note that, in general, if $\hat{u}(x)$ satisfies $A(\tau, \mathrm{D}_x, \eta)\hat{u}(x) = 0$ and tends to zero as $x \to +\infty$ it can be expressed as

$$\hat{u}(x) = \sum_{k=1}^{m_+} c_k \frac{1}{2\pi\mathrm{i}} \oint \frac{\mathrm{e}^{\mathrm{i}x\xi}\xi^{k-1}}{A_+(\tau,\xi,\eta)} \mathrm{d}\xi,$$

where $c_1, \ldots, c_m$ are all arbitrary constants, and integration is understood as taken along a simple closed curve in the complex ξ-plane which encloses all the roots of $A_+(\tau, \xi, \eta) = 0$.

The necessary and sufficient condition for $\hat{u}$ satisfying the boundary condition

$$B_j(\tau, \mathrm{D}_x, \eta)\hat{u}(0) = \hat{g}_j \quad (j = 1, \ldots, \mu)$$

is that $(c_1, \ldots, c_m)$ has the property

$$\sum_{k=1}^{m_+} \frac{1}{2\pi \mathrm{i}} \oint \frac{B_j(\tau, \xi, \eta)\xi^{k-1}}{A_+(\tau, \xi, \eta)} \mathrm{d}\xi c_k = \hat{g}_j \quad (j = 1, \ldots, \mu).$$

Therefore, when $\mu = m_+$ and

$$R(\tau, \eta) = \det\left(\frac{1}{2\pi \mathrm{i}} \oint \frac{B_j(\tau, \xi, \eta)\xi^{k-1}}{A_+(\tau, \xi, \eta)} \mathrm{d}\xi\right)_{j,k=1,\ldots,m_+} \neq 0,$$

a unique $(c_1, \ldots, c_{m_+})$ exists. The converse of this statement is also true.

Let us call $R(\tau, \eta)$ the *Lopatinski determinant* for $\{A, B_j (j = 1, \ldots, m_+)\}$, and write $R_{kj}(\tau, \eta)$ for the (j, k)-cofactor of

$$\left(\frac{1}{2\pi \mathrm{i}} \oint \frac{B_j(\tau, \xi, \eta)\xi^{k-1}}{A_+(\tau, \xi, \eta)} \mathrm{d}\xi\right)_{j,k=1,\ldots,m_+}$$

Since

$$\left(\frac{1}{2\pi \mathrm{i}} \oint \frac{B_j(\tau, \xi, \eta)\xi^{k-1}}{A_+(\tau, \xi, \eta)} \mathrm{d}\xi\right)^{-1}_{j,k=1,\ldots,m_+} = \left(\frac{R_{jk}(\tau, \eta)}{R(\tau, \eta)}\right)_{j,k=1,\ldots,m_+}$$

we have

$$c_j = \sum_{k=1}^{m_+} \frac{R_{jk}(\tau, \eta)}{R(\tau, \eta)} g_k.$$

Hence, in the case of $\hat{f} \equiv 0$ the solution $\hat{u}$ of $(\hat{\mathrm{P}})$ becomes

$$\begin{aligned} \hat{u}(x) &= \sum_{j=1}^{m_+} \frac{1}{2\pi \mathrm{i}} \oint \frac{\mathrm{e}^{\mathrm{i}x\xi}\xi^{j-1}}{A_+(\tau, \xi, \eta)} \mathrm{d}\xi \sum_{k=1}^{m_+} \frac{R_{jk}(\tau, \eta)}{R(\tau, \eta)} \hat{g}_k \\ &= \sum_{k=1}^{m_+} \frac{1}{2\pi} \oint \mathrm{e}^{\mathrm{i}x\xi} \left\{ \frac{\sum_{j=1}^{m_+} \xi^{j-1} \frac{R_{jk}(\tau, \eta)}{R(\tau, \eta)}}{\mathrm{i}A_+(\tau, \xi, \eta)} \right\} \mathrm{d}\xi \hat{g}_k \\ &= \sum_{k=1}^{m_+} \frac{1}{2\pi} \oint \mathrm{e}^{\mathrm{i}x\xi} P_k(\tau, \xi, \eta) \mathrm{d}\xi \hat{g}_k = \sum_{k=1}^{m_+} Q_k(x; \tau, \eta) \hat{g}_k. \end{aligned}$$

At the same time, for $x > 0$,

$$\hat{u}_0(x) = \frac{1}{2\pi}\int_{-\infty}^{\infty} \frac{e^{ix\xi}\tilde{f}(\zeta)}{A(\tau,\xi,\eta)}d\xi, \quad \tilde{f}(\xi) = \int_0^{\infty} e^{-ix\xi}\hat{f}(x)\,dx$$

satisfies the equation $A(\tau, D_x, \eta)\hat{u}_0(x) = \hat{f}(x)$. Therefore

$$\hat{u}(x) = \hat{u}_0(x) - \sum_{i=1}^{m_+} Q_k(x;\tau,\eta)B_k(\tau, D_x, \eta)\hat{u}_0(0)$$

$$= \frac{1}{2\pi}\int_{-\infty}^{\infty} \frac{e^{ix\xi}\tilde{f}(\xi)}{A(\tau,\xi,\eta)}d\xi$$

$$-\sum_{k=1}^{m_+} Q_k(x;\tau,\eta)\frac{1}{2\pi}\int_{-\infty}^{\infty} \frac{B_k(\tau,\xi,\eta)\tilde{f}(\xi)}{A(\tau,\xi,\eta)}d\xi$$

satisfies the condition

$$A(\tau, D_x, \eta)\hat{u}(x) = \hat{f}(x), \quad x > 0,$$
$$B_j(\tau, D_x, \eta)\hat{u}(0) = 0 \quad (j = 1, \ldots, m_+).$$

2.4

In §2.3 we singled out the necessary and sufficient condition

$$\mu = m_+ \quad \text{and} \quad R(\tau,\eta) \neq 0$$

for $(\hat{\mathrm{P}})$ being solvable when we fixed the parameter (τ,η) in $\{\operatorname{Im}\tau < -\gamma_0, \eta \in \mathbb{R}^{n-1}\}$. The question arises: is there any relation between this and the $\mathscr{E}$-well-posedness of (P)? In the rest of this chapter, we deal with this problem, assuming $\mu = m_+$. As a conclusion we shall find that one of the necessary (but not sufficient) conditions of (P) being $\mathscr{E}$-well-posed is that there exists $\gamma_1 (\geqq \gamma_0)$ such that $R(\tau,\eta) \neq 0$ in $\{\operatorname{Im}\tau < -\gamma_1, \eta \in \mathbb{R}^{n-1}\}$.

To begin with, we establish the following lemma to facilitate our subsequent arguments.

Lemma 2.1

If (P) is $\mathscr{E}$-well-posed, then there exists $p > 0$ such that $R(\tau,\eta) \neq 0$ in

$$\operatorname{Im}\tau < -p\{\log(1 + |\tau| + |\eta|) + 1\}, \quad \eta \in \mathbb{R}^{n-1}.$$

Proof

We divide the proof into three parts.

Part 1. If the above statement is not true, then for $p = 1, 2, 3, \ldots$ there exists $(\tau^{(p)}, \eta^{(p)})$ such that

$$\operatorname{Im}\tau^{(p)} < -p\{\log(1 + |\tau^{(p)}| + |\eta^{(p)}|) + 1\}, \quad \eta^{(p)} \in \mathbb{R}^{n-1}.$$

and $R(\tau^{(p)}, \eta^{(p)}) = 0$. Therefore there exists a unit vector $(c_1{}^{(p)}, \ldots, c_{m_+}{}^{(p)})$

such that

$$\sum_{k=1}^{m_+} \frac{1}{2\pi i} \oint \frac{B_j(\tau^{(p)}, \xi, \eta^{(p)})\xi^{k-1}}{A_+(\tau^{(p)}, \xi, \eta^{(p)})} d\xi r_p^{m_+ - k} c_k^{(p)} = 0,$$

$$r_p = (|\tau^{(p)}|^2 + |\eta^{(p)}|^2)^{1/2} \to \infty, \quad \text{as } p \to +\infty$$

Letting

$$v_p(x) = \sum_{k=1}^{m_+} \left\{ \frac{1}{2\pi i} \oint \frac{e^{ix\xi}\xi^{k-1}}{A_+(\tau^{(p)}, \xi, \eta^{(p)})} d\xi r_p^{m_+ - k} \right\} c_k^{(p)},$$

we see that

$$A(\tau^{(p)}, D_x, \eta^{(p)}) v_p(x) = 0, x > 0,$$

$$B_j(\tau^{(p)}, D_x, \eta^{(p)}) v_p(0) = 0 \quad (j = 1, \ldots, m_+)$$

and

$$r_p^{-k} \sup_{x>0} |D_x^k v_p(x)| \leqq C_k,$$

where C_k is a constant not dependent on p. Therefore, if we put

$$u_p(t, x, y) = v_p(x) e^{it\tau(p) + iy\eta(p)}$$

we see that

$$A(D_t, D_x, D_y) u_p = 0 \quad (t, x, y) \in \mathbb{R}^{n+1},$$

$$B_j(D_t, D_x, D_y) u_p|_{x=0} = 0 \quad (j = 1, \ldots, m_+), \quad (t, y) \in \mathbb{R}^n$$

and

$$r_p^{-(j+k+|\nu|)} \sup_{(x,y)\in\mathbb{R}_+^n} |D_t^j D_x^k D_y^\nu u_p(0, x, y)| \leqq C_{jk\nu}$$

where $C_{jk\nu}$ is a constant not dependent on p.

Part 2. We now observe the behaviour of $v_p(x)$ at $x = 0$ to obtain the relation

$$\begin{bmatrix} v_p(0) \\ r_p^{-1} D_x v_p(0) \\ \vdots \\ r_p^{-m_+ + 1} D_x^{m_+ - 1} v_p(0) \end{bmatrix} = \begin{bmatrix} 0 & & 1 \\ & \cdot^{\cdot^{\cdot}} & \alpha_1^{(p)} \\ & \cdot^{\cdot^{\cdot}} & \vdots \\ 1 & \alpha_1^{(p)} \cdots & \alpha_{m_+ - 1}^{(p)} \end{bmatrix} \begin{bmatrix} c_1^{(p)} \\ c_2^{(p)} \\ \vdots \\ c_{m_+}^{(p)} \end{bmatrix},$$

where

$$|\alpha_j^{(p)}| = \left| \frac{1}{2\pi i} \oint \frac{\xi^{j+m_+ - 1}}{A_+(\tau^{(p)}, \xi, \eta^{(p)})} d\xi r_p^{-j} \right| \leqq C_j.$$

From this we have

$$1 = \sum_{j=1}^{m_+} |c_j^{(p)}|^2 \leqq C \sum_{j=0}^{m_+ - 1} r_p^{-2j} |D_x^j v_p(0)|^2,$$

so that we end up with

$$\sum_{j=0}^{m_+-1} r_p^{-2j}|D_x^j u_p(t,0,y)|^2 = \sum_{j=0}^{m-1} r_p^{-2j}|D_x^j v_p(0)|^2 e^{-2t\,\mathrm{Im}\,\tau^{(p)}}$$
$$\geqq ce^{-2t\,\mathrm{Im}\,\tau^{(p)}}$$
$$\geqq ce^{2tp(\log(1+|\tau^{(p)}|+|\eta^{(p)}|)+1)}$$
$$\geqq c(1+r_p)^{2tp}.$$

Part 3. Since we assume that (P) is $\mathscr{E}$-well-posed, for a fixed $t_0 > 0$, there exist constants C and N such that for $p = 1, 2, 3, \dots$ we must have

$$\sum_{j=0}^{m_+-1} |D_x^j u_p(t_0, 0, 0)|$$
$$\leqq C \sum_{j=0}^{m-1} \sum_{k+|\nu|\leqq N} \sup_{(x,y)\in\mathbb{R}_+^n} |D_t^j D_x^k D_y^\nu u_p(0,x,y)|.$$

But from the result established in part 1 of the proof we get that the right hand term $\leqq C' r_p^{m-1+N}$, and from the result established in part 2, the left hand term $\geqq c r_p^{2t_0 p}$. Consequently, we have

$$c r_p^{2t_0 p} \leqq C' r_p^{m-1+N}.$$

Letting $p \to +\infty$ we obtain a contradiction. ∎

2.5

In this subsection we shall digress from the main line of argument in order to establish some elementary facts from classical algebraic geometry which are used for the elimination of the logarithmic function that appeared in Lemma 2.1. First, we need some new concepts for this purpose.

$M \subset \mathbb{R}^m$ is said to be an *algebraic set* iff there exists a polynomial $p(\xi_1, \dots, \xi_m)$ with real coefficients such that

$$M = \{(\xi_1, \dots, \xi_m) \in \mathbb{R}^m \mid P(\xi_1, \dots, \xi_m) = 0\}.$$

$M \subset \mathbb{R}^m$ is said to be a *quasi-algebraic set* iff there exists an algebraic set

$$N = \{(\xi_1, \dots, \xi_n) \in \mathbb{R}^n \mid P(\xi_1, \dots, \xi_n) = 0\} \text{ in } \mathbb{R}^n \ (n \geqq m)$$

such that M coincides with the image of the orthogonal projection of N on its m-dimensional subspace, i.e. M is representable as

$$M = \{(\xi_1, \dots, \xi_m) \in \mathbb{R}^m \mid (\xi_1, \dots, \xi_m, \xi_{m+1}, \dots, \xi_n) \in N\}.$$

In particular, consider real polynomials

$$P_j(\xi)(j = 1, \dots, p), \quad Q_j(\xi)(j = 1, \dots, q), \quad R_j(\xi)(j = 1, \dots, r).$$

Then, if we know that

$$M = \{\xi \in \mathbb{R}^m | P_j(\xi) < 0 (j = 1, \ldots, p), \quad Q_j(\xi) \leqq 0 \, (j = 1, \ldots, q),$$
$$R_j(\xi) = 0 (j = 1, \ldots, r)\}$$

we can see that M is a quasi-algebraic set. To see this, consider the algebraic set

$$N = \Big\{ (\xi_1, \ldots, \xi_m, \xi_{m+1}, \ldots, \xi_{m+p}, \xi_{m+p+1}, \ldots, \xi_{m+p+q}) \in \mathbb{R}^{m+p+q} \Big|$$
$$\sum_{j=1}^{p} (\xi_{m+j}{}^2 P_j(\xi_1, \ldots, \xi_m) + 1)^2 + \sum_{j=1}^{q} (Q_j(\xi_1, \ldots, \xi_m) + \xi_{m+p+j}{}^2)^2$$
$$+ \sum_{j=1}^{r} R_j(\xi_1, \ldots, \xi_m)^2 = 0 \Big\}.$$

Then, obviously M can be recovered as the image of the orthogonal projection of N on its m-dimensional subspace.

Seidenberg's theorem† says that, given an arbitrary quasi-algebraic set, it can be represented as a finite sum of the special type of quasi-algebraic sets mentioned above. Using this fact, we present the following two lemmas concerning quasi-algebraic sets.

Seidenberg's lemma (I)

Let M be a quasi-algebraic set of $\mathbb{R}^2$. If the section of M at s,

$$M_s = \{t | (s, t) \in M\},$$

is not empty when s is made large, then $\mu(s) = \sup_{t \in M_s} t$ becomes either $\mu(s) = +\infty$ for $s \geqq s_0$ or

$$\mu(s) = As^\alpha(1 + o(1)), \quad s \to +\infty,$$

where A is a real number and α is a rational number.

Proof

In the case of an algebraic set, this is obvious from the fact that an algebraic function of a single variable can be expressed in the form of a Puiseux series.‡ In the general case, from Seidenberg's theorem, we see that there exist special quasi-algebraic sets $M^{(i)}$ such that $M = \bigcup_{i=1}^{N} M^{(i)}$.

† This theorem is not proved in this book. The interested reader should refer to A. Seidenberg, A new decision method for elementary algebra. *Ann. Math.* **60** (1954).

‡ Translator's note: For a fixed natural number k, $\sum_{i=-\infty}^{\infty} c_i t^{i/k}$ is called a Puiseux series, where the c_i are elements of a field, and t is a local canonical parameter. A detailed discussion of algebraic functions in relation to Puiseux series can be found in C. L. Siegel, *Topics in complex function theory*, vol. 1. Wiley–Interscience, New York (1969).

There exists a polynomial $P^{(i)}(s,t)$ in the defining expression of the special quasi-algebraic set $M^{(i)}$ such that

$$\tilde{M}^{(i)} = \{(s,t) \mid P^{(i)}(s,t) = 0\},$$

and

$$\mu(s) = \sup_{t \in M_s} t = \max_{1 \leqq i \leqq N} \sup_{t \in M_s^{(i)}} t = \max_{1 \leqq i \leqq N} \sup_{t \in \tilde{M}_s^{(i)}} t.$$

This proves the lemma. ■

We recast this lemma in a convenient form for future use.

Seidenberg's lemma (II)

Let M be a quasi-algebraic set of $\mathbb{R}^n$, and $P_j(\xi)$, $Q_j(\xi)$ $(j = 1,2)$ be polynomials with real coefficients over $\mathbb{R}^m$. If $Q_j(\xi) \neq 0$ $(j = 1,2)$ on M, and for large s,

$$M_s = \left\{ \xi \in M \,\middle|\, \frac{P_1(\xi)}{Q_1(\xi)} = s \right\}$$

is not empty, then

$$\mu(s) = \sup_{\xi \in M_s} \frac{P_2(\xi)}{Q_2(\xi)}$$

becomes $\mu(s) = +\infty$ for $s \geqq$ some s_0 or

$$\mu(s) = As^{\alpha}(1 + o(1)), \quad s \to +\infty$$

where A is a real number and α is a rational number.

Proof

Putting

$$M' = \left\{ (\eta_1, \eta_2) \in \mathbb{R}^2 \,\middle|\, \eta_j = \frac{P_j(\xi_1, \ldots, \xi_m)}{Q_j(\xi_1, \ldots, \xi_m)} \ (j = 1,2), (\xi_1, \ldots, \xi_m) \in M \right\}.$$

we can show that M' is a quasi-algebraic set. To see this, suppose that for $n, m (n \geqq m)$; M can be expressed as

$$M = \{(\xi_1, \ldots, \xi_m) \in \mathbb{R}^m \mid (\xi_1, \ldots, \xi_n) \in N\},$$
$$N = \{(\xi_1, \ldots, \xi_n) \in \mathbb{R}^n \mid P(\xi_1, \ldots, \xi_n) = 0\}.$$

Write

$$\begin{aligned} N' = \{(\eta_1, \eta_2, \xi_1, \ldots, \xi_n) \in \mathbb{R}^{n+2} \mid & P_1(\xi_1, \ldots, \xi_m) - \eta_1 Q_1(\xi_1, \ldots, \xi_m) = 0, \\ & P_2(\xi_1, \ldots, \xi_m) - \eta_2 Q_2(\xi_1, \ldots, \xi_m) = 0, P(\xi_1, \ldots, \xi_n) = 0\} \\ = \{(\eta_1, \eta_2, \xi_1, \ldots, \xi_n) \in \mathbb{R}^{n+2} \mid & Q(\eta_1, \eta_2, \xi_1, \ldots, \xi_n) = 0\}, \end{aligned}$$

where

$$Q(\eta_1, \eta_2, \xi_1, \ldots \Lambda \xi_n) = \{P_1(\xi_1, \ldots, \xi_m) - \eta_1 Q_1(\xi_1, \ldots, \xi_m)\}^2 + \{P_2(\xi_1, \ldots, \xi_m) - \eta_2 Q_2(\xi_1, \ldots, \xi_m)\}^2 + \{P(\xi_1, \ldots, \xi_n)\}^2.$$

Then N' is an algebraic set, and M' is the image of the orthogonal projection of N' on its 2-dimensional subspace; therefore, the first form of Seidenberg's lemma is applicable to M'. In fact, by the hypothesis, we see that

$$M_s' = M' \cap \{\eta_1 = s\} = \left\{(\eta_1, \eta_2) \middle| \eta_1 = \frac{P_1(\xi)}{Q_1(\xi)} = s,\ \eta_2 = \frac{P_2(\xi)}{Q_2(\xi)},\ \xi \in M\right\} \neq \varnothing \text{ as } s \to +\infty.$$

Then we can write

$$\mu(s) = \sup_{\eta_2 \in M_s'} \eta_2 = \sup_{\xi \in M, P_1(\xi)/Q_1(\xi) = s} \frac{P_2(\xi)}{Q_2(\xi)}.$$

The proof is complete. ■

Note. In the assumption of the second form of Seidenberg's lemma, we can write either

$$M_s = \left\{\xi \in M \middle| \frac{P_1(\xi)}{Q_1(\xi)} < s\right\},$$

or

$$M_s = \left\{\xi \in M \middle| \frac{P_1(\xi)}{Q_1(\xi)} \leqq s\right\}.$$

If the first condition holds, let

$$M' = M \times \mathbb{R}^1, \quad M_s' = \left\{(\xi, t) \in M' \middle| \frac{P_1(\xi)}{Q_1(\xi)} + \frac{1}{t^2} = s\right\}.$$

Since the projection of M_s' on the ξ-space is M_s,

$$\mu(s) = \sup_{\xi \in M_s} \frac{P_2(\xi)}{Q_2(\xi)} = \sup_{(\xi, t) \in M_s'} \frac{P_2(\xi)}{Q_2(\xi)}.$$

Now the second form of Seidenberg's lemma can be applied.

The same applies to the second condition.

Finally, we show how to define an algebraic set and quasi-algebraic set for the complex space $\mathbb{C}^m$.

$M \subset \mathbb{C}^m$ is said to be an *algebraic set* iff there exists a polynomial $P(\zeta_1, \dots, \zeta_m)$, with complex coefficients, such that

$$M = \{(\zeta_1, \dots, \zeta_m) \in \mathbb{C}^m \mid P(\zeta_1, \dots, \zeta_m) = 0\}.$$

In this case, by embedding $\mathbb{C}^m$ in $\mathbb{R}^{2m}$ we set

$$\tilde{M} = \{(\xi_1, \eta_1, \dots, \xi_m, \eta_m) \in \mathbb{R}^{2m} \mid P(\xi_1 + i\eta_1, \dots, \xi_m + i\eta_m) = 0\}.$$

If we write

$$\operatorname{Re}\{P(\xi_1 + i\eta_1, \dots, \xi_m + i\eta_m)\} = P_1(\xi_1, \eta_1, \dots, \xi_m, \eta_m),$$

$$\operatorname{Im}\{P(\xi_1 + i\eta_1, \dots, \xi_m + i\eta_m)\} = P_2(\xi_1, \eta_1, \dots, \xi_m, \eta_m),$$

$$Q(\xi_1, \eta_1, \dots, \xi_m, \eta_m) = \{P_1(\xi_1, \eta_1, \dots, \xi_m, \eta_m)\}^2 + \{P_2(\xi_1, \eta_1, \dots, \xi_m, \eta_m)\}^2,$$

we see that

$$\tilde{M} = \{(\xi_1, \eta_1, \dots, \xi_m, \eta_m) \in \mathbb{R}^{2m} \mid Q(\xi_1, \eta_1, \dots, \xi_m, \eta_m) = 0\}.$$

This shows that $\tilde{M}$ is an algebraic set of $\mathbb{R}^{2m}$.

$M \subset \mathbb{C}^m$ is said to be a *quasi-algebraic set* iff $N \subset \mathbb{C}^m$ is the image of the orthogonal projection of an algebraic set $N \subset \mathbb{C}^n (n \geqq m)$ on its m-dimensional complex subspace. Since $\mathbb{C}^m$ is embedded in $\mathbb{R}^{2m}$ we can write

$$\begin{aligned}\tilde{M} &= \{(\xi_1, \eta_1, \dots, \xi_m, \eta_m) \in \mathbb{R}^{2m} \mid (\xi_1 + i\eta_1, \dots, \xi_n + i\eta_n) \in N\} \\ &= \{(\xi_1, \eta_1, \dots, \xi_m, \eta_m) \in \mathbb{R}^{2m} \mid (\xi_1, \eta_1, \dots, \xi_n, \eta_n) \in \tilde{N}\}.\end{aligned}$$

Hence $\tilde{M}$ is a quasi-algebraic set of $\mathbb{R}^{2m}$.

2.6

In this subsection we shall demonstrate that, using the second form of Seidenberg's lemma, we can drop the log function in the inequality which appeared in Lemma 2.1. To this end, we prove

Lemma 2.2

$$M = \{(\tau, \eta) \in \mathbb{C}^n \mid \operatorname{Im} \tau < -\gamma_0, \eta \in \mathbb{R}^{n-1}, R(\tau, \eta) = 0\}$$

is a quasi-algebraic set of $\mathbb{C}^n$.

Proof

We divide the proof into two parts.

Part 1. First, we wish to derive an alternative representation of the Lopatinski determinant. Since

$$R(\tau, \eta) = \det\left(\frac{1}{2\pi i} \oint \frac{B_j(\tau, \xi, \eta)\xi^{k-1}}{\prod_{i=1}^{m_+} (\xi - \xi_i^+(\tau, \eta))} d\xi\right)_{j,k=1,\dots,m_+}$$

letting

$$\mathscr{R}(\tau,\eta;\xi_1,\ldots,\xi_{m_+}) = \det\left(\frac{1}{2\pi i}\oint\frac{B_j(\tau,\xi,\eta)\xi^{k-1}}{\prod_{i=1}^{m_+}(\xi-\xi_i)}\mathrm{d}\xi\right)_{j,k=1,\ldots,m_+}$$

we see that

$$R(\tau,\eta) = \mathscr{R}(\tau,\eta;\xi_1{}^+(\tau,\eta),\ldots,\xi_{m_+}^+(\tau,\eta)).$$

On the other hand, setting

$$B_j(\tau,\xi,\eta) = \sum_{l=0}^{r_j} b_{jl}(\tau,\eta)\xi^{r_j-l},$$

we have

$$\frac{1}{2\pi i}\oint\frac{B_j(\tau,\xi,\eta)\xi^{k-1}}{\prod_{i=1}^{m_+}(\xi-\xi_i)}\mathrm{d}\xi = \sum_{l=0}^{r_j} b_{jl}(\tau,\eta)\frac{1}{2\pi i}\oint\frac{\xi^{r_j-l+k-1}}{\prod_{i=1}^{m_+}(\xi-\xi_i)}\mathrm{d}\xi.$$

Since

$$\frac{1}{2\pi i}\oint\frac{\xi^{s-1}}{\prod_{i=1}^{m_+}(\xi-\xi_i)}\mathrm{d}\xi$$

is a function which is *everywhere continuous in terms of the degree of homogeneity* $s-m_+$ with respect to the vector $(\xi_1,\ldots,\xi_{m_+})$,[†] it is zero if $s-m_+<0$; otherwise it is a homogeneous polynomial of degree $s-m_+$. Moreover, it is symmetric with respect to the vector $(\xi_1,\ldots,\xi_{m_+})$. Therefore $\mathscr{R}(\tau,\eta;\xi_1,\ldots,\xi_{m_+})$ is a polynomial with respect to $(\tau,\eta,\xi_1,\ldots,\xi_{m_+})$ and symmetric with respect to $(\xi_1,\ldots,\xi_{m_+})$.

Part 2. We write

$$\begin{aligned}A(\tau,\xi,\eta) &= \xi^m + a_1(\tau,\eta)\xi^{m-1} + \ldots + a_m(\tau,\eta)\\ &= \prod_{i=1}^m(\xi-\xi_i(\tau,\eta)).\end{aligned}$$

Then, we have

$$(*)\qquad \begin{cases}\sum\xi_j = -a_1(\tau,\eta),\\ \sum_{i<j}\xi_i\cdot\xi_j = a_2(\tau,\eta),\\ \quad\vdots\\ \xi_1\ldots\xi_m = (-1)^m a_m(\tau,\eta)\end{cases}$$

This is equivalent to saying that as a set $\{\xi_1,\ldots,\xi_m\}$ is identical with $\{\xi_1(\tau,\eta),\ldots,\xi_m(\tau,\eta)\}$.

† This means that, if we regard f as a function defined over $\mathbb{R}^{m_+}$, f is everywhere continuous over $\mathbb{R}^{m_+}$, and satisfies the following condition: for an arbitrary $\lambda>0$, $f(\lambda\xi_1,\ldots,\lambda\xi_{m_+}) = \lambda^{s-m_+}f(\xi_1,\ldots,\xi_{m_+})$.

On the other hand, for $\operatorname{Im}\tau < -\gamma_0$, $\eta\in\mathbb{R}^{n-1}$, we have

$$A(\tau,\xi,\eta) = \prod_{i=1}^{m+}(\xi - \xi_i^+(\tau,\eta)) \prod_{i=1}^{m-}(\xi - \xi_i^-(\tau,\eta)),$$

$$\operatorname{Im}\xi_i^+(\tau,\eta) > 0, \quad \operatorname{Im}\xi_i^-(\tau,\eta) < 0.$$

Therefore, letting

$$(**) \qquad \operatorname{Im}\xi_1 \geqq \operatorname{Im}\xi_2 \geqq \ldots \geqq \operatorname{Im}\xi_m$$

we see that as a set $\{\xi_1,\ldots,\xi_m\}$ which satisfies $(*)$ and $(**)$ becomes

$$\{\xi_1,\ldots,\xi_{m+}\} = \{\xi_1^+(\tau,\eta),\ldots,\xi_{m+}^+(\tau,\eta)\},$$
$$\{\xi_{m_+ +1},\ldots,\xi_m\} = \{\xi_1^-(\tau,\eta),\ldots,\xi_{m-}^-(\tau,\eta)\},$$

therefore, if we set

$$N = \{(\tau,\eta,\xi_1,\ldots,\xi_m)\in\mathbb{C}^{n+m} | \operatorname{Im}\tau < -\gamma_0,\ \operatorname{Im}\eta = 0; (*),(**);$$
$$\mathscr{R}(\tau,\eta;\xi_1,\ldots\xi_{m+}) = 0\},$$

then N becomes a quasi-algebraic set. Hence

$$M = \{(\tau,\eta)\in\mathbb{C}^n | (\tau,\eta,\xi_1,\ldots,\xi_m)\in N\}$$

is also a quasi-algebraic set. ■

Theorem 2 *(the first necessary condition for $\mathscr{E}$-well-posedness)*
If (P) is $\mathscr{E}$-well-posed, then there exists a positive number $\gamma_1 \geqq \gamma_0$ such that $R(\tau,\eta) \neq 0$ for $\{\operatorname{Im}\tau < -\gamma_1, \eta\in\mathbb{R}^{n-1}\}$.

Proof
By Lemma 2.2,

$$M = \{(\tau,\eta)\in\mathbb{C}^n | \operatorname{Im}\tau < -\gamma_0,\ \eta\in\mathbb{R}^{n-1},\ R(\tau,\eta) = 0\}$$

is a quasi-algebraic set. Letting $M \neq \varnothing$, we see that for $s \geqq$ some s_0,

$$M_s = \{(\tau,\eta)\in M \,|\, |\tau|^2 + |\eta|^2 < s\} \neq \varnothing.$$

We write

$$\mu(s) = \sup_{(\tau,\eta)\in M_s} \{-\operatorname{Im}\tau\}$$

and then apply the second form of Seidenberg's lemma. Obviously

$$\gamma_0 \leqq \mu(s) \leqq \sqrt{s}.$$

Therefore, there exist A and α such that

$$\mu(s) = As^\alpha(1 + o(1)), \quad s \to +\infty.$$

At the same time, by Lemma 2.1 we see that

$$M \subset \{-\operatorname{Im}\tau \leqq p\{\log(1 + |\tau| + |\eta|) + 1\},\ \eta\in\mathbb{R}^{n-1}\}.$$

This means that

$$A \geqq \gamma_0, \quad \alpha = 0.$$

That is,

$$\mu(s) = A + o(1), \quad s \to +\infty.$$

Since $\mu(s)$ is an increasing function, we see that

$$\sup_{(\tau,\eta)\in M} \{-\operatorname{Im}\tau\} = A.$$

Setting $A = \gamma_1$ we see that the statement is true. ■

3. Another necessary condition for $\mathscr{E}$-well-posedness

In §2, we obtained one of the necessary conditions for the initial value problem being $\mathscr{E}$-well-posed. In this section we shall find the remaining necessary condition. As far as $R(\tau, \eta)$ is concerned, we shall extend the domain of η to a complex domain, and, having done so, we shall study the behaviour of $R(\tau, \eta)$ within the conceptual framework of 'a function of hyperbolic type'.

3.1

In this subsection, we begin by giving the definition of a function of hyperbolic type in general terms, and establish two lemmas which are useful in the subsequent argument. As we observed in the previous subsection, if the support of a function f defined in $\mathbb{R}^n$ is contained in $B_a - \Gamma'$, then, by a Fourier–Laplace transform $\mathscr{L}f = g$, $g(\zeta)$ becomes regular (or holomorphic) in the domain $\mathbb{R}^n + i\Omega$ and Ω stretches in the direction of Γ, i.e. $\Omega + \Gamma = \Omega$. If $g(\zeta) \neq 0$, then $g(\zeta)^{-1}$ is also regular. What about the inverse Fourier–Laplace transform of such $g(\zeta)^{-1}$? Is the support of $\mathscr{L}^{-1}[g^{-1}]$ contained in $B_b - \Gamma'$? The idea of hyperbolicity will be the key for the answers to these problems.

Let $\Gamma \subset \mathbb{R}^n \backslash \{0\}$ be a cone. Then $g(\zeta)$ is said to be *hyperbolic in the direction of* $-\Gamma$ iff it satisfies the following conditions (a)–(d):

(a) $g(\zeta)$ is regular in $\mathbb{R}^n + i\Omega$, where Ω is a domain of $\mathbb{R}^n$ stretching in the direction of Γ, i.e. $\Omega + \Gamma \subset \Omega$.

(b) There exists $h \in \mathbb{Z}$ such that for any $\zeta^0 \in \mathbb{R}^n + i\Omega$, and any $\zeta^1 \in \mathbb{R}^n + i\Gamma$,

$$\lambda^{-h} g(\zeta^0 + \lambda\zeta) \to g_0(\zeta) \not\equiv 0, \quad \text{where } \lambda\zeta \in \mathbb{R}^n + i\Gamma, \quad \text{as } |\lambda| \to +\infty,$$

is uniformly convergent for $\zeta \in U$ (a neighbourhood of ζ^1).

(c) There exists $\zeta^0 \in \mathbb{R}^n + i\Omega$ such that

$$g(\zeta^0 + (\mathbb{R}^n + i\Gamma)) \neq 0.$$

(d) $g_0(\Gamma) \neq 0$.

Note that from condition (b), $g_0(\zeta)$ is regular in $\mathbb{R}^n + i\Gamma$, and does not depend on $\zeta^0 \in \mathbb{R}^n + i\Omega$. To see this, we observe that for $\zeta^0 \in \mathbb{R}^n + i\Omega$, $\zeta \in U$,

$$\lambda^{-h} g(\zeta^0 + \lambda\zeta) - g_0(\zeta) = \left\{ \lambda^{-h} g\left(\zeta^0 + \lambda\left(\frac{\zeta^0 - \zeta^0}{\lambda} + \zeta\right)\right) - g_0\left(\frac{\zeta^0 - \zeta^0}{\lambda} + \zeta\right)\right\} + \left\{ g_0\left(\frac{\zeta^0 - \zeta^0}{\lambda} + \zeta\right) - g_0(\zeta)\right\}.$$

Then, letting $|\lambda| \to +\infty$, we see that the first term on the right hand side of the above equality tends to zero owing to the uniform convergence of g, and the second one tends to zero owing to the continuity of g_0.

Let us call $g_0(\zeta)$ the *principal part* of $g(\zeta)$. We now see that

$$g_0(\mu\zeta) = \mu^h g_0(\zeta), \quad \mu > 0, \quad \zeta \in \mathbb{R}^n + i\Gamma.$$

In fact we observe that

$$\lambda^{-h} g(\zeta^0 + \lambda\mu\zeta) \to g_0(\mu\zeta) \text{ as } |\lambda| \to +\infty,$$

$$\lambda^{-h} g(\zeta^0 + \lambda\mu\zeta) = \mu^h (\lambda\mu)^{-h} g(\zeta^o + (\lambda\mu)\zeta) \to \mu^h g_0(\zeta) \quad \text{as } |\lambda| \to +\infty.$$

Since $h \in \mathbb{Z}$, we see that the right hand side of

$$g_0(\mu\zeta) = \mu^h g_0(\zeta), \quad \mu > 0$$

has a single-valued and regular analytic continuation to $\mathbb{C}^1 \backslash \{0\}$ with respect to μ. Therefore, $g_0(\zeta)$ also has a regular analytical continuation to

$$\bigcup_{\mu \in \mathbb{C}^1 \backslash \{0\}} \mu(\mathbb{R}^n + i\Gamma).$$

Then clearly

$$g_0(\lambda\zeta) = \lambda^h g_0(\zeta), \quad \lambda \in \mathbb{C}^1 \backslash \{0\}, \quad \zeta \in \bigcup_{\mu \in \mathbb{C}^1 \backslash \{0\}} \mu(\mathbb{R}^n + i\Gamma).$$

Lemma 3.1

If $g(\zeta)$ is hyperbolic in the direction of $-\Gamma$, then

$$g_0(\mathbb{R}^n \pm i\Gamma) \neq 0.$$

Proof

Since $g_0(\mathbb{R}^n - i\Gamma) = (-1)^h g_0(\mathbb{R}^n + i\Gamma)$, it is enough to prove $g_0(\mathbb{R}^n + i\Gamma) \neq 0$. To do this we derive a contradiction by assuming that there exists

$$\zeta^1 = \xi^1 + i\eta^1 \in \mathbb{R}^n + i\Gamma$$

such that $g_0(\zeta^1) = 0$. We divide the proof into three parts.

Part 1. For $\operatorname{Re}\lambda > 0$,

$$\xi^1 + i\lambda\eta^1 = (\xi^1 - \operatorname{Im}\lambda\eta^1) + i\operatorname{Re}\lambda\eta^1 \in \mathbb{R}^n + i\Gamma.$$

From this, by writing

$$\varphi_0(\lambda) = g_0(\xi^1 + \mathrm{i}\lambda\eta^1)$$

we see that $\varphi_0(\lambda)$ is regular at $\mathrm{Re}\,\lambda > 0$. Also, from the fact that $g_0(\Gamma) \neq 0$, we have

$$\lambda^{-h}\varphi_0(\lambda) = \lambda^{-h}g_0(\xi^1 + \mathrm{i}\lambda\eta^1) = g_0(\xi^1/\lambda + \mathrm{i}\eta^1)$$
$$\to g_0(\mathrm{i}\eta^1) = \mathrm{i}^h g_0(\eta^1) \neq 0 \quad \text{as } 0 < \lambda \to +\infty,$$

therefore $\varphi_0(\lambda) \not\equiv 0$. Since $\varphi_0(1) = 0$, $\lambda = 1$ becomes a zero point of $\varphi_0(\lambda)$ of degree $l(\geqq 1)$. That is, in a neighbourhood of $\lambda = 1$, we can write

$$\varphi_0(\lambda) = (\lambda - 1)^l c(\lambda), \quad c(\lambda) \neq 0.$$

Therefore, for $0 < \delta < \delta_0$ we see that

$$\inf_{|\lambda - 1| = \delta} |\varphi_0(\lambda)| = c_\delta > 0.$$

Part 2. On the other hand, for $\zeta^0 = \xi^0 + \mathrm{i}\eta^0 \in \mathbb{R}^n + \mathrm{i}\Omega$ we have

$$g(\zeta^0 + \mathbb{R}^n + \mathrm{i}\Gamma) \neq 0, \quad \text{for any } \zeta^0.$$

Since, for $\mathrm{Re}\,\lambda > 0$, $\mu > 0$ we have

$$\mu(\xi^1 + \mathrm{i}\lambda\eta^1) \in \mathbb{R}^n + \mathrm{i}\Gamma,$$

it follows that

$$\varphi_\mu(\lambda) = g(\zeta^0 + \mu(\xi^1 + \mathrm{i}\lambda\eta^1))$$

is regular in $\mathrm{Re}\,\lambda > 0$ and is never zero.

Part 3. We see that when $|\lambda - 1| \leqq \delta_1 (< \delta_0)$,

$$\mu^{-h}\varphi_\mu(\lambda) = \mu^{-h}g(\zeta^0 + \mu(\xi^1 + \mathrm{i}\lambda\eta^1)) \to g_0(\xi^1 + \mathrm{i}\lambda\eta^1) = \varphi_0(\lambda) \quad \text{as } \mu \to +\infty,$$

where the convergence is uniform. From this, for $\mu \geqq \mu_0$, we have

$$\sup_{|\lambda - 1| = \delta_1} |\mu^{-h}\varphi_\mu(\lambda) - \varphi_0(\lambda)| < c_{\delta_1}.$$

From the result obtained in part 1, for $\mu \geq \mu_0$ we see that

$$|\mu^{-h}\varphi_\mu(\lambda) - \varphi_0(\lambda)| < |\varphi_0(\lambda)|, \quad |\lambda - 1| = \delta_1.$$

This implies that, by Rouché's theorem,† the number of roots of $\varphi_\mu(\lambda) = 0$ in $|\lambda - 1| < \delta_1$ is equal to that of $\varphi_0(\lambda) = 0$ That is, the number is exactly equal to $l(\geqq 1)$.

This contradicts the result obtained in part 2. ■

Lemma 3.2

Let $g(\zeta)$ satisfy the conditions (a), (b) and (d) in the definition of a hyperbolic function. Assume that there exists $\eta^* \in \Gamma$ such that Γ stretches in the

† Translator's note: See E. C. Titchmarsh, *The theory of functions* (second edition). Oxford University Press (1939) p. 116.

direction of η^*, i.e. $\Gamma^* = \{\eta = t\eta^* | t > 0\}$. Then the following condition holds.

(c′) If there exists $\zeta^0 \in \mathbb{R}^n + i\Omega$ and

$$g(\zeta^0 + (\mathbb{R}^n + i\Gamma^*)) \neq 0,$$

then (c) is true.

Proof

We assume that there exists $\zeta^1 = \xi^1 + i\eta^1 \in \mathbb{R}^n + i\Gamma$, and $g(\zeta^0 + \zeta^1) = 0$, and then derive a contradiction.

Part 1. Let Re $\lambda \geq 0$, Re $\nu \geq 0$. Since

$$\begin{aligned}(\xi^0 + i\eta^0) &+ (\xi^1 + i\lambda\eta^1) + i\nu\eta^* \\ &= (\xi^0 + \xi^1 - \operatorname{Im}\lambda\eta^1 - \operatorname{Im}\nu\eta^*) + i(\eta^0 + \operatorname{Re}\lambda\eta^1 + \operatorname{Re}\nu\eta^*) \\ &\in \mathbb{R}^n + i(\eta^0 + \Gamma\cup\{0\} + \Gamma^*\cup\{0\}) \\ &= \mathbb{R}^n + i(\eta^0 + \Gamma\cup\{0\}) \subset \mathbb{R}^n + i\Omega,\end{aligned}$$

putting

$$\varphi(\lambda, \nu) = g(\zeta^0 + (\xi^1 + i\lambda\eta^1) + i\nu\eta^*)$$

we see from (a) that $\varphi(\lambda, \nu)$ is regular in this domain. By fixing $\nu = \nu_0$, we find from (b), (d) that

$$\lambda^{-h}\varphi(\lambda, \nu_0) = \lambda^{-h}g(\zeta^0 + (\xi^1 + i\lambda\eta^1) + i\nu_0\eta^*) \to g_0(i\eta^1) \neq 0 \text{ as } 0 < \lambda \to +\infty,$$

therefore $\varphi(\lambda, \nu_0) \not\equiv 0$.

Since $\varphi(1, 0) = g(\zeta^0 + \zeta^1) = 0$, by *Weierstrass' preparation theorem* there exists a continuous $\lambda = \lambda(\nu)$ in the neighbourhood of $\nu = 0$ and $\lambda(0) = 1$, $\varphi(\lambda(\nu), \nu) = 0$ (see Fig. 6).

Let Re $\lambda = 0$, Re $\nu > 0$. We have

$$(\xi^1 + i\lambda\eta^1) + i\nu\eta^* \in \mathbb{R}^n + i\Gamma^*.$$

Then, from (c′) we conclude that

$$\varphi(\lambda, \nu) = g(\zeta^0 + (\xi^1 + i\lambda\eta^1) + i\nu\eta^*) \neq 0.$$

This shows that if Re $\nu > 0$ then Re $\lambda(\nu) > 0$.

Part 2. Next, we wish to see that $\lambda(\nu)$ can be infinitely extended along the positive axis of ν. If there exists a cluster point λ_0 of $\lambda(\nu)$ as $\nu \to \nu_0$, then Re$\lambda_0 > 0$, $\varphi(\lambda_0, \nu_0) = 0$. By using the same Weierstrass' preparation theorem at $(\lambda, \nu) = (\lambda_0, \nu_0)$, $\lambda(\nu)$ has a continuous extension beyond $\nu = \nu_0$. Letting $|\lambda(\nu)| \to +\infty$ as $\nu \to \nu_0$, we have

$$0 = \lambda(\nu)^{-h}\varphi(\lambda(\nu), \nu) = \lambda(\nu)^{-h}g(\xi^0 + (\xi^1 + i\lambda(\nu)\eta^1) + i\nu\eta^*) \\ \to g_0(i\eta^1) \neq 0,$$

a contradiction.

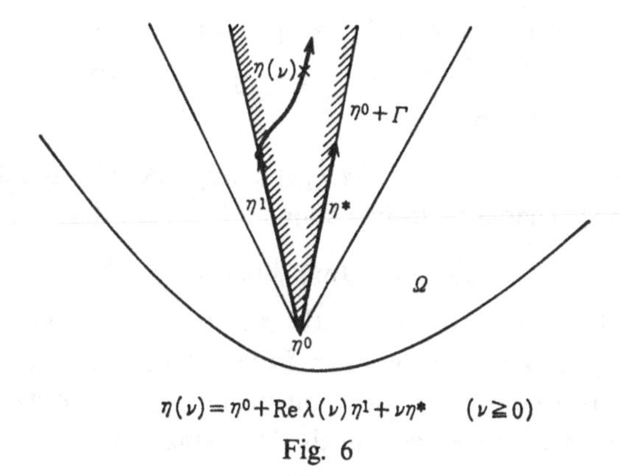

$\eta(\nu) = \eta^0 + \text{Re}\,\lambda(\nu)\eta^1 + \nu\eta^* \quad (\nu \geqq 0)$

Fig. 6

Therefore, there exists a continuous function $\lambda(\nu)$ defined over $\nu \geqq 0$ such that $\text{Re}\,\lambda(\nu) > 0$, $\varphi(\lambda(\nu), \nu) = 0$.

Part 3. Now, consider the case that there exists $\nu_k > 0$ such that

$$\frac{\nu_k}{\lambda(\nu_k)} = \nu_k' \to \nu_0' \quad \text{as } \nu_k \to \infty.$$

Since $\text{Re}\,\nu_k' > 0$, we find $\text{Re}\,\nu_0' \geqq 0$, therefore

$$\eta^1 + \text{Re}\,\nu_0'\eta^* \in \Gamma.$$

Also, from (b) we see that

$$\begin{aligned} 0 &= \lambda(\nu_k)^{-h}\varphi(\lambda(\nu_k), \nu_k) \\ &= \lambda(\nu_k)^{-h} g(\zeta^0 + \xi^1 + \lambda(\nu_k)(\mathrm{i}\eta^1 + \mathrm{i}\nu_k'\eta^*)) \\ &\to g_0(\mathrm{i}\eta^1 + \mathrm{i}\nu_0'\eta^*) \neq 0, \end{aligned}$$

which is a contradiction.

Next, consider the case that there exists $\nu_k > 0$ such that

$$\frac{\lambda(\nu_k)}{\nu_k} = \lambda_k' \to \lambda_0' \quad \text{as } \nu_k \to \infty.$$

From this we find $\text{Re}\,\lambda_k' > 0$, so that $\text{Re}\,\lambda_0' \geqq 0$. Hence $\text{Re}\,\lambda_0'\eta^1 + \eta^* \in \Gamma$. From (b) we see that

$$\begin{aligned} 0 &= \nu_k^{-h}\varphi(\lambda(\nu_k), \nu_k) \\ &= \nu_k^{-h} g(\zeta^0 + \xi^1 + \nu_k(\mathrm{i}\lambda_k'\eta^1 + \mathrm{i}\eta^*)) \\ &\to g_0(\mathrm{i}\lambda_0'\eta^1 + \mathrm{i}\eta^*) \neq 0. \end{aligned}$$

We have arrived at a contradiction again. ■

3.2

In this subsection we use the general knowledge of hyperbolic functions acquired in §3.1 to deal with the polynomial $A(\tau, \xi, \eta)$ where we assume the condition (i) of Hypothesis (A) (see p. 59) i.e.

$$A(\tau,\xi,\eta)\neq 0,\quad \operatorname{Im}\tau<-\gamma_0,\quad (\xi,\eta)\in\mathbb{R}^n,$$
$$A_0(1,0,0)\neq 0.$$

This is the same as saying that $A(\tau,\xi,\eta)$ is hyperbolic in the direction of $(1,0,0)$ (see §3.1). Therefore, from Lemma 3.1, we see that

$$A_0(\tau,\xi,\eta)\neq 0,\quad \operatorname{Im}\tau\neq 0,\quad (\xi,\eta)\in\mathbb{R}^n.$$

This implies that for $(\xi,\eta)\in\mathbb{R}^n$ all the roots of $A_0(\tau,\xi,\eta)=0$ are real numbers. We take the largest number from them and call it $\tau(\xi,\eta)$. Now, we see that τ is a continuous function defined in the domain $(\xi,\eta)\in\mathbb{R}^n$, and is homogeneous of degree 1 in the following sense:

$$\tau(\lambda\xi,\lambda\eta)=\lambda\tau(\xi,\eta),\quad \lambda>0.$$

Setting

$$\Gamma=\{(\tau,\xi,\eta)\in\mathbb{R}^{n+1}\,|\,\tau>\tau(\xi,\eta)\},$$

we see that Γ is a conical domain which stretches in the direction of $(1,0,0)$ such that

$$A_0(\tau,\xi,\eta)\neq 0,\quad (\tau,\xi,\eta)\in\Gamma.$$

Therefore, we see that Γ is a connected component of the set

$$\{A_0(\tau,\xi,\eta)\in\mathbb{R}^{n+1}\,|\,A_0(\tau,\xi,\eta)\neq 0\}$$

containing the point $(1,0,0)$.

Here we have that Γ is a conical domain stretching in the direction of $(1,0,0)$ such that

$$A(\tau,\xi,\eta)\neq 0,\quad \operatorname{Im}\tau<-\gamma_0,\quad (\xi,\eta)\in\mathbb{R}^n,$$
$$A_0(\tau,\xi,\eta)\neq 0,\quad (\tau,\xi,\eta)\in\Gamma.$$

From this and Lemma 3.2 we see that

$$A(\tau,\xi,\eta)\neq 0,\quad (\tau,\xi,\eta)\in\mathbb{R}^{n+1}-\mathrm{i}(\gamma_0(1,0,0)+\Gamma),$$

i.e. $A(\tau,\xi,\eta)$ is hyperbolic in the direction of Γ. Therefore, from Lemma 3.1 again, we have

$$A_0(\tau,\xi,\eta)\neq 0,\quad (\tau,\xi,\eta)\in\mathbb{R}^{n+1}-\mathrm{i}\Gamma.$$

We regard this as a condition which guarantees the convexity of Γ.

Lemma 3.3
Γ is convex.

Proof
Take an arbitrary $e_0\in\Gamma$ and basis $\{e_0,e_1,\ldots,e_n\}$ of $\mathbb{R}^{n+1}$ containing e_0.

Then we have a representation

$$(\tau, \xi, \eta) = \tau' e_0 + \xi' e_1 + \eta_1' e_2 + \ldots + \eta_{n-1}' e_n.$$

Let

$$A_0(\tau' e_0 + \xi' e_1 + \eta_1' e_2 + \ldots + \eta_{n-1}' e_n) = A_0'(\tau', \xi', \eta'),$$

then we see that

$$A_0'(\tau', \xi', \eta') \neq 0, \quad \mathrm{Im}\,\tau' \neq 0, \quad (\xi', \eta') \in \mathbb{R}^n,$$
$$A_0'(1,0,0) \neq 0.$$

This shows that we can define Γ' from A_0' by the same procedure that we employed when we defined Γ from A_0. We see that Γ' is a conical domain stretching in the direction of $(1,0,0)$ in (τ', ξ', η')-space. Γ' is also a connected component of the set

$$\{(\tau', \xi', \eta') \in \mathbb{R}^{n+1} \mid A_0'(\tau', \xi', \eta') \neq 0\}$$

containing $(1,0,0)$. Since Γ is a connected component of the set

$$\{(\tau, \xi, \eta) \in \mathbb{R}^{n+1} | A_0(\tau, \xi, \eta) \neq 0\}$$
$$= \{(\tau, \xi, \eta) \in \mathbb{R}^{n+1} | (\tau, \xi, \eta) = \tau' e_0 + \xi' e_1 + \eta'_1 e_2 + \ldots + \eta_{n-1}' e_n, A_0'(\tau', \xi', \eta') \neq 0\}$$

containing e_0, we can say that

$$\Gamma = \{(\tau, \xi, \eta) \in \mathbb{R}^{n+1} | (\tau, \xi, \eta) =$$
$$\tau' e_0 + \xi' e_1 + \eta_1' e_2 + \ldots + \eta_{n-1}' e_n, (\tau', \xi', \eta') \in \Gamma'\}$$

From the fact that

$$\lambda(1,0,0) + \Gamma' \subset \Gamma', \quad \lambda > 0,$$

i.e. Γ' stretches in the direction of $(1,0,0)$, we see that

$$\lambda e_0 + \Gamma$$
$$= \{(\tau, \xi, \eta) \in \mathbb{R}^{n+1} | (\tau, \xi, \eta) = \tau' e_0 + \ldots + \eta_{n-1}' e_n, (\tau' \xi', \eta') \in \lambda(1,0,0) + \Gamma'\}$$
$$\subset \{(\tau, \xi, \eta) \in \mathbb{R}^{n+1} | (\tau, \xi, \eta) = \tau' e_0 + \ldots + \eta_{n-1}' e_n, (\tau', \xi', \eta') \in \Gamma'\}$$
$$= \Gamma,$$

therefore Γ stretches in the direction of e_0. Since $e_0 \in \Gamma$ was arbitrary, Γ stretches in the direction of Γ itself. That is, Γ is convex. ∎

Let us write $\dot{\Gamma}$ for the orthogonal projection of Γ on to the (τ, η)-space. This means

$$\dot{\Gamma} = \{(\tau, \eta) \in \mathbb{R}^n | (\tau, \xi, \eta) \in \Gamma\}.$$

It is obvious that $\dot{\Gamma}$ is a convex conical domain of $\mathbb{R}^n$. Recall that Γ can be expressed as

$$\Gamma = \{(\tau, \xi, \eta) | \tau > \tau(\xi, \eta)\}.$$

by using a continuous function $\tau(\xi, \eta)$ defined in $\mathbb{R}^n$. From this we find

$$\dot{\Gamma} = \{(\tau, \eta) | \tau > \tau(\eta)\}, \quad \tau(\eta) = \inf_{\xi \in \mathbb{R}^1} \tau(\xi, \eta).$$

Because $\dot{\Gamma}$ is convex, $\tau(\eta)$ becomes a continuous function in $\mathbb{R}^{n-1}$.

On the other hand, from the convexity of Γ we see that for $(\tau,\eta)\in\dot{\Gamma}$, the set

$$\{\xi|(\tau,\xi,\eta)\in\Gamma\}$$

must be an open interval. We write

$$\xi_{\min}(\tau,\eta)=\inf_{(\tau,\xi,\eta)\in\Gamma}\xi,\quad \xi_{\max}(\tau,\eta)=\sup_{(\tau,\xi,\eta)\in\Gamma}\xi$$

for both ends of the interval. Then, obviously these are continuous functions of $(\tau,\eta)\in\mathbb{R}^n$, and

$$\Gamma=\{(\tau,\xi,\eta)\in\mathbb{R}^{n+1}|(\tau,\eta)\in\dot{\Gamma},\quad \xi_{\min}(\tau,\eta)<\xi<\xi_{\max}(\tau,\eta)\}.$$

Using the knowledge acquired so far, we examine the roots $A(\tau,\xi,\eta)=0$ with respect to ξ. We see that

$$A(\tau,\xi,\eta)\neq 0,\quad (\tau,\xi,\eta)\in\mathbb{R}^{n+1}-\mathrm{i}(\gamma_0(1,0,0)+\Gamma),$$

which implies

$$A(\tau,\xi,\eta)\neq 0,\quad -(\operatorname{Im}\tau+\gamma_0,\operatorname{Im}\xi,\operatorname{Im}\eta)\in\Gamma,$$

that is, we have $A(\tau,\xi,\eta)\neq 0$ for

$$-(\operatorname{Im}\tau-\gamma_0,-\operatorname{Im}\eta)\in\dot{\Gamma},$$

$$\xi_{\min}(-\operatorname{Im}\tau-\gamma_0,-\operatorname{Im}\eta)<-\operatorname{Im}\xi<\xi_{\max}(-\operatorname{Im}\tau-\gamma_0,-\operatorname{Im}\eta).$$

Since the roots of $A(\tau,\xi,\eta)=0$ with respect to ξ are continuous with respect to (τ,η), there are two distinct families of roots, namely, ξ satisfying

$$-\operatorname{Im}\xi\leqq\xi_{\min}(-\operatorname{Im}\tau-\gamma_0,-\operatorname{Im}\eta),$$

and ξ satisfying

$$-\operatorname{Im}\xi\geqq\xi_{\max}(-\operatorname{Im}\tau-\gamma_0,-\operatorname{Im}\eta),$$

provided

$$(-\operatorname{Im}\tau-\gamma_0,-\operatorname{Im}\eta)\in\dot{\Gamma}$$

In particular, for $-\operatorname{Im}\tau-\gamma_0>0,\ -\operatorname{Im}\eta=0$, the first family represents the roots whose imaginary parts are all positive, and the second family represents the roots whose imaginary parts are negative. Write these families as

$$\{\xi_j^{+}(\tau,\eta)\}_{j=1,\ldots,m_+},\ \{\xi_j^{-}(\tau,\eta)\}_{j=1,\ldots,m_-},$$

respectively. Now, consider a simple closed curve C_+ on the complex ξ-plane such that the interior domain of C_+ contains $\{\xi_j^{+}(\tau,\eta)\}_{j=1,\ldots,m_+}$ and the exterior domain of C_+ contains $\{\xi_j^{-}(\tau,\eta)\}_{j=1,\ldots,m_-}$. Then, we see that

$$S_k^{+}(\tau,\eta)=\sum_{j=1}^{m_+}(\xi_j^{+}(\tau,\eta))^k=\frac{1}{2\pi\mathrm{i}}\oint_{C_+}\frac{(\partial/\partial\xi)A(\tau,\xi,\eta)\xi^k}{A(\tau,\xi,\eta)}\mathrm{d}\xi$$

is regular for $(\tau, \eta) \in \mathbb{R}^n - i(\gamma_0(1,0) + \dot{\Gamma})$. Therefore, for an arbitrary symmetric polynomial $P(\xi_1, \ldots, \xi_{m_+})$, $P(\xi_1^+(\tau,\eta), \ldots, \xi_{m_+}^+(\tau,\eta))$ is also regular in the same domain.

Recall that a Lopatinski determinant $R(\tau, \eta)$ defined at $\operatorname{Im}\tau < -\gamma_0$, $\eta \in \mathbb{R}^{n-1}$ (see §2) can be expressed as

$$R(\tau,\eta) = \mathscr{R}(\tau,\eta;\xi_1^+(\tau,\eta), \ldots, \xi_{m_+}^+(\tau,\eta)),$$

where $\mathscr{R}$ is a polynomial of $(\tau, \eta, \xi_1^+, \ldots, \xi_{m_+}^+)$ and is symmetric with respect to $(\xi_1^+, \ldots, \xi_{m_+}^+)$. Therefore, $R(\tau, \eta)$ has a regular analytic continuation to $\mathbb{R}^n - i(\gamma_0(1,0) + \dot{\Gamma})$. For the same reason, if we put

$$\begin{aligned} A_+(\tau,\xi,\eta) &= \prod_{j=1}^{m_+} (\xi - \xi_j^+(\tau,\eta)) \\ &= \xi^{m_+} + a_1^+(\tau,\eta)\xi^{m_+-1} + \ldots + a_{m_+}^+(\tau,\eta) \end{aligned}$$

then $\{a_j^+(\tau,\eta)\}_{j=1,\ldots,m_+}$ also has a regular analytical continuation to $\mathbb{R}^n - i\{\gamma_0(1,0) + \dot{\Gamma}\}$.

3.3

In this subsection, by an argument similar to the one in the previous subsection, but with the addition of an extra parameter v, we demonstrate that the principal part $R_0(\tau, \eta)$ of $R(\tau, \eta)$ can be defined. To see this, we write $A_k(\tau, \xi, \eta)$ for the sum of the homogeneous terms of degree $(m-k)$ appearing in the polynomial $A(\tau, \xi, \eta)$ of degree m. Using them, we can write

$$A(\tau,\xi,\eta) = A_0(\tau,\xi,\eta) + A_1(\tau,\xi,\eta) + \ldots + A_m.$$

Set

$$v^m A\left(\frac{\tau}{v}, \frac{\xi}{v}, \frac{\eta}{v}\right) = A_0(\tau,\xi,\eta) + vA_1(\tau,\xi,\eta) + \ldots + v^m A_m = \tilde{A}(\tau,\xi,\eta;v),$$

where $\tilde{A}(\tau,\xi,\eta;v)$ is a homogeneous polynomial of degree m with respect to (τ, ξ, η, v). We see that, in $0 \leqq v \leqq 1$,

$$\tilde{A}(\tau,\xi,\eta;v) \neq 0, \quad (\tau,\xi,\eta) \in \mathbb{R}^{n+1} - i(\gamma_0(1,0,0) + \Gamma).$$

In fact, if $0 < v \leqq 1$ then

$$\frac{1}{v}(\tau,\xi,\eta) \in \mathbb{R}^{n+1} - i(\gamma_0(1,0,0) + \Gamma),$$

therefore

$$\tilde{A}(\tau,\xi,\eta;v) = v^m A\left(\frac{\tau}{v}, \frac{\xi}{v}, \frac{\eta}{v}\right) \neq 0,$$

and if $v = 0$, then

$$\tilde{A}(\tau,\xi,\eta;0) = A_0(\tau,\xi,\eta) \neq 0.$$

From this we find that for $0 \leqq \nu \leqq 1$, $(\tau, \eta) \in \mathbb{R}^n - \mathrm{i}(\gamma_0(1,0) + \dot{\Gamma})$ the roots of the equation $\tilde{A}(\tau, \xi, \eta; \nu) = 0$ with respect to ξ can be classified into two families, namely $\{\tilde{\xi}_j^+(\tau, \eta; \nu)\}_{j=1,\ldots,m_+}$ and $\{\tilde{\xi}_j^-(\tau, \eta; \nu)\}_{j=1,\ldots,m_-}$. Note that from the definition of $\tilde{A}$ at $0 < \nu \leqq 1$ we have

$$\xi_j^+(\tau, \eta; \nu) = \nu \xi_j^+\left(\frac{\tau}{\nu}, \frac{\eta}{\nu}\right).$$

Setting

$$\tilde{B}_j(\tau, \xi, \eta; \nu) = \nu^{r_j} B_j\left(\frac{\tau}{\nu}, \frac{\xi}{\nu}, \frac{\eta}{\nu}\right),$$

$$\tilde{\mathscr{R}}(\tau, \eta, \nu; \xi_1, \ldots, \xi_{m_+}) = \det\left(\frac{1}{2\pi \mathrm{i}} \oint \frac{\tilde{B}_j(\tau, \xi, \eta; \nu)\xi^{k-1}}{\prod_{i=1}^{m_+}(\xi - \xi_i)} \mathrm{d}\xi\right)_{j,k=1,\ldots,m_+},$$

$$\tilde{R}(\tau, \eta; \nu) = \tilde{\mathscr{R}}(\tau, \eta, \nu; \tilde{\xi}_1^+(\tau, \eta; \nu), \ldots, \tilde{\xi}_{m_+}^+(\tau, \eta; \nu)),$$

we see that $\tilde{B}_j$ is a homogeneous polynomial of degree r_j, $\tilde{\mathscr{R}}$ is a homogeneous polynomial of degree $\sum_{j=1}^{m_+}(r_j - j + 1)$ (which we call h_0) with respect to $(\tau, \eta, \nu, \xi_1, \ldots, \xi_{m_+})$; and $\tilde{R}$ is a homogeneous polynomial of degree h_0 with respect to (τ, η, ν), and is regular at $\{(\tau, \eta) \in \mathbb{R}^n - \mathrm{i}(\gamma_0(1,0) + \dot{\Gamma}), 0 \leqq \nu \leqq 1\}$.

At the same time, we see that for $0 < \nu \leqq 1$,

$$\begin{aligned}\tilde{R}(\tau, \eta; \nu) &= \det\left(\frac{1}{2\pi \mathrm{i}} \oint \frac{\tilde{B}_j(\tau, \xi, \eta; \nu)\xi^{k-1}}{\prod_{i=1}^{m_+}(\xi - \tilde{\xi}_i^+(\tau, \eta; \nu))} \mathrm{d}\xi\right)_{j,k} \\ &= \det\left(\frac{1}{2\pi \mathrm{i}} \oint \frac{\nu^{r_j} B_j(\tau/\nu, \xi/\nu, \eta/\nu)\xi^{k-1}}{\prod_{i=1}^{m_+}(\xi - \nu\xi_i^+(\tau/\nu, \eta/\nu))} \mathrm{d}\xi\right)_{j,k} \\ &= \nu^{\sum_{j=1}^{m_+}(r_j - j + 1)} \det\left(\frac{1}{2\pi \mathrm{i}} \oint \frac{B_j(\tau/\nu, \xi, \eta/\nu)\xi^{k-1}}{\prod_{i=1}^{m_+}(\xi - \xi_i^+(\tau/\nu, \eta/\nu))} \mathrm{d}\xi\right)_{j,k} \\ &= \nu^{\sum_{j=1}^{m_+}(r_j - j + 1)} R\left(\frac{\tau}{\nu}, \frac{\eta}{\nu}\right) = \nu^{h_0} R\left(\frac{\tau}{\nu}, \frac{\eta}{\nu}\right).\end{aligned}$$

Since $\tilde{R}$ is regular at $\{(\tau, \eta) \in \mathbb{R}^n - \mathrm{i}(\gamma_0(1,0) + \dot{\Gamma}), \quad 0 \leqq \nu \leqq 1\}$ we have

$$\tilde{R}(\tau, \eta; \nu) = \tilde{R}(\tau, \eta; 0) + \frac{\partial \tilde{R}}{\partial \nu}(\tau, \eta; 0)\nu + \frac{1}{2!}\frac{\partial^2 \tilde{R}}{\partial \nu^2}(\tau, \eta; 0)\nu^2 + \ldots.$$

Note that if we take a compact set $\mathscr{K} \subset \mathbb{R}^n - \mathrm{i}(\gamma_0(1,0) + \dot{\Gamma})$ then there exists $\delta_{\mathscr{K}}$ such that in

$$\{(\tau, \eta) \in \mathscr{K}, \quad |\nu| < \delta_{\mathscr{K}}\}$$

the above series becomes uniformly convergent. Letting $\tilde{R}(\tau,\eta;\nu)\not\equiv 0$, we see that there exists at least one coefficient of ν^k which is not identically zero. Let us take such k and write h_1 for the largest one among them. By setting $h=h_0-h_1$, from the previous argument we obtain

Lemma 3.4

$R(\tau,\eta)$ is regular at $\mathbb{R}^n-\mathrm{i}(\gamma_0(1,0)+\dot{\Gamma})$. In particular, if $R(\tau,\eta)\not\equiv 0$ then there exists $h\in\mathbb{Z}$ such that,

$$\nu^h R\left(\frac{\tau}{\nu},\frac{\eta}{\nu}\right)=R_0(\tau,\eta)+R_1(\tau,\eta)\nu+\ \dots,\qquad R_0\not\equiv 0.$$

If we take a compact set $\mathscr{K}\subset\mathbb{R}^n-\mathrm{i}(\gamma_0(1,0)+\dot{\Gamma})$ then we have $\delta_{\mathscr{K}}>0$ such that the series on the right hand side of the above equality becomes uniformly convergent in

$$\{(\tau,\eta)\in\mathscr{K},\quad |\nu|<\delta_{\mathscr{K}}\},$$

and in

$$\{(\tau,\eta)\in\mathscr{K},\quad 0<\nu<\delta_{\mathscr{K}}\}$$

it coincides with the term on the left hand side of the equality. Moreover $R_j(\tau,\eta)$ is regular at $\mathbb{R}^n-\mathrm{i}\dot{\Gamma}$ and is homogeneous of degree $h-j$.

Using Lemma 3.4, we see that the principal part of $R(\tau,\eta)(\not\equiv 0)$ is, in fact, $R_0(\tau,\eta)$. To see this let $(\tau_0,\eta_0)\in\mathbb{R}^n-\mathrm{i}(\gamma_0(1,0)+\dot{\Gamma})$, and $\mathscr{K}$ be a compact set of $\mathbb{R}^n-\mathrm{i}\dot{\Gamma}$. For $(\tau,\eta)\in\mathscr{K}$, $0<\nu<\delta_{\mathscr{K}}$ we have

$$\begin{aligned}\nu^h R\left(\tau_0+\frac{\tau}{\nu},\eta_0+\frac{\eta}{\nu}\right)&=\nu^h R\left(\frac{\nu\tau_0+\tau}{\nu},\frac{\nu\eta_0+\eta}{\nu}\right)\\&=R_0(\nu\tau_0+\tau,\nu\eta_0+\eta)+R_1(\nu\tau_0+\tau,\nu\eta_0+\eta)\nu+\ \dots,\end{aligned}$$

where the series on the right is uniformly convergent. Also, we know that $R_j(\tau,\eta)$ is regular at $\mathbb{R}^n-\mathrm{i}\dot{\Gamma}$. From this

$$R_j(\nu\tau_0+\tau,\nu\eta_0+\eta)\to R_j(\tau,\eta)\quad\text{as }\nu\to 0$$

is uniformly convergent at $(\tau,\eta)\in\mathscr{K}$. Therefore

$$\nu^h R\left(\tau_0+\frac{\tau}{\nu},\eta_0+\frac{\eta}{\nu}\right)\to R_0(\tau,\eta)\quad\text{as }\nu\to +0$$

is also uniformly convergent at $(\tau,\eta)\in\mathscr{K}$.

3.4

We are now able to use the Lopatinski determinant $R(\tau,\eta)$ in a complex domain with respect to (τ,η). Recall that in §2 we obtained Theorem 2 giving a necessary condition for $\mathscr{E}$-well-posedness, but the argument there was confined to the case of η being a real vector. Note that the same argument is valid for η in a complex domain.

More precisely, we can prove

Theorem 2′
If (P) is $\mathscr{E}$-well-posed, then there exists a positive $\gamma_1(\geqq \gamma_0)$ such that $R(\tau, \eta) \neq 0$ when

$$\operatorname{Im}\tau < -\gamma_1(|\operatorname{Im}\eta| + 1), \quad \eta \in \mathbb{C}^{n-1}.$$

A sketch of the proof.

The same argument as was used in the proof of Theorem 2 can be used for this purpose. That is, after proving statements similar to Lemmas 2.1 and 2.2, we apply Seidenberg's lemma.

For the statement corresponding to Lemma 2.1 we obtain, by a similar argument:

Lemma 2.1′
If (P) is $\mathscr{E}$-well-posed, then there exists p such that $R(\tau, \eta) \neq 0$ when

$$\operatorname{Im}\tau < -p\{|\operatorname{Im}\eta| + \log(1 + |\tau| + |\eta|) + 1\}, \quad \eta \in \mathbb{C}^{n-1}.$$

In the proof of Lemma 2.1′ note that if p is sufficiently large, then not only is the range of (τ, η) contained in $\mathbb{R}^n - i(\gamma_0(1, 0) + \dot{\Gamma})$, but also the inequalities $\operatorname{Im}\xi_j^+(\tau, \eta) > 0 \quad (j = 1, \ldots, m_+)$ become valid.

We also have the statement corresponding to Lemma 2.2:

Lemma 2.2′

$$M' = \{(\tau, \eta) \in \mathbb{C}^n | \operatorname{Im}\tau < -\gamma_0(|\operatorname{Im}\eta|^2 + 1)^{1/2}, \quad R(\tau, \eta) = 0\}$$

is a quasi-algebraic set.

The proof of the lemma is similar to that of Lemma 2.2. That is, using the same symbols as there, we can prove that

$$N' = \{(\tau, \eta, \xi_1, \ldots, \xi_m) \in \mathbb{C}^{n+m} | \operatorname{Im}\tau < 0, (\operatorname{Im}\tau)^2 > \gamma_0{}^2(|\operatorname{Im}\eta|^2 + 1),$$
$$(*), (**), \mathscr{R}(\tau, \eta; \xi_1, \ldots, \xi_m) = 0\}$$

is a quasi-algebraic set, so that we obtain the expression

$$M' = \{(\tau, \eta) \in \mathbb{C}^n | (\tau, \eta, \xi_1, \ldots, \xi_m) \in N'\}.$$

Setting

$$\mu'(s) = \sup_{\substack{(\tau,\eta)\in M' \\ |\tau|^2 + |\eta|^2 < s}} \frac{(\operatorname{Im}\tau)^2}{|\operatorname{Im}\eta|^2 + 1},$$

from Lemma 2.1′ we obtain

$$\gamma_0{}^2 \leqq \mu'(s) \leqq C(\log s)^2.$$

We then apply the second form of Seidenberg's lemma to obtain the desired result. ■

With this preparation, we can state

Theorem 3 *(the second necessary condition for $\mathscr{E}$-well-posedness)*
If (P) is $\mathscr{E}$-well-posed, then $R_0(1,0) \neq 0$ (where R_0 is the principal part of R).

Proof
The proof is by contradiction. Suppose that $R_0(1,0) = 0$, there exists $(\tau_0, \eta_0) \in \mathbb{R}^n - i\dot{\Gamma}$, and $R_0(\tau_0, \eta_0) \neq 0$. In Lemma 3.4 we obtained the series expansion of R as

$$v^h R\left(\frac{\tau}{v}, \frac{\eta}{v}\right) = R_0(\tau, \eta) + R_1(\tau, \eta)v + \dots.$$

We set

$$(\tau, \eta) = 2i\gamma_0(1,0) + \mu(\tau_0, \eta_0),$$

where we assume $\mu \in \mathbb{C}^1$ and $|\mu| < \delta_1$ which can be made sufficiently small. That is,

$$v^h R\left(\frac{-2i\gamma_0 + \mu\tau_0}{v}, \frac{\mu\eta_0}{v}\right)$$
$$= R_0(-2i\gamma_0 + \mu\tau_0, \mu\eta_0) + R_1(-2i\gamma_0 + \mu\tau_0, \mu\eta_0)v + \dots,$$

and the sum on the right hand side converges uniformly for $|\mu| < \delta_1, |v| < \delta_2$.

Let us write the sum as $F(\mu, v)$. Then $F(\mu, v)$ is regular for $|\mu| < \delta_1, |v| < \delta_2$. Since

$$F(\mu, 0) = R_0(-2i\gamma_0 + \mu\tau_0, \mu\eta_0)$$

is also regular for $0 \leqq \mu < +\infty$, and since

$$\mu^{-h}F(\mu, 0) = R_0\left(-\frac{2i\gamma_0}{\mu} + \tau_0, \eta_0\right) \to R_0(\tau_0, \eta_0) \neq 0 \quad \text{as } \mu \to +\infty,$$

$F(\mu, 0) \not\equiv 0$ in a neighbourhood of $\mu = 0$. Now we see that

$$F(0,0) = R_0(-2i\gamma_0, 0) = (-2i\gamma_0)^h R_0(1,0) = 0.$$

Therefore, by Weierstrass' preparation theorem, which we used before, there exists a continuous function $\mu = \mu(v)$ in a neighbourhood of $v = 0$ such that $\mu(0) = 0, F(\mu(v), v) \equiv 0$. For $0 < v < \delta$, write

$$\tau(v) = \frac{-2i\gamma_0 + \mu(v)\tau_0}{v}, \quad \eta(v) = \frac{\mu(v)\eta_0}{v}.$$

Then, using the series expansion of R we see that $R(\tau(v), \eta(v)) \equiv 0$. On the other hand we have

$$-v \operatorname{Im} \tau(v) = 2\gamma_0 + o(1),$$
$$-v \operatorname{Im} \eta(v) = o(1)$$

as $v \to +0$. This is a contradiction of what we claimed in Theorem 2'. ■

With the knowledge acquired in §§2 and 3, we can now see that if (P) is $\mathscr{E}$-well-posed, then the Lopatinski determinant $R(\tau, \eta)$ must be hyperbolic in the direction of $(1,0)$.

In other words, considering Lemma 3.4, we have

(a) $R(\tau, \eta)$ is regular in $\mathbb{R}^n - \mathrm{i}(\gamma_0(1,0) + \dot{\Gamma})$.

(b) For any $(\tau_0, \eta_0) \in \mathbb{R}^n - \mathrm{i}(\gamma_0(1,0) + \dot{\Gamma})$,

$$\lambda^{-h} R(\tau_0 + \lambda\tau, \eta_0 + \lambda\eta) \to R_0(\tau, \eta) \not\equiv 0, \lambda(\tau, \eta) \in \mathbb{R}^n - \mathrm{i}\dot{\Gamma} \quad \text{as } |\lambda| \to +\infty,$$

is uniformly convergent in every compact set of $\mathbb{R}^n - \mathrm{i}\dot{\Gamma}$.

Also, from Theorem 2' we see that

(c) $R(\tau, \eta) \neq 0, \quad \operatorname{Im} \tau < -\gamma_0, \quad \eta \in \mathbb{R}^{n-1}$.

And from Theorem 3, we have

(d) $R_0(1,0) \neq 0$.

(Compare conditions (a)–(d) stated in §3.1.)

Moreover, by Theorem 2', we see that there exists a small cone which contains $(1,0)$ as an interior point, and R is hyperbolic in the direction of this cone. A question arises: to what extent can we enlarge such cones? In the next subsection we study this problem.

3.5

Let us assume that $R(\tau, \eta)$ is hyperbolic in the direction of $(1,0)$ i.e. $R_0(1,0) \neq 0$ and

$$R(\tau, \eta) \neq 0, \quad \operatorname{Im} \tau < -\gamma_1, \quad \eta \in \mathbb{R}^{n-1}.$$

By Lemma 3.1, this implies

$$R_0(\tau, \eta) \neq 0, \quad \operatorname{Im} \tau \neq 0, \quad \eta \in \mathbb{R}^{n-1}.$$

Write $\dot{\Sigma}$ for the connected component of the open cone

$$\{(\tau, \eta) \in \dot{\Gamma} \,|\, R_0(\tau, \eta) \neq 0\}$$

containing $(1,0)$. In this case we can prove

Lemma 3.5

There exists a continuous function $\tau_0(\eta)$ defined on $\mathbb{R}^{n-1}$ such that

$$\dot{\Sigma} = \{(\tau, \eta) \in \mathbb{R}^n \,|\, \tau > \tau_0(\eta)\}.$$

Proof
Recall that, for $\dot{\Gamma}$, there exists a continuous function $\tau(\eta)$ such that

$$\dot{\Gamma} = \{(\tau, \eta) \in \mathbb{R}^n \mid \tau > \tau(\eta)\}.$$

Also, since $R_0(1,0) \neq 0$, we see that $R_0(\tau, \eta) \neq 0$ for

$$\{(\tau, \eta) \in \mathbb{R}^n \mid \tau > C|\eta|\}.$$

Then, for a fixed $\eta \in \mathbb{R}^{n-1}$ one of the following two situations arises.

Case 1. For all $\tau > \tau(\eta)$, $\quad R_0(\tau, \eta) \neq 0$,
Case 2. For some $\tau > \tau(\eta)$, $\quad R_0(\tau, \eta) = 0$.

Let

$$\tau_0(\eta) = \begin{cases} \tau(\eta) & \text{in case 1,} \\ \sup\limits_{R_0(\tau,\eta)=0} \tau & \text{in case 2.} \end{cases}$$

We prove that τ_0 is a continuous function. First, we wish to establish upper semi-continuity at $\eta = \eta^0$. To see this we observe the fact that $R_0 \neq 0$ for

$$\{\tau > \tau_1 = C(|\eta^0| + 1), \quad |\eta - \eta^0| < 1\}.$$

For any $\varepsilon > 0$, we have $R_0 \neq 0$ for

$$\{\tau_0(\eta^0) + \varepsilon \leqq \tau \leqq \tau_1, \quad \eta = \eta^0\}.$$

Since R_0 is continuous, $R_0 \neq 0$ for

$$\{\tau_0(\eta^0) + \varepsilon \leqq \tau \leqq \tau_1, \quad |\eta - \eta^0| < \delta\} \quad (0 < \delta < 1).$$

Therefore, for

$$\{\tau_0(\eta^0) + \varepsilon \leqq \tau < +\infty, \quad |\eta - \eta^0| < \delta\}$$

we find $R_0 \neq 0$, and

$$\tau_0(\eta) < \tau_0(\eta^0) + \varepsilon, \quad |\eta - \eta^0| < \delta.$$

This proves upper semi-continuity.

Next, we prove lower semi-continuity at $\eta = \eta^0$. If case 1 holds at $\eta = \eta^0$, then

$$\tau_0(\eta) \geqq \tau(\eta), \quad \tau_0(\eta^0) = \tau(\eta^0).$$

Therefore, from the continuity of τ, we see that τ_0 is lower semi-continuous at $\eta = \eta^0$.

If case 2 holds at $\eta = \eta^0$, then $(\tau_0(\eta^0), \eta^0) \in \dot{\Gamma}$ and $R_0(\tau_0(\eta^0), \eta^0) = 0$. But $R_0(\tau, \eta^0) \not\equiv 0$, so that, by Weierstrass' preparation theorem, which we mentioned before, there exists a continuous function $\tau = \varphi(\eta)$ in a neighbourhood of $\eta = \eta^0$ such that $\varphi(\eta^0) = \tau_0(\eta^0)$, and $R_0(\varphi(\eta), \eta) \equiv 0$. Recall our assumption that

$$R_0(\tau, \eta) \neq 0, \quad \operatorname{Im} \tau \neq 0, \quad \eta \in \mathbb{R}^{n-1}.$$

Then, we see that the function $\varphi(\eta)$ takes real values as long as η is within a real neighbourhood of η_0. Hence $\tau_0(\eta) \geqq \varphi(\eta)$, $\tau_0(\eta^0) = \varphi(\eta^0)$ so that from the continuity of $\varphi(\eta)$ we see that τ_0 is lower semi-continuous at $\eta = \eta^0$.

With the continuity of $\tau_0(\eta)$ we see that

$$\Omega = \{(\tau, \eta) \in \mathbb{R}^n \mid \tau > \tau_0(\eta)\}$$

is, in fact, a domain, and its boundary can be written

$$\partial\Omega = \{(\tau, \eta) \in \mathbb{R}^n \mid \tau = \tau_0(\eta)\}.$$

Thus, from the definition of Γ_0, we have

$$\Omega \subset \dot{\Sigma}, \quad \partial\Omega \subset (\dot{\Sigma})^c,$$

where c indicates complement. Hence $\Omega = \dot{\Sigma}$. ■

Lemma 3.5 says that $\dot{\Sigma}(\subset \dot{\Gamma})$ is an open cone stretching in the direction of $(1, 0)$, and $R_0(\dot{\Sigma}) \neq 0$. By Lemma 3.2, we see that

$$R(\tau, \eta) \neq 0, \quad (\tau, \eta) \in \mathbb{R}^n - \mathrm{i}(\gamma_1(1, 0) + \dot{\Sigma}).$$

That is, $R(\tau, \eta))$ is hyperbolic in the direction of $\dot{\Sigma}$. Therefore, by Lemma 3.1, we have

$$R_0(\tau, \eta) \neq 0, \quad (\tau, \eta) \in \mathbb{R}^n \pm \mathrm{i}\dot{\Sigma}.$$

Also, using the same method in the proof of Lemma 3.3, we can see that, by a rotation of the coordinate axes, $\dot{\Sigma}$ stretches in the direction of $\dot{\Sigma}$ itself, which implies that $\dot{\Sigma}$ is convex. To see that $\dot{\Sigma}$ stretches in any direction of $\dot{\Sigma}$ itself, we proceed as follows: Consider the condition

$$R_0(\tau, \eta) \neq 0, \quad (\tau, \eta) \in \mathbb{R}^n \pm \mathrm{i}\dot{\Sigma}.$$

By rotation of the coordinates axis we can move the $(1, 0)$-axis to a given direction. Then we obtain an expression of $\dot{\Sigma}$ similar to the one in Lemma 3.5.

Summing up, we have

Proposition 3.6
If the Lopatinski determinant $R(\tau, \eta)$ is hyperbolic in the direction of $(1, 0)$, then if we write $\dot{\Sigma}$ for the connected component of

$$\{(\tau, \eta) \in \mathbb{R}^n \mid R_0(\tau, \eta) \neq 0\}$$

containing $(1, 0)$, $\dot{\Sigma}$ becomes a convex open cone, and $R(\tau, \eta)$ is hyperbolic in the direction of $\dot{\Sigma}$.

4. The construction of a solution for a typical hyperbolic boundary value problem

From the argument of §§2 and 3, we have so far observed that, given an initial value problem of type (P), the corresponding Lopatinski determinant

$R(\tau,\eta)$ is hyperbolic in the direction of the vector $(1,0)$. The argument of the remaining sections will lead us to conclude that the converse of this statement is also true. To this end, in the present section we deal with a special problem; we first construct a concrete solution for a 'pure' boundary value problem†, called (P″), by means of the Fourier–Laplace transform. First recall our statement at the end of §3 that if $R(\tau,\eta)$ is hyperbolic in the direction of $(1,0)$, then $R(\tau,\eta)$ is also hyperbolic in the direction of $\dot{\Sigma}$. In what follows, our argument is based on this fact.

4.1

In general, in order to perform an inverse Fourier–Laplace transform on the inverse of a hyperbolic function $g(\zeta)$, we need some sort of 'estimation from below' for $|g(\zeta)|$. We shall obtain this type of estimation of $|R(\tau,\eta)|$ and $|A_+(\tau,\xi,\eta)|$.

Lemma 4.1

Let K be an arbitrary compact set which is contained in $\dot{\Sigma}$.

(i) There exists $\lambda_0 > 0$ such that

$$\tilde{K} = \bigcup_{\lambda \geqq \lambda_0} \lambda K \subset \gamma_1(1,0) + \dot{\Sigma}.$$

(ii) There exist a positive number c and a real number α such that

$$|R(\tau,\eta)| \geqq c(|\tau| + |\eta|)^\alpha, \quad (\tau,\eta) \in \mathbb{R}^n - \mathrm{i}\tilde{K}.$$

Proof

(i) is obvious.

(ii) Choose an arbitrary $(\tau_0,\eta_0) \in \dot{\Sigma}$ and an ε-neighbourhood U_ε of (τ_0,η_0) ($U_\varepsilon \subset \dot{\Sigma}$). It is sufficient to prove the statement for $\bar{U}_\varepsilon = K$. Set

$$\tilde{U}_\varepsilon = \bigcup_{\lambda \geqq \lambda_0} \lambda U_\varepsilon \subset \gamma_1(1,0) + \dot{\Sigma}.$$

We can see that this is a quasi-algebraic set. In fact, $\tilde{U}_\varepsilon$ is the image of the orthogonal projection of the quasi-algebraic set

$$\{(\tau,\eta,\lambda) \in \mathbb{R}^{n+1} \,|\, \lambda \geqq \lambda_0, |\tau - \lambda\tau_0|^2 + |\eta - \lambda\eta_0|^2 < \lambda^2\varepsilon^2\}$$

on (τ,η)-space.

Therefore,

$$M = \{(\tau,\eta) \in \mathbb{R}^n - \mathrm{i}\tilde{U}_\varepsilon, (\xi_1,\ldots,\xi_m) \in \mathbb{C}^m; \textstyle\sum \xi_i = -a_1(\tau,\eta),\ldots,$$
$$\textstyle\prod \xi_i = (-1)^m a_m(\tau,\eta), \mathrm{Im}\,\xi_1 \geqq \mathrm{Im}\,\xi_2 \geqq \ldots \geqq \mathrm{Im}\,\xi_m\}$$

† This means the exclusion of initial conditions.

is also a quasi-algebraic set. We see that if $(\tau,\eta,\xi_1,\dots,\xi_m)\in M$, then

$$(\tau,\eta)\in\mathbb{R}^n - i(\gamma_1(1,0)+\dot{\Sigma}),\quad (\xi_1,\dots,\xi_{m_+})=\{\xi_1{}^+(\tau,\eta),\dots,\xi_{m_+}{}^+(\tau,\eta)\},$$

therefore

$$\mathcal{R}(\tau,\eta;\xi_1,\dots,\xi_{m_+})=R(\tau,\eta)\neq 0.$$

Set

$$\mu(s)=\sup_{\substack{(\tau,\eta,\xi_1,\dots,\xi_m)\in M\\ |\tau|^2+|\eta|^2\leq s}}\frac{1}{|\mathcal{R}(\tau,\eta;\xi_1,\dots,\xi_{m_+})|^2}$$

$$=\sup_{\substack{(\tau,\eta)\in\mathbb{R}^n-i\tilde{U}_\varepsilon\\ |\tau|^2+|\eta|^2\leq s}}\frac{1}{|R(\tau,\eta)|^2}<+\infty.$$

By the second form of Seidenberg's lemma we have

$$\mu(s)=As^\alpha(1+o(1)),\quad s\to+\infty.$$

Therefore, for $(\tau,\eta)\in\mathbb{R}^n-i\tilde{U}_\varepsilon$
we have

$$\frac{1}{|R(\tau,\eta)|^2}\leqq C(|\tau|^2+|\eta|^2)^\alpha. \quad \blacksquare$$

We now seek an estimation for

$$A_+(\tau,\xi,\eta)=\prod_{j=1}^{m_+}(\xi-\xi_j{}^+(\tau,\eta)).$$

Notice that this is symmetric with respect to $\{\xi_1{}^+,\dots,\xi_{m_+}{}^+\}$. Therefore, it becomes a polynomial of ξ with coefficients regular for

$$(\tau,\eta)\in\mathbb{R}^n-i(\gamma_0(1,0)+\dot{\Gamma}).$$

Furthermore, since $\xi_j{}^+(\tau,\eta)$ satisfies

$$-\operatorname{Im}\xi_j{}^+(\tau,\eta)\leqq\xi_{\min}(-\operatorname{Im}\tau-\gamma_0,-\operatorname{Im}\eta)$$

we see that

$$A_+(\tau,\xi,\eta)\neq 0,\quad (\tau,\xi,\eta)\in\mathbb{R}^{n+1}-i(\gamma_0(1,0,0)+\Gamma_+),$$

where

$$\Gamma_+=\{(\tau,\xi,\eta)|(\tau,\eta)\in\dot{\Gamma},\xi>\xi_{\min}(\tau,\eta)\}.$$

Using the same notation we prove

Lemma 4.2
Let K be an arbitrary compact set contained in Γ_+. There exists $\lambda_0>0$ such that

$$\tilde{K}=\bigcup_{\lambda\geqq\lambda_0}(\lambda K)\subset\gamma_0(1,0,0)+\Gamma_+,$$

and

$$|A_+(\tau,\xi,\eta)| \geqq \gamma_0{}^{m_+}, \quad (\tau,\xi,\eta)\in\mathbb{R}^{n+1} - \mathrm{i}\tilde{K}.$$

Proof

For $(\tau,\xi,\eta)\in\mathbb{R}^{n+1} - \mathrm{i}\tilde{K}$, we can write

$$-(\operatorname{Im}\tau, \operatorname{Im}\xi, \operatorname{Im}\eta) = \lambda(\tau',\xi',\eta'), \quad \lambda \geqq \lambda_0, \quad (\tau',\xi',\eta')\in K.$$

We therefore have

$$\begin{aligned} -\operatorname{Im}\xi_j{}^+(\tau,\eta) &\leqq \xi_{\min}(-\operatorname{Im}\tau - \gamma_0, -\operatorname{Im}\eta) \\ &= \lambda\xi_{\min}\left(\tau' - \frac{\gamma_0}{\lambda}, \eta'\right). \end{aligned}$$

Setting $\operatorname{dis}(K,\partial\Gamma_+) = \delta(>0)$ we see that for $\gamma_0/\lambda \leqq \frac{1}{2}\delta$,

$$\xi' < \xi_{\min}\left(\tau' - \frac{\gamma_0}{\lambda}, \eta'\right) + \tfrac{1}{2}\delta.$$

From this we find

$$\begin{aligned} -\operatorname{Im}\xi = \lambda\xi' &> \lambda\{\xi_{\min}(\tau' - \gamma_0/\lambda, \eta') + \tfrac{1}{2}\delta\} \\ &\geqq -\operatorname{Im}\xi_j{}^+(\tau,\eta) + \tfrac{1}{2}\lambda\delta. \end{aligned}$$

Therefore

$$-\operatorname{Im}\xi + \operatorname{Im}\xi_j{}^+(\tau,\eta) \geqq \gamma_0.$$

Let

$$\lambda_1 = \max(\lambda_0, 2\gamma_0/\delta).$$

Then, for

$$(\tau,\xi,\eta)\in\mathbb{R}^{n+1} - \mathrm{i}\bigcup_{\lambda\geqq\lambda_1}(\lambda K)$$

we have

$$\begin{aligned} |A_+(\tau,\xi,\eta)| &= \sum_{j=1}^{m_+} |\xi - \xi_j{}^+(\tau,\eta)| \\ &\geqq \prod_{j=1}^{m_+} |\operatorname{Im}\xi - \operatorname{Im}\xi_j{}^+(\tau,\eta)| \geqq \gamma_0{}^{m_+}. \end{aligned}$$

■

4.2

Recall our argument for the problem ($\hat{\mathrm{P}}$) in §2. We said that a solution $\hat{u}(x)$ of

$$\begin{cases} A(\tau, \mathrm{D}_x, \eta)\hat{u}(x) = 0, & x > 0, \\ B_j(\tau, \mathrm{D}_x, \eta)\hat{u}(0) = \hat{g}_j & (j = 1, \ldots, m_+), \\ \hat{u}(x) \to 0 \text{ as } x \to +\infty, \end{cases}$$

can be represented as

$$\hat{u}(x) = \sum_{k=1}^{m_+} \frac{1}{2\pi} \oint e^{ix\xi} P_k(\tau, \xi, \eta)\, d\xi\, \hat{g}_k, \quad x > 0$$

for $(\tau, \eta) \in \mathbb{R}^n - i\{\gamma_1(1,0) + \dot{\Sigma}\}$, where P_k is defined as

$$P_k = (\tau, \xi, \eta) = \frac{1}{i} \sum_{j=1}^{m_+} \frac{\xi^{j-1}}{A_+(\tau, \xi, \eta)} \cdot \frac{R_{jk}(\tau, \eta)}{R(\tau, \eta)}.$$

In this subsection we observe the inverse Fourier–Laplace transform of such P_k.

By Lemma 4.2, we see that for a compact set K of Γ_+, the following inequality holds:

$$\sum_{j=1}^{m_+} \left| \frac{\xi^{j-1}}{A_+(\tau, \xi, \eta)} \right| \leqq C|\xi|^{m_+ - 1}, \quad (\tau, \xi, \eta) \in \mathbb{R}^{n+1} - i \bigcup_{\lambda \geqq \lambda_0} (\lambda K).$$

More precisely, starting from the inequality

$$|\xi_j^+(\tau, \eta)| \leqq C(|\tau| + |\eta|),$$

by looking at the two cases $|\xi| \geq 2C(|\tau| + |\eta|)$, $|\xi| < 2C(|\tau| + |\eta|)$ separately, we obtain the inequality

$$\sum_{j=1}^{m_+} \left| \frac{\xi^{j-1}}{A_+(\tau, \xi, \eta)} \right| \leqq C \frac{(|\tau| + |\eta|)^{m_+}}{|\xi| + |\tau| + |\eta|}, \quad (\tau, \xi, \eta) \in \mathbb{R}^{n+1} - i \bigcup_{\lambda \geqq \lambda_0} (\lambda K).$$

For $R_{jk}(\tau, \eta)$ we have an expression

$$R_{jk}(\tau, \eta) = \mathscr{R}_{jk}(\tau, \eta; \xi_1^+(\tau, \eta), \ldots, \xi_{m_+}^+(\tau, \eta)),$$

where $\mathscr{R}_{jk}(\tau, \eta; \xi_1, \ldots, \xi_{m_+})$ is the (k,j)-cofactor of

$$\left(\frac{1}{2\pi i} \oint \frac{B_j(\tau, \xi, \eta)\xi^{k-1}}{\prod_{i=1}^{m_+} (\xi - \xi_i)} d\xi \right)_{j,k=1,\ldots,m_+}$$

and is a polynomial in $(\tau, \eta, \xi_1, \ldots, \xi_{m_+})$, which is symmetric with respect to $(\xi_1, \ldots, \xi_{m_+})$. Therefore, as in the case of $R(\tau, \eta)$ we see that $R_{jk}(\tau, \eta)$ is regular in $\mathbb{R}^n - i(\gamma_0(1,0) + \dot{\Sigma})$. Since

$$|\xi_j^+(\tau, \eta)| \leqq C(|\tau| + |\eta|),$$

we obtain

$$|R_{jk}(\tau, \eta)| \leqq C'(|\tau| + |\eta|)^l.$$

Then, the application of the estimation of $R(\tau, \eta)$ (see Lemma 4.1) yields

$$\left| \frac{R_{jk}(\tau, \eta)}{R(\tau, \eta)} \right| \leqq C(|\tau| + |\eta|)^N, \quad (\tau, \eta) \in \mathbb{R}^n - i \bigcup_{\lambda \geqq \lambda_0} (\lambda K)$$

for a compact set K in $\dot{\Sigma}$.

Write

$$\Sigma = \{(\tau, \xi, \eta) \in \mathbb{R}^{n+1} | (\tau, \eta) \in \dot{\Sigma}\}, \quad \Sigma_+ = \Sigma \cap \Gamma_+.$$

For an arbitrary compact set K in Σ_+ we see that

$$|P_k(\tau, \xi, \eta)| \leqq C \frac{(|\tau| + |\eta|)^N}{|\xi| + |\tau| + |\eta|}, \quad (\tau, \xi, \eta) \in \mathbb{R}^{n+1} - \mathrm{i} \bigcup_{\lambda \geqq \lambda_0} (\lambda K)$$

(where P_k is as defined at the beginning of this subsection). Therefore, by Theorem 1(a), we can apply an inverse Fourier–Laplace transform to $P_k(\tau, \xi, \eta)$. Also, using Theorem 1(c) we can examine the nature of the support of $\mathscr{L}^{-1}[P_k]$. Summing up, we have

Proposition 4.3
Let

$$G_k(t, x, y) = \mathscr{L}^{-1}_{(\tau, \xi, \eta) \to (t, x, y)}[P_k(\tau, \xi, \eta)],$$
$$(\operatorname{Im} \tau, \operatorname{Im} \xi, \operatorname{Im} \eta) \in -\{\gamma_1(1, 0, 0) + \Sigma_+\}$$

then

$$G_k(t, x, y) \in \mathscr{S}'_{-\{\gamma_1(1,0,0) + \Sigma_+\}},$$
$$\operatorname{supp}[G_k(t, x, y)] \subset (\Sigma_+)'$$

where $(\Sigma_+)'$ is the adjoint cone of Σ_+.
Note. $G_k(t, x, y)$ is called a *Poisson function.*

Now we examine $(\Sigma_+)'$, which is an adjoint cone (see §1.3 for the definition). But, in passing, a word about cones. In general, any cone Γ and its adjoint cone Γ' have the following properties:

(i) $(\Gamma_1 \cup \Gamma_2)' = \Gamma_1' \cap \Gamma_2'$,
(ii) $C[\Gamma_1] \cap C[\Gamma_2] = C[\Gamma_1' \cup \Gamma_2']$,
(iii) $(C[\Gamma])' = \Gamma'$,

where $C(\Gamma)$ is the *convex hull* (or convex closure)† of Γ.

Now in our case we have

$$\Sigma_+ = \Gamma_+ \cap \Sigma,$$
$$\Gamma_+ = C[\Gamma \cup L_+], \quad L_+ = \{(0, \xi, 0); \xi > 0\}.$$

From this we see that

$$(\Sigma_+)' = (\Gamma_+ \cap \Sigma)' = C[\Gamma_+' \cup \Sigma'],$$
$$\Gamma_+' = (C[\Gamma \cup L_+])' = (\Gamma \cup L_+)' = \Gamma' \cap L_+'$$
$$= \Gamma' \cap \{x \geqq 0\}.$$

† Translator's note: $C(\Gamma)$ is the minimal convex set that contains Γ.

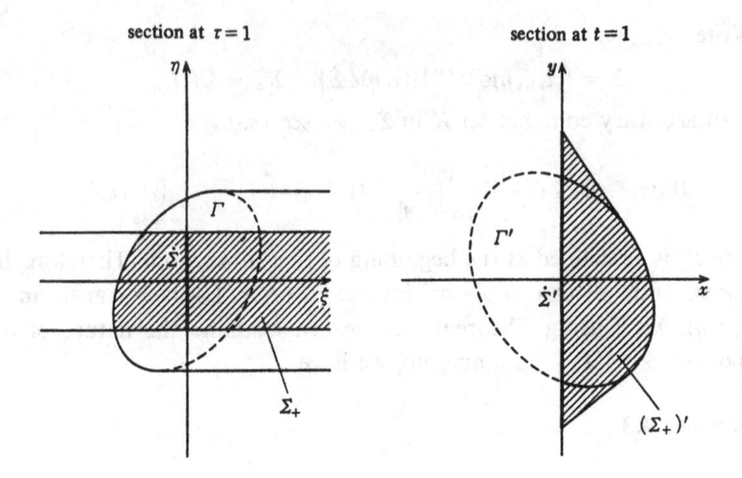

Fig. 7

On the other hand, we have

$$\Sigma = \{(\tau, \xi, \eta) \in \mathbb{R}^{n+1} | (\tau, \eta) \in \dot{\Sigma}\}.$$

From this we see that

$$\Sigma' = \{(t, 0, y) \in \mathbb{R}^{n+1} | (t, y) \in (\dot{\Sigma})'\},$$

where $(\dot{\Sigma})'$ is the adjoint cone of $\dot{\Sigma}$ in $\mathbb{R}^n$. (See Fig. 7.)

Let G_k be a Poisson function. For each

$$f(t, y) \in \mathscr{S}'_{-\{\gamma_1(1,0)+\dot{\Sigma}\}},$$

we define

$$(G_k f)(t, x, y) = G_k(t, x, y) * \{f(t, y) \otimes \delta_x\}$$

where δ_x is the *Dirac distribution*† of $x \in \mathbb{R}$, and $\otimes$ indicates tensor product. We call $G_k f$ the *Poisson operator* operating G_k from the left as a convolution. Since

$$f(t, y) \otimes \delta_x \in \mathscr{S}'_{-\{\gamma_1(1,0,0)+\Sigma\}} \subset \mathscr{S}'_{-\{\gamma_1(1,0,0)+\Sigma_+\}},$$

and

$$G_k(t, x, y) \in \mathscr{S}'_{-\{\gamma_1(1,0,0)+\Sigma_+\}}$$

it is easy to see that

$$(G_k f)(t, x, y) \in \mathscr{S}'_{-\{\gamma_1(1,0,0)+\Sigma_+\}}.$$

From Proposition 1.5 (i), we see that if

$$f_j(t, y) \to f(t, y) \quad (\mathscr{S}'_{-\{\gamma_1(1,0)+\dot{\Sigma}\}}) \text{ as } j \to \infty,$$

† Translator's note: See Schwartz, *Mathematics for the physical sciences*.

then

$$(G_k f_j)(t,x,y) \to (G_k f)(t,x,y) \quad (\mathscr{S}'_{-\{\gamma_1(1,0,0)+\Sigma_+\}}) \text{ as } j \to \infty.$$

Using (ii) of the same Proposition we also see that

$$\begin{aligned}\operatorname{supp}[G_k f] &= \operatorname{supp}[G_k] + \operatorname{supp}[f \otimes \delta_x] + (\Sigma_+)' \\ &\subset \operatorname{supp}[f \otimes \delta_x] + (\Sigma_+)' \\ &= \operatorname{supp}[f] \times \{x=0\} + (\Sigma_+)'.\end{aligned}$$

Furthermore, we can prove

Proposition 4.4

If $f \in \mathscr{S}'_{-\{\gamma_1(1,0)+\dot{\Sigma}\}}$ then

$$G_k f \in C^\infty([0,\infty); \mathscr{S}'_{-\{\gamma_1(1,0)+\dot{\Sigma}\}}),$$

and $G_k f$ satisfies the following equations for A, B_j, B_k

$$\begin{aligned}&A(G_k f) = 0, \quad x > 0, \\ &B_j(G_k f)|_{x=+0} = 0 \quad (j = 1, \ldots, m_+, j \neq k). \\ &B_k(G_k f)|_{x=+0} = f.\end{aligned}$$

In particular, if $f \in \mathscr{S}_{-\{\gamma_1(1,0)+\Sigma\}}$, then

$$G_k f \in \mathscr{S}_{-\{\gamma_1(1,0,0)+\Sigma_+\}}(\overline{\mathbb{R}_+^{n+1}}).$$

That is, there exists an extension

$$\widetilde{G_k f} \in \mathscr{S}_{-\{\gamma_1(1,0,0)+\Sigma_+\}}(\mathbb{R}^{n+1}).$$

Note. We write $\mathscr{F}_0, \mathscr{L}_0$ for a Fourier transform and a Fourier–Laplace transform in (t,y)-space, respectively (corresponding to $\mathscr{F}$, $\mathscr{L}$ in (t,x,y)-space).

Proof

For fixed $(\tau_0, \xi_0, \eta_0) \in \{\gamma_1(1,0,0) + \Sigma_+\}$ and $(\tau', \xi', \eta') \in \mathbb{R}^{n+1}$, we have

$$P_k(\tau' - i\tau_0, \xi' - i\xi_0, \eta' - i\eta_0) \in L^2(-\infty, +\infty)$$

with respect to ξ'. Setting

$$\begin{aligned}&g_k(x; \tau' - i\tau_0, \eta' - i\eta_0) \\ &= \underset{N \to +\infty}{\text{l.i.m.}} \frac{1}{2\pi} \int_{-N}^{N} e^{ix\xi'} P_k(\tau' - i\tau_0, \xi' - i\xi_0, \eta' - i\eta_0) d\xi',\end{aligned}$$

we find

$$g_k(x; \tau' - i\tau_0, \eta' - i\eta_0)$$

$$= \begin{cases} \dfrac{1}{2\pi} \displaystyle\oint_{C+} e^{ix\xi'} P_k(\tau' - i\tau_0, \xi' - i\xi_0, \eta' - i\eta_0) d\xi', & x > 0, \\ 0, & x < 0, \end{cases}$$

where C_+ is a closed curve in the upper half ξ'-space.† Therefore, for any $s(\geq 0)$, we see that

$$\sup_{x>0}(|x|+1)^s \sum_{l+j+|\nu|\leqq s}\left|\left(\frac{\partial}{\partial x}\right)^l\left(\frac{\partial}{\partial \tau'}\right)^j\left(\frac{\partial}{\partial \eta'}\right)^\nu g_k(x;\tau'-i\tau_0,\eta'-i\eta_0)\right|$$
$$\leqq C_s(\tau')+|\eta'|)^{\alpha_s}.$$

Write

$$G_k'(x;t,y) = \mathscr{L}^{-1}{}_{0(\tau,\eta)\to(t,y)}[e^{x\xi_0}g_k(x;\tau,\eta)]$$
$$= e^{t\tau_0+x\xi_0+y\cdot\eta_0}\mathscr{F}_{0(\tau',\eta')\to(t,y)}[g_k(x;\tau'-i\tau_0,\eta'-i\eta_0)].$$

Then, for $f\in\mathscr{S}'_{-\{\gamma_1(1,0)+\Sigma\}}$ we have

$$(G_k'f)(t,x,y) = G_k'(x;t,y)\underset{(t,y)}{*}f(t,y)$$
$$= e^{t\tau_0+x\xi_0+y\cdot\eta_0}$$
$$\times\mathscr{F}_{0(\tau',\eta)\to(t,y)}[g_k(x;\tau'-i\tau_0,\eta'-i\eta_0)$$
$$\times(\mathscr{L}_0f)(\tau'-i\tau_0,\eta'-i\eta_0)],$$

and, consequently, we see that

$$(G_k'f)(t,x,y)\in C^\infty([0,\infty);\mathscr{S}'_{-\{\gamma_1(1,0)+\Sigma\}}).$$

In particular, if $f\in\mathscr{S}_{-\{\gamma_1(1,0)+\Sigma\}}$ we have

$$(G_k'f)(t,x,y)\in\mathscr{S}_{-\{\gamma_1(1,0,0)+\Sigma_+\}}(\overline{\mathbb{R}_+^{n+1}}).$$

We now wish to prove

$$\langle(G_kf)(t,x,y),\varphi(t,x,y)\rangle = \langle(G_k'f)(t,x,y),\varphi(t,x,y)\rangle$$

for an arbitrary $\varphi(t,x,y)\in\mathscr{D}$. To this end, first, notice that

$$(G_kf)(t,x,y) = \mathscr{L}^{-1}_{(\tau,\xi,\eta)\to(t,x,y)}[P_k(\tau,\xi,\eta)(\mathscr{L}_0f)(\tau,\eta)]$$
$$= e^{t\tau_0+x\xi_0+y\cdot\eta_0}$$
$$\times\mathscr{F}_{(\tau',\xi',\eta')\to(t,x,y)}[P_k(\tau'-i\tau_0,\xi'-i\xi_0,\eta'-i\eta_0)$$
$$\times(\mathscr{L}_0f)(\tau'-i\tau_0,\eta'-i\eta_0)].$$

From this we have

$$\langle G_kf,\varphi\rangle_{t,x,y}$$
$$= \int_{\operatorname{Im}(\tau,\xi,\eta)=-(\tau_0\xi_0,\eta_0)} P_k(\tau,\xi,\eta)(\mathscr{L}_0f)(\tau,\eta)(\bar{\mathscr{L}}\varphi)(\tau,\xi,\eta)\,d\tau\, d\xi\, d\eta,$$

$$(\mathscr{L}\varphi)(\tau'-i\tau_0,\xi'-i\xi_0,\eta'-i\eta_0)$$
$$= \bar{\mathscr{F}}_{(t,x,y)\to(\tau',\xi',\eta')}[e^{t\tau_0+x\xi_0+y\cdot\eta_0}\varphi(t,x,y)]$$
$$= \frac{1}{(2\pi)^{n+1}}\int e^{i\{t(\tau'-i\tau_0)+x(\xi'-i\xi_0)+y\cdot(\eta'-i\eta_0)\}}\varphi(t,x,y)\,dt\,dx\,dy.$$

† Translator's note: 1. i.m. (limit in the *mean*) means the limit in $L^2(\mathbb{R})$-space.

On the other hand, we have

$$\int P_k(\tau' - i\tau_0, \xi' - i\xi_0, \eta - i\eta_0)(\mathscr{L}\varphi)(\tau' - i\tau_0, \xi' - i\xi_0, \eta' - i\eta_0)\,d\xi'$$

$$= \int e^{x\xi_0} g_k(x;\tau' - i\tau_0, \eta' - i\eta_0)(\mathscr{L}_0\varphi)(x;\tau' - i\tau_0, \xi' - i\xi_0)\,dx$$

so we see that

$$\langle G_k f, \varphi\rangle = \int_0^\infty \left\{ \int e^{x\xi_0} g_k(x;\tau' - i\tau_0, \eta' - i\eta_0)(\mathscr{L}_0 f)(\tau' - i\tau_0, \eta' - i\eta_0) \right.$$

$$\left. \times (\bar{\mathscr{L}}_0\varphi)(x;\tau' - i\tau_0, \eta' - i\eta_0)\,d\tau'\,d\eta' \right\} dx$$

$$= \langle G_k' f, \varphi\rangle. \quad ■$$

4.3

In this subsection, we shall establish some facts about Green's operators. To do this, we write

$$E(t,x,y) = \mathscr{L}^{-1}_{(\tau,\xi,\eta)\to(t,x,y)}\left[\frac{1}{A(\tau,\xi,\eta)}\right]$$

for the fundamental solution of the *differential operator* $A = A(D_t, D_x, D_y)$ where A satisfies condition (i) of Hypothesis (A) (see §2). Then we see that

$$E(t,x,y) \in \mathscr{S}'_{-\{\gamma_0(1,0,0)+\Gamma\}},$$

$$\operatorname{supp}[E(t,x,y)] \subset \Gamma'.$$

If we consider $(E*f)(t,x,y)$, for $f(t,x,y) \in \mathscr{S}'_{\{\gamma_0(1,0,0)+\Gamma\}}$ we find

$$(E*f)(t,x,y) \in \mathscr{S}'_{-\{\gamma_0(1,0,0)+\Gamma\}},$$

$$\operatorname{supp}[(E*f)(t,x,y)] \subset \operatorname{supp}[f(t,x,y)] + \Gamma'.$$

and $E*f$ satisfies

$$A(D_t, D_x, D_y)(E*f)(t,x,y) = f(t,x,y).$$

In particular, if $f \in \mathscr{S}_{-\{\gamma_0(1,0,0)+\Gamma\}}$ then $E*f \in \mathscr{S}_{-\{\gamma_0(1,0,0)+\Gamma\}}$.

Lemma 4.5

Let $f(t,x,y) \in \mathscr{S}'_{-\{\gamma_0(1,0,0)+\Gamma\}}$ and let the support of f be contained in $\{x \geqq a, (t,y) \in \mathbb{R}^n\}$. Then

$$(E*f)(t,x,y) \in C^\infty((-\infty, a]; \mathscr{S}'_{-\{\gamma_0(1,0)+\dot{\Gamma}\}})$$

Proof

We wish to obtain an expression for $E*f$ for $x < a$ which leads us to the conclusion. Take $\varphi(t,x,y) \in \mathscr{D}$ with its support contained in

$\{x \leqq b, (t, y) \in \mathbb{R}^n\}$ $(b < a)$. For $(\tau_0, \xi_0, \eta_0) \in \gamma_1(1,0,0) + \Gamma$ we have

$$\begin{aligned}(E * f)(t, x, y) &= \mathscr{L}^{-1}_{(\tau,\xi,\eta)\to(t,x,y)}\left[\frac{(\mathscr{L}f)(\tau,\xi,\eta)}{A(\tau,\xi,\eta)}\right]\\ &= e^{t\tau_0 + x\xi_0 + y\cdot\eta_0}\bar{\mathscr{F}}_{(\tau',\xi',\eta')\to(t,x,y)}\left[\frac{(\mathscr{L}f)(\tau' - i\tau_0, \xi' - i\xi_0, \eta' - i\eta_0)}{A(\tau' - i\tau_0, \xi' - i\xi_0, \eta' - i\eta_0)}\right].\end{aligned}$$

From this we have

$$\langle E * f, \varphi\rangle_{t,x,y} =$$

$$\left\langle \frac{(\mathscr{L}f)(\tau' - i\tau_0, \xi' - i\xi_0, \eta' - i\eta_0)}{A(\tau' - i\tau_0, \xi' - i\xi_0, \eta' - i\eta_0)}, \mathscr{F}_{(t,x,y)\to(\tau',\xi',\eta')}[e^{t\tau_0 + x\xi_0 + y\cdot\eta_0}\varphi(t,x,y)]\right\rangle_{\tau',\xi',\eta'}$$

$$= \int_{\operatorname{Im}(\tau,\xi,\eta) = -(\tau_0,\xi_0,\eta_0)} \frac{(\mathscr{L}f)(\tau,\xi,\eta)}{A(\tau,\xi,\eta)}(\bar{\mathscr{L}}\varphi)(\tau,\xi,\eta)\,d\tau\,d\xi\,d\eta,$$

where

$$(\bar{\mathscr{L}}\varphi)(\tau,\xi,\eta) = \frac{1}{(2\pi)^{n+1}}\int e^{i(t\tau + x\xi + y\cdot\eta)}\varphi(t,x,y)\,dt\,dx\,dy.$$

Since $\operatorname{supp}[\varphi] \subset \{x \leqq b, (t, y) \in \mathbb{R}^n\}$, from Theorem 1(c) (see §1) for any $\varepsilon > 0$ and any $N > 0$, we have

$$\begin{aligned}&|(\bar{\mathscr{L}}\varphi)(\tau' - i\tau_0, \xi' - i(\xi_0 + \lambda), \eta' - i\eta_0)|\\ &\leqq Ce^{(b+\varepsilon)\lambda}(1 + |\tau'| + |\xi'| + |\eta'|)^{-N}, \quad \lambda > 0.\end{aligned}$$

On the other hand, since $f \in \mathscr{S}'_{-\{\gamma_0(1,0,0)+\Gamma\}}$, and $\operatorname{supp}[f] \subset \{x \geq a, (t, y) \in \mathbb{R}^n\}$ we see that $f \in \mathscr{S}'_{-\{\gamma_0(1,0,0)+\Gamma_+\}}$. Consequently, we have

$$\begin{aligned}&|(\mathscr{L}f)(\tau' - i\tau_0, \xi' - i(\xi_0 + \lambda), \eta' - i\eta_0)|\\ &\leqq Ce^{(-a+\varepsilon)\lambda}(1 + |\tau'| + |\xi'| + |\eta'|)^l, \quad \lambda > 0\end{aligned}$$

for any arbitrary $\varepsilon > 0$. Therefore

$$\langle E * f, \varphi\rangle_{t,x,y} = \int_{\operatorname{Im}(\tau,\eta) = -(\tau_0,\eta_0)} d\tau\,d\eta \oint_{C_-}\frac{(\mathscr{L}f)(\tau,\xi,\eta)}{A(\tau,\xi,\eta)}(\bar{\mathscr{L}}\varphi)(\tau,\xi,\eta)\,d\xi,$$

where the closed curve C_- lies in $\operatorname{Im}\xi < -\xi_0$ and contains all the roots $\{\xi_j^-(\tau,\eta)\}_{j=1,\ldots,m_-}$ of $A(\tau,\xi,\eta) = 0$ with respect to ξ that lie in the region $\operatorname{Im}\xi < -\xi_0$.

Letting

$$(\bar{\mathscr{L}}\varphi)(\tau,\xi,\eta) = \frac{1}{2\pi}\int_{-\infty}^{\infty} e^{ix\xi}(\bar{\mathscr{L}}_0\varphi)(x;\tau,\eta)\,dx,$$

we have

$$\oint_{C_-} \frac{(\mathscr{L}f)(\tau,\xi,\eta)}{A(\tau,\xi,\eta)}(\bar{\mathscr{L}}\varphi)(\tau,\xi,\eta)\,\mathrm{d}\xi$$

$$= \oint_{-\infty}^{\infty}\left(\frac{1}{2\pi}\int_{C_-} \mathrm{e}^{\mathrm{i}x\xi}\frac{(\mathscr{L}f)(\tau,\xi,\eta)}{A(\tau,\xi,\eta)}\mathrm{d}\xi\right)(\bar{\mathscr{L}}_0\varphi)(x;\tau,\eta)\,\mathrm{d}x.$$

From this we see that

$$\langle E*f,\varphi\rangle = \int_{-\infty}^{\infty}\left\{\int\!\!\int_{\mathrm{Im}(\tau,\eta)=-(\tau_0,\eta_0)}\left(\frac{1}{2\pi}\oint_{C_-} \mathrm{e}^{\mathrm{i}x\xi}\frac{(\mathscr{L}f)(\tau,\xi,\eta)}{A(\tau,\xi,\eta)}\mathrm{d}\xi\right)\right.$$

$$\left.(\bar{\mathscr{L}}_0\varphi)(x;\tau,\eta)\,\mathrm{d}\tau\,\mathrm{d}\eta\right\}\mathrm{d}x$$

$$= \left\langle \mathscr{L}^{-1}_{0(\tau,\eta)\to(t,y)}\left[\frac{1}{2\pi}\oint_{C_-} \mathrm{e}^{\mathrm{i}x\xi}\frac{(\mathscr{L}f)(\tau,\xi,\eta)}{A(\tau,\xi,\eta)}\mathrm{d}\xi\right],\varphi\right\rangle.$$

That is, for $x < a$, we have the expression

$$(E*f)(t,x,y) = \mathscr{L}^{-1}_{0(\tau,\eta)\to(t,y)}\left[\frac{1}{2\pi}\oint_{C_-} \mathrm{e}^{\mathrm{i}x\xi}\frac{(\mathscr{L}f)(\tau,\xi,\eta)}{A(\tau,\xi,\eta)}\mathrm{d}\xi\right].$$

Hence, obviously

$$(E*f)(t,x,y)\in C^\infty((-\infty,a];\mathscr{S}'_{-\{\gamma_0(1,0)+\dot{\Gamma}\}}). \qquad \blacksquare$$

Now, let $f(t,x,y)\in\mathscr{S}'_{-\{\gamma_0(1,0)+\Gamma\}}$. Assume that the support of f is in $\{x>0,(t,y)\in\mathbb{R}^n\}$ and

$$\mathrm{dis}(\mathrm{supp}[f],\{x=0,(t,y)\in\mathbb{R}^n\}) = d > 0.$$

From Lemma 4.5 we see that

$$(E*f)(t,x,y)\in C^\infty((-\infty,d];\mathscr{S}'_{-\{\gamma_0(1,0)+\dot{\Gamma}\}}),$$

therefore

$$B_j(\mathrm{D}_t,\mathrm{D}_x,\mathrm{D}_y)(E*f)(t,x,y)|_{x=+0}\in\mathscr{S}'_{-\{\gamma_0(1,0)+\dot{\Gamma}\}}.$$

In general, given a function f such that

$$f(t,x,y)\in\mathscr{S}'_{-\{\gamma_0(1,0,0)+I\}},$$
$$B_j(\mathrm{D}_t,\mathrm{D}_x,\mathrm{D}_y)(E*f)(t,x,y)|_{x=+0}\in\mathscr{S}'_{-\{\gamma_0(1,0)+\dot{\Gamma}\}},$$

we can define a *Green's operator* G as satisfying

$$Gf = E*\mathrm{f} - \sum_{j=1}^{m_+} G_j[B_j(E*f)|_{x=+0}]\in\mathscr{S}'_{-\{\gamma_1(1,0,0)+\Sigma_+\}}.$$

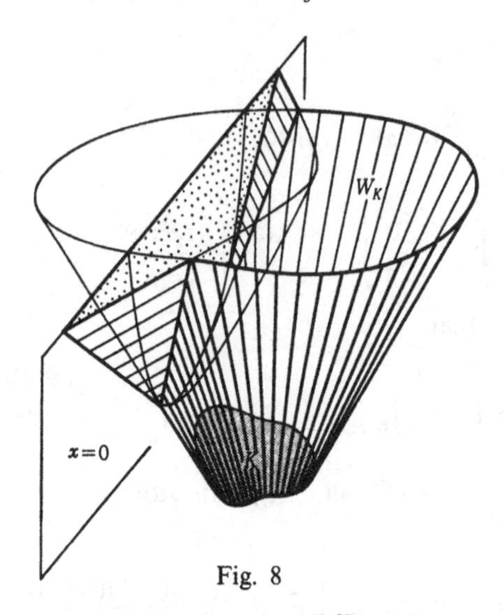

Fig. 8

Since

$$\operatorname{supp}[E*f] \subset \operatorname{supp}[f] + \Gamma',$$

$$\operatorname{supp}[G_j[B_j(E*f)|_{x=+0}]] \subset \operatorname{supp}[B_j(E*f)|_{x=+0}] + \Sigma_x{}'$$
$$\subset \{\operatorname{supp}[f] + \Gamma'\} \cap \{x = 0, (t, y) \in \mathbb{R}^n\} + \Sigma_+{}',$$

setting

$$W_K = [(K + \Gamma') \cap \{x \geqq 0\}] \cap [(K + \Gamma') \cap \{x = 0, (t, y) \in \mathbb{R}^n\} + \Sigma_+{}']$$

for any set K (see Fig. 8) we obtain

$$\operatorname{supp}[Gf] \cap \{x \geqq 0, (t, y) \in \mathbb{R}^n\} \subset W_{\operatorname{supp}[f]}.$$

Therefore, the following proposition is true.

Proposition 4.6
Assume $f \in \mathscr{S}'_{-\{\gamma_1(1,0,0)+\Sigma_+\}}$, $\operatorname{supp}[f] \subset \{x > 0, (t, y) \in \mathbb{R}^n\}$ and

$$\operatorname{dis}(\operatorname{supp}[f], \{x = 0, (t, y) \in \mathbb{R}^n\}) > 0.$$

Then,

$$Gf \in \mathscr{S}'_{-\{\gamma_1(1,0,0)+\Sigma_+\}},$$
$$\operatorname{supp}[Gf] \cap \{x \geqq 0, (t, y) \in \mathbb{R}^n\} \subset W_{\operatorname{supp}[f]}$$

and Gf gives a solution for the boundary value problem:

$$A(\mathrm{D}_t, \mathrm{D}_x, \mathrm{D}_y)(Gf) = f, \quad (t, x, y) \in \mathbb{R}_+{}^{n+1},$$
$$B_j(\mathrm{D}_t, \mathrm{D}_x, \mathrm{D}_y)(Gf)|_{x=+0} = 0 \quad (j = 1, \ldots, m_+), (t, y) \in \mathbb{R}^n.$$

If $f \in \mathscr{S}_{-\{\gamma_1(1,0,0)+\Sigma_+\}}$, then irrespective of the location of the support of f, a Green's operator can be defined, and it will satisfy the condition

$$Gf \in \mathscr{S}_{-\{\gamma_1(1,0,0)+\Sigma_+\}}(\overline{\mathbb{R}_+{}^{n+1}}).$$

Gf gives a solution for the same boundary value problem.

The main results in §4 are Propositions 4.4 and 4.6. Using these results we can construct a concrete solution in $\mathscr{E}$ for an initial value problem. In other words, if the data in $\mathscr{E}$-domains satisfy the compatibility condition of infinite order, then there exists a solution belonging to an $\mathscr{E}$-domain. In the case of the initial value being zero, we can explicitly state this as

Theorem 4

Let $R(\tau, \eta)$ be hyperbolic in the direction of $(1,0)$ and assume

$$f(t,x,y)\in\mathscr{E}(\overline{\mathbb{R}_+{}^{n+1}}),\quad \operatorname{supp}[f]\subset\{t\geqq 0,(x,y)\in\mathbb{R}\},$$
$$g_j(t,y)\in\mathscr{E}(\mathbb{R}^n),\quad \operatorname{supp}[g_j]\subset\{t\geqq 0,(x,y)\in\mathbb{R}^n\}\quad (j=1,\ldots,m_+).$$

Then there exists

$$u(t,x,y)\in\mathscr{E}(\overline{\mathbb{R}_+{}^{n+1}}),\quad \operatorname{supp}[u]\subset\{t\geq 0,(x,y)\in\mathbb{R}^n\}$$

such that

$$A(\mathrm{D}_t,\mathrm{D}_x,\mathrm{D}_y)u=f,\quad t>0,\quad x>0,\quad y\in\mathbb{R}^{n-1},$$
$$B_j(\mathrm{D}_t,\mathrm{D}_x,\mathrm{D}_y)u|_{x=+0}=g_j\quad (j=1,\ldots,m_+),\quad t>0,\quad y\in\mathbb{R}^{n-1}$$

and if $f\in\mathscr{S}_{-\{\gamma_1(1,0,0)+\Sigma_+\}}$ then, irrespective of the location of $\operatorname{supp}[f]$, the following inclusion property holds:

$$\operatorname{supp}[u]\subset W_{\operatorname{supp}[f]}\cup\left[\bigcup_{j=1}^{m_+}\operatorname{supp}[g_j]+\Sigma_+{}'\right].$$

Proof

First, we verify the above statements for $\{f(t,x,y)\in\mathscr{D}(\mathbb{R}^{n+1}), g_j(t,y)\in\mathscr{D}(\mathbb{R}^n)$ $(j=1,\ldots,m_+)\}$. From Propositions 4.4 and 4.6, we see that

$$u=Gf+\sum_{j=1}^{m_+}G_jg_j\in\mathscr{E}(\overline{\mathbb{R}_+{}^{n+1}})$$

satisfies

$$A(\mathrm{D}_t,\mathrm{D}_x,\mathrm{D}_y)u=f,\quad (t,x,y)\in\mathbb{R}_+{}^{n+1},$$
$$B_j(\mathrm{D}_t,\mathrm{D}_x,\mathrm{D}_y)u|_{x=+0}=g_j\quad (j=1,\ldots,m_+),\quad (t,y)\in\mathbb{R}^n,$$

and

$$\operatorname{supp}[u]\cap\{x\geqq 0,(t,y)\in\mathbb{R}^n\}\subset W_{\operatorname{supp}[f]}\cup\left[\bigcup_{j=1}^{m_+}\operatorname{supp}[g_j]+\Sigma_+{}'\right].$$

Next, we deal with the case of $\{f(t,x,y)\in\mathscr{E}(\mathbb{R}^{n+1}), g_j(t,y)\in\mathscr{E}(\mathbb{R}^n)$ $(j=1,..,m_+)\}$. Consider the locally finite† partition of unity in $\mathbb{R}^{n+1}$:

$$\sum_{i=1}^{\infty}\alpha_i(t,x,y)=1,\quad \alpha_i(t,x,y)\in\mathscr{D}(\mathbb{R}^{n+1}).$$

† Translator's note: This means $\{\operatorname{supp}[\alpha_i]\}$ is locally finite. 'Partition of unity' is defined under this condition.

By using this, the data can be also partitioned as

$$Au = \alpha_i f, \quad (t, x, y) \in \mathbb{R}_+{}^{n+1},$$
$$B_j u|_{x=0} = \alpha_i g_j|_{x=0} \quad (j = 1, \ldots, m_+), \quad (t, y) \in \mathbb{R}^n.$$

Then there exists a solution $u \in \mathscr{E}(\overline{\mathbb{R}_+{}^{n+1}})$ satisfying

$$\operatorname{supp}[u] \cap \{x \geqq 0, (t, y) \in \mathbb{R}^n\} \subset W_{\operatorname{supp}[\alpha_i f]} \cup \left[\bigcup_{j=1}^{m_+} \operatorname{supp}[\alpha_i g_j|_{x=0}] + \Sigma_+{}' \right].$$

Write u_i for such u. Since the support of f is in $\{t \geqq 0, (x, y) \in \overline{\mathbb{R}_+{}^n}\}$ and the support of g_j, is in $\{t \geqq 0, y \in \mathbb{R}^{n-1}\}$, given an arbitrary compact set K of $\mathbb{R}^{n+1}$ the number of indexes i for which u_i does not vanish on K is finite. Write

$$u = \sum_{i=1}^{\infty} u_i,$$

then $u \in \mathscr{E}(\overline{\mathbb{R}_+{}^{n+1}})$ and satisfies

$$A(\mathrm{D}_t, \mathrm{D}_x, \mathrm{D}_y)u = f, \quad (t, x, y) \in \mathbb{R}_+{}^n,$$
$$B_j(\mathrm{D}_t, \mathrm{D}_x, \mathrm{D}_y)u|_{x=+0} = g_j \quad (j = 1, \ldots, m_+), \quad (t, y) \in \mathbb{R}^n.$$

The inclusion property

$$\operatorname{supp}[u] \cap \{x \geqq 0, (t, y) \in \mathbb{R}^n\} \subset W_{\operatorname{supp}[f]} \cup \left[\bigcup_{j=1}^{m_+} \operatorname{supp}[g_j] + \Sigma_+{}' \right]$$

is obviously true. ■

5. Adjoint problems and the uniqueness of solutions

In general terms, it is probably true to say that the proofs of the existence and uniqueness of solutions of partial differential equations require basically different approaches. However, given an initial value problem, if we are able to derive its 'adjoint' problem, and find that the new problem is within the same family of problem type, then the proofs of the existence and uniqueness of solutions are closely linked to each other. In such cases we can consider just one of them and dispense with the other because, roughly speaking, the existence of solutions for the original problem guarantees uniqueness for the adjoint problem and vice versa.

In §4, our argument was centred on the existence of a solution for a hyperbolic boundary value problem itself. In the present section we are mainly concerned with the adjoint problem; we shall begin by deriving the adjoint of the original hyperbolic problem, and then prove that the new problem is also hyperbolic. Having done this, we shall see that the existence theorem given in §4 is, in fact, applicable to a solution for the adjoint problem, and, consequently, the uniqueness of the corresponding solution for the original problem can be established from the guaranteed existence of the solution in the adjoint case.

5.1

First, we see that for a partial differential operator with constant coefficients

$$(*) \qquad A(\mathrm{D}_t, \mathrm{D}_x, \mathrm{D}_y) = \sum_{j+k+|\nu| \leqq m} a_{jk\nu} \mathrm{D}_t^{\,j} \mathrm{D}_x^{\,k} \mathrm{D}_y^{\,\nu}$$

we can obtain the *formal adjoint operator*

$$A^*(\mathrm{D}_t, \mathrm{D}_x, \mathrm{D}_y) = \sum_{j+k+|\nu| \leqq m} \overline{a_{jk\nu}} \mathrm{D}_t^{\,j} \mathrm{D}_x^{\,k} \mathrm{D}_y^{\,\nu}$$

with respect to the inner product defined in L^2-space. We consider this in the half-space $\mathbb{R}_+^{\,n+1}$ where $x > 0$. Let

$$(u, v) = \int_{\mathbb{R}_+^{\,n+1}} u(t, x, y)\overline{v(t, x, y)}\, \mathrm{d}t\, \mathrm{d}x\, \mathrm{d}y,$$

$$\langle u, v \rangle = \int_{\mathbb{R}^n} u(t, 0, y)\overline{v(t, 0, y)}\, \mathrm{d}t\, \mathrm{d}y.$$

For $u, v \in \mathscr{D}(\mathbb{R}_+^{\,n+1})$ we see that

$$(*) \qquad (A(\mathrm{D}_t, \mathrm{D}_x, \mathrm{D}_y)u, v) - (u, A^*(\mathrm{D}_t, \mathrm{D}_x, \mathrm{D}_y)v) = 0.$$

But, if $u, v \in \mathscr{D}(\overline{\mathbb{R}_+^{\,n+1}})$ then on the right hand side of the above equation $(*)$ there remain some integrals defined over $x = 0$, i.e. if we write

$$A(\mathrm{D}_t, \mathrm{D}_x, \mathrm{D}_y) = a_0 \mathrm{D}_x^{\,m} + a_1(\mathrm{D}_t, \mathrm{D}_y)\mathrm{D}_x^{\,m-1} + \ldots + a_m(\mathrm{D}_t, \mathrm{D}_y)$$

then

$$\begin{aligned}
&(A(\mathrm{D}_t, \mathrm{D}_x, \mathrm{D}_y)u, v) - (u, A^*(\mathrm{D}_t, \mathrm{D}_x, \mathrm{D}_y)v) \\
&\quad = \{(a_0 \mathrm{D}_x^{\,m} u, v) - (u, \bar{a}_0 \mathrm{D}_x^{\,m} v)\} \\
&\qquad + \{(a_1 \mathrm{D}_x^{\,m-1} u, v) - (u, a_1^* \mathrm{D}_x^{\,m-1} v)\} \\
&\qquad + \ldots + \{(a_{m-1} \mathrm{D}_x u, v) - (u, a_{m-1}^* \mathrm{D}_x v)\} \\
&\qquad + \{(a_m u, v) - (u, a_m^* v)\} \\
&\quad = \mathrm{i}\{\langle a_0 \mathrm{D}_x^{\,m-1} u, v \rangle + \langle a_0 \mathrm{D}_x^{\,m-2} u, \mathrm{D}_x v \rangle + \ldots + \langle a_0 u, \mathrm{D}_x^{\,m-1} v \rangle\} \\
&\qquad + \mathrm{i}\{\langle a_1 \mathrm{D}_x^{\,m-2} u, v \rangle + \langle a_1 \mathrm{D}_x^{\,m-2} u, \mathrm{D}_x v \rangle \\
&\qquad + \ldots + \langle a_1 u, \mathrm{D}_x^{\,m-2} v \rangle\} \\
&\qquad + \ldots + \mathrm{i}\langle a_{m-1} u, v \rangle.
\end{aligned}$$

Therefore, setting

$$\mathscr{A}(D_t, D_y) = \begin{bmatrix} 0 & & & a_0 \\ & & \cdot^{\cdot^{\cdot}} & a_1(D_t, D_y) \\ & \cdot^{\cdot^{\cdot}} & & \vdots \\ a_0 & a_1(D_t, D_y) & \dots & a_{m-1}(D_t, D_y) \end{bmatrix}$$

we have

$$(A(D_t, D_x, D_y)u, v) - (u, A^*(D_t, D_x, D_y)v)$$
$$= i\left\langle \mathscr{A}(D_t, D_y) \begin{bmatrix} D_x^{\ m-1} \\ \vdots \\ 1 \end{bmatrix} u, \begin{bmatrix} D_x^{\ m-1} \\ \vdots \\ 1 \end{bmatrix} v \right\rangle.$$

Let us look again at conditions (ii) and (iii) of Hypothesis (A), which was given at the beginning of §2 (p. 59). We see that $A_0(0,1,0) \neq 0$, $B_{j0}(0,1,0) \neq 0$ and, for the degree r_j of B_j we have

$$0 \leqq r_j \leqq m-1, \quad r_i \neq r_j \quad (i \neq j).$$

To simplify things, let us set

$$A_0(0,1,0) = B_{10}(0,1,0) = \dots = B_{m+0}(0,1,0) = 1.$$

We supplement $\{r_1, \dots, r_{m+}\}$ with $\{r_{m_+ +1}, \dots, r_m\}$ to obtain

$$\{r_1, \dots, r_{m+}\} \cup \{r_{m_+ +1}, \dots, r_m\} = \{0, 1, \dots, m-1\}.$$

Then we let

$$B_j = D_x^{\ r_j} \quad (j = m_+ + 1, \dots, m)$$

and derive

$$\begin{bmatrix} B_1(D_t, D_x, D_y) \\ \vdots \\ B_m(D_t, D_x, D_y) \end{bmatrix} = \begin{bmatrix} b_{11}(D_t, D_y) \dots b_{1m}(D_t, D_y) \\ \vdots \\ b_{m1}(D_t, D_y) \dots b_{mm}(D_t, D_y) \end{bmatrix} \begin{bmatrix} D_x^{\ m-1} \\ \vdots \\ 1 \end{bmatrix}$$

$$= \mathscr{B}(D_t, D_y) \begin{bmatrix} D_x^{\ m-1} \\ \vdots \\ 1 \end{bmatrix}.$$

Since

$$\det(\mathscr{B}(\tau, \eta)) = \pm 1,$$

we see that, in the inverse matrix $\mathscr{B}^{-1}(\tau, \eta)$ of $\mathscr{B}(\tau, \eta)$, every element becomes a polynomial of (τ, η), and

$$\mathscr{B}^{-1}(D_t, D_y)\mathscr{B}(D_t, D_y) = 1.$$

Therefore

$$\begin{bmatrix} D_x^{m-1} \\ \vdots \\ 1 \end{bmatrix} = \mathscr{B}^{-1}(D_t, D_y) \begin{bmatrix} B_1(D_t, D_x, D_y) \\ \vdots \\ B_m(D_t, D_x, D_y) \end{bmatrix}.$$

Hence

$$\left\langle \mathscr{A}(D_t, D_y) \begin{bmatrix} D_x^{m-1} \\ \vdots \\ 1 \end{bmatrix} u, \begin{bmatrix} D_x^{m-1} \\ \vdots \\ 1 \end{bmatrix} v \right\rangle$$

$$= \left\langle \mathscr{A}(D_t, D_y)\mathscr{B}^{-1}(D_t, D_y) \begin{bmatrix} B_1(D_t, D_x, D_y) \\ \vdots \\ B_m(D_t, D_x, D_y) \end{bmatrix} u, \begin{bmatrix} D_x^{m-1} \\ \vdots \\ 1 \end{bmatrix} v \right\rangle$$

$$= \left\langle \begin{bmatrix} B_1(D_t, D_x, D_y) \\ \vdots \\ B_m(D_t, D_x, D_y) \end{bmatrix} u, (\mathscr{B}^{-1})^*(D_t, D_y)\mathscr{A}^*(D_t, D_y) \begin{bmatrix} D_x^{m-1} \\ \vdots \\ 1 \end{bmatrix} v \right\rangle.$$

Setting

$$\begin{aligned}(D_x^{m-1}, \ldots, 1)\mathscr{A}(D_t, D_y)\mathscr{B}^{-1}(D_t, D_y) \\ = (B_1'(D_t, D_x, D_y), \ldots, B_m'(D_t, D_x, D_y))\end{aligned}$$

we see that B_j' becomes a polynomial of degree $r_j' = m - 1 - r_j$.

Summing up, we have

Lemma 5.1 *(Green's formula)*

For $u, v \in \mathscr{D}(\overline{\mathbb{R}_+^{n+1}})$,

$$\begin{aligned}(A(D_t, D_x, D_y)u, v) - (u, A^*(D_t, D_x, D_y)v) \\ = i \sum_{j=1}^{m} \langle B_j(D_t, D_x, D_y)u, B_j'^*(D_t, D_x, D_y)v \rangle.\end{aligned}$$

Now, let

$$u \in \mathscr{E}(\overline{\mathbb{R}_+^{n+1}}), \quad \operatorname{supp}[u] \subset \{t \geqq 0, (x, y) \in \mathbb{R}^n\}$$

such that

$$B_j(D_t, D_x, D_y)u|_{x=0} = 0 \quad (j = 1, \ldots, m_+), \quad (t, y) \in (0, T) \times \mathbb{R}^{n-1}.$$

We write $\mathscr{D}(A)$ for the entire set of such u. Consider v such that

$$v \in \mathscr{E}(\overline{\mathbb{R}_+^{n+1}}), \quad \operatorname{supp}[v] \subset \{t \leqq T, (x, y) \in \mathbb{R}^n\}.$$

Then, for any $u \in \mathscr{D}(A)$, the necessary and sufficient conditions for v

satisfying

$$(A(\mathrm{D}_t, \mathrm{D}_x, \mathrm{D}_y)u, v) = (u, A^*(\mathrm{D}_t, \mathrm{D}_x, \mathrm{D}_y)v)$$

are

$$B_j'^*(\mathrm{D}_t, \mathrm{D}_x, \mathrm{D}_y)v|_{x=0} = 0 \quad (j = m_+ + 1, \ldots, m), \quad (t, y) \in (0, T) \times \mathbb{R}^{n-1}.$$

To see that they are necessary, we choose $j_0 \geqq m_+ + 1$ and $\varphi(t, y) \in \mathscr{D}((0, T) \times \mathbb{R}^{n-1})$, and construct $u \in \mathscr{D}((0, T) \times \overline{\mathbb{R}_+{}^n})$ satisfying the condition

$$B_j(\mathrm{D}_t, \mathrm{D}_x, \mathrm{D}_y)u|_{x=0} = 0 \quad (1 \leqq j \leqq m, j \neq j_0),$$
$$B_{j_0}(\mathrm{D}_t, \mathrm{D}_x, \mathrm{D}_y)u|_{x=0} = \varphi(t, y).$$

From this, we see that, in fact, $u \in \mathscr{D}(A)$. Given a v, we can apply Green's formula to obtain

$$(A(\mathrm{D}_t, \mathrm{D}_x, \mathrm{D}_y)u, v) - (u, A^*(\mathrm{D}_t, \mathrm{D}_x, \mathrm{D}_y)v)$$
$$= \mathrm{i}\langle \varphi, (B_{j_0}')^*(\mathrm{D}_t, \mathrm{D}_x, \mathrm{D}_y)v \rangle.$$

By our assumption we see that the term on the left hand side is zero. Since φ is arbitrary, we have

$$B_{j_0}'^*(\mathrm{D}_t, \mathrm{D}_x, \mathrm{D}_y)v|_{x=0} = 0.$$

This shows that the conditions are necessary. Bearing these facts in mind we proceed to the main subject. To begin with, we set up an initial boundary value problem as follows:

$$(\mathrm{P}) \quad \begin{cases} A(\mathrm{D}_t, \mathrm{D}_x, \mathrm{D}_y)u = f, \quad (t, x, y) \in (0, T) \times \mathbb{R}_+{}^n, \\ B_j(\mathrm{D}_t, \mathrm{D}_x, \mathrm{D}_y)u|_{x=0} = g_j \quad (j = 1, \ldots, m_+), \quad (t, y) \in (0, T) \times \mathbb{R}^{n-1}, \\ \mathrm{D}_t{}^j u|_{t=0} = u_j \quad (j = 0, \ldots, m-1), \quad (x, y) \in \mathbb{R}_+{}^n. \end{cases}$$

Then we define the *adjoint problem* (P*) as

$$(\mathrm{P}^*) \quad \begin{cases} A^*(\mathrm{D}_t, \mathrm{D}_x, \mathrm{D}_y)v = \varphi, & (t, x, y) \in (0, T) \times \mathbb{R}_+{}^n, \\ B_j'^*(\mathrm{D}_t, \mathrm{D}_x, \mathrm{D}_y)v|_{x=0} = \psi_j & (j = m_+ + 1, \ldots, m), \\ & (t, y) \in (0, T) \times \mathbb{R}^{n-1}, \\ \mathrm{D}_t{}^j v|_{t=T} = v_j \quad (j = 0, \ldots, m-1), & (x, y) \in \mathbb{R}_+{}^n. \end{cases}$$

First, we see that the following statement is true: if (P) satisfies Hypothesis (A), then (P*) satisfies the conditions:

(i) $A^*(\tau, \xi, \eta)$ is hyperbolic in the direction of $(-1, 0, 0)$,

(ii) $A_0{}^*(0, 1, 0) \neq 0$,

(iii) $B_{j_0}'(0, 1, 0) \neq 0, \quad 0 \leqq r_j' \leqq m - 1, \quad r_i' \neq r_j' \quad (i \neq j)$.

To see this, notice that from the definition of A^* (formal adjoint) we have

$$A^*(\tau, \xi, \eta) = \bar{A}(\tau, \xi, \eta) = \overline{A(\bar{\tau}, \bar{\xi}, \bar{\eta})}.$$

Since $A_0(\tau, \xi, \eta)$ has real coefficients we see that

$$A_0{}^*(\tau, \xi, \eta) = A_0(\tau, \xi, \eta),$$

therefore, from this, the conditions (i), (ii) follow. At the same time we have

$$(B_1{}'(\tau, \xi, \eta), \ldots, B_m{}'(\tau, \xi, \eta)) = (\xi^{m-1}, \ldots, 1)\mathscr{A}(\tau, \eta)\mathscr{B}^{-1}(\tau, \eta),$$
$$(B_{10}{}'(\tau, \xi, \eta), \ldots, B_{m0}{}'(\tau, \xi, \eta)) = (\xi^{m-1}, \ldots, 1)\mathscr{A}_0(\tau, \eta)\mathscr{B}_0{}^{-1}(\tau, \eta),$$

so we see that

$$[B_{10}{}'(0,1,0), \ldots, B_{m0}{}'(0,1,0)] = [1, \ldots, 1]\begin{bmatrix} 0 & & 1 \\ & \cdot^{\cdot^{\cdot}} & \\ 1 & & 0 \end{bmatrix}\mathscr{B}_0{}^{-1}(0,0)$$
$$= [1, \ldots, 1]\mathscr{B}_0{}^{-1}(0,0) = [1, \ldots, 1].$$

This shows that condition (iii) is satisfied.

Thus, we see that there is an essential similarity between the two problems. Therefore, we can expect that for the adjoint problem (P*), we can construct a solution as we did in §4. In order to investigate this we first study the behaviour of the Lopatinski determinant for the adjoint problem. Letting

$$(\tau, \eta) \in \mathbb{R}^n + \mathrm{i}\{\gamma_0(1,0) + \dot{\Gamma}\},$$

we see that

$$(\bar{\tau}, \bar{\eta}) \in \mathbb{R}^n - \mathrm{i}\{\gamma_0(1,0) + \dot{\Gamma}\}.$$

From this, using the same notation as before, we see that

$$A^*(\tau, \xi, \eta) = \overline{A(\bar{\tau}, \bar{\xi}, \bar{\eta})} = \overline{\prod_{j=1}^{m_+} (\bar{\xi} - \xi_j{}^+(\bar{\tau}, \bar{\eta})) \prod_{j=1}^{m_-} (\bar{\xi} - \xi_j{}^-(\bar{\tau}, \bar{\eta}))}$$
$$= \prod_{j=1}^{m_+} (\xi - \overline{\xi_j{}^+(\bar{\tau}, \bar{\eta})}) \prod_{j=1}^{m_-} (\xi - \overline{\xi_j{}^-(\bar{\tau}, \bar{\eta})}).$$

Therefore, we obtain the Lopatinski determinant for the adjoint problem as

$$R^*(\tau, \eta) = \det\left(\frac{1}{2\pi\mathrm{i}} \oint \frac{B'^*_{m_+ + j}(\tau, \xi, \eta)\xi^{k-1}}{\prod_{i=1}^{m_-} (\xi - \overline{\xi_i{}^-(\bar{\tau}, \bar{\eta})})} \mathrm{d}\xi\right)_{j,k=1,\ldots,m_-}$$

We rewrite $R^*(\tau, \eta)$ as

$$R^*(\tau, \eta) = \det\left(\frac{1}{2\pi\mathrm{i}} \oint \frac{\overline{B'_{m_+ + j}(\bar{\tau}, \bar{\xi}, \bar{\eta})}\xi^{k-1}}{\prod_{i=1}^{m_-} (\xi - \overline{\xi_i{}^-(\bar{\tau}, \bar{\eta})})} \mathrm{d}\xi\right)_{j,k=1,\ldots,m_-}$$
$$= \overline{\det\left(\frac{1}{2\pi\mathrm{i}} \oint \frac{B'_{m_+ + j}(\bar{\tau}, \bar{\xi}, \bar{\eta})\bar{\xi}^{k-1}}{\prod_{i=1}^{m_-} (\bar{\xi} - \xi_i{}^-(\bar{\tau}, \bar{\eta}))} \mathrm{d}\bar{\xi}\right)_{j,k=1,\ldots,m_-}}$$
$$= \overline{R'(\bar{\tau}, \bar{\eta})}.$$

Our goal is to prove

$$R'(\tau,\eta) = \pm R(\tau,\eta),$$

but to do this we need precise information about how $\{B_1(\tau,\xi,\eta),\ldots, B_m(\tau,\xi,\eta)\}$, $\{B_1'(\tau,\xi,\eta),\ldots,B_m'(\tau,\xi,\eta)\}$, are related to $A(\tau,\xi,\eta)$. We shall clarify the matter in the following subsection.

5.2

We have already come across a relation of this kind (see §5.1). Expressed as a polynomial in ξ, it was

$$\sum_{j=1}^{m} B_j(\xi)B_j'(\xi') = [\xi'^{m-1},\ldots,1]\,\mathscr{A}\begin{bmatrix}\xi^{m-1}\\ \vdots\\ 1\end{bmatrix}.$$

That is, if we write

$$\begin{bmatrix}B_1(\xi)\\ \vdots\\ B_m(\xi)\end{bmatrix} = \mathscr{B}\begin{bmatrix}\xi^{m-1}\\ \vdots\\ 1\end{bmatrix}, \quad \begin{bmatrix}B_1'(\xi)\\ \vdots\\ B_m'(\xi)\end{bmatrix} = \mathscr{B}'\begin{bmatrix}\xi^{m-1}\\ \vdots\\ 1\end{bmatrix},$$

then, we have

$${}^t\mathscr{B}'\mathscr{B} = \mathscr{A}.$$

At the same time, we have

$$\begin{aligned}
&[\xi'^{m-1},\ldots,1)\mathscr{A}\begin{bmatrix}\xi^{m-1}\\ \vdots\\ 1\end{bmatrix}\\
&= (\xi^{m-1} + \xi^{m-2}\xi' + \ldots + \xi'^{m-1}) + a_1(\xi^{m-2} + \xi^{m-3}\xi'\\
&\quad + \ldots + \xi'^{m-2}) + \ldots + a_{m-1}\\
&= \frac{\xi^m - \xi'^m}{\xi - \xi'} + a_1\frac{\xi^{m-1} - \xi'^{m-1}}{\xi - \xi'} + \ldots + a_{m-1}\frac{\xi - \xi'}{\xi - \xi'}\\
&= \frac{A(\xi) - A(\xi')}{\xi - \xi'}.
\end{aligned}$$

Using this we rewrite the above relation as

$$\sum_{j=1}^{m} B_j(\xi)B_j'(\xi') = \frac{A(\xi) - A(\xi')}{\xi - \xi'}.$$

If $\{B_1(\xi),\ldots,B_m(\xi)\}$, $\{B_1'(\xi),\ldots,B_m'(\xi)\}$ satisfy this condition, they are said to be *adjoint with respect to* $A(\xi)$. Using this terminology we state the following lemma, which deals with a typical case.

Lemma 5.2
Let

$$A(\xi)=\prod_{j=1}^{m_+}(\xi-\xi_j^{\,+})\prod_{j=1}^{m_-}(\xi-\xi_j^{\,-})=A_+(\xi)A_-(\xi)$$

be a decomposition of $A(\xi)$ where $\xi_i^{\,+}\neq\xi_j^{\,-}(i=1,\dots,m_+,j=1,\dots,m_-)$. If the relations

$$A_\pm(\xi)=\xi^{m_\pm}+a_1^{\,\pm}\xi^{m_\pm-1}+\dots+a_{m_\pm}^{\,\pm},$$

$$A_\pm^{\,(j)}(\xi)=\begin{cases}\xi^j+a_1^{\,\pm}\xi^{j-1}+\dots+a_j^{\,\pm} & (j\leqq m_\pm-1),\\ \xi^{j-m_\pm}A_\pm(\xi) & (j\geqq m_\pm)\end{cases}$$

hold, then

$$\{A_+^{\,(0)}(\xi),\dots,A_+^{\,(m-1)}(\xi)\},\quad \{A_-^{\,(m-1)}(\xi),\dots,A_-^{\,(0)}(\xi)\}$$

are adjoint with respect to $A(\xi)$.

Proof
The following calculation verifies the lemma.

$$\frac{A(\xi)-A(\xi')}{\xi-\xi'}=\frac{A_+(\xi)A_-(\xi)-A_+(\xi')A_-(\xi')}{\xi-\xi'}$$

$$=\frac{A_+(\xi)-A_+(\xi')}{\xi-\xi'}A_-(\xi)+A_+(\xi')\frac{A_-(\xi)-A_-(\xi')}{\xi-\xi'}$$

$$=[\xi'^{m_+-1},\dots,1]\begin{bmatrix} & 0 & & 1\\ & & \cdot^{\cdot^{\cdot}} & a_1^{\,+}\\ & \cdot^{\cdot^{\cdot}} & & \vdots\\ 1 & a_1^{\,+} & \cdots & a_{m_+-1}^{\,+}\end{bmatrix}\begin{bmatrix}\xi^{m_+-1}\\ \vdots\\ 1\end{bmatrix}A_-(\xi)$$

$$+A_+(\xi')[\xi'^{m_--1},\dots,1]\begin{bmatrix} & 0 & & 1\\ & & \cdot^{\cdot^{\cdot}} & a_1^{\,-}\\ & \cdot^{\cdot^{\cdot}} & & \vdots\\ 1 & a_1^{\,-} & \cdots & a_{m_--1}^{\,-}\end{bmatrix}\begin{bmatrix}\xi^{m_--1}\\ \vdots\\ 1\end{bmatrix}$$

$$=[A_+^{\,(0)}(\xi'),\dots,A_+^{\,(m_+-1)}(\xi')]\begin{bmatrix}A_-^{\,(m-1)}(\xi)\\ \vdots\\ A_-^{\,(m_-)}(\xi)\end{bmatrix}$$

$$+[A_+^{\,(m-1)}(\xi'),\dots,A_+^{\,(m_+)}(\xi')]\begin{bmatrix}A_-^{\,(0)}(\xi)\\ \vdots\\ A_-^{\,(m_--1)}(\xi)\end{bmatrix}. \quad \blacksquare$$

With arbitrary s-, t-tuples of polynomials $\{P_1(\xi),\dots,P_s(\xi)\}$, $\{Q_1(\xi),\dots,Q_t(\xi)]$ we associate an (s, t)-matrix

$$\left\langle \begin{bmatrix} P \\ \vdots \\ P_s \end{bmatrix}, \begin{bmatrix} Q_1 \\ \vdots \\ Q_t \end{bmatrix} \right\rangle_A = \left[\frac{1}{2\pi i} \oint \frac{P_j(\xi) Q_k(\xi)}{A(\xi)} d\xi \right]_{\substack{j=1,\dots,s \\ k=1,\dots,t}}$$

where the integral sign in the term on the right hand side means integration along a closed curve enclosing all the roots of $A(\xi) = 0$ in its interior. Then we have

Lemma 5.3

$$\{B_1(\xi),\dots,B_m(\xi)\}, \quad \{B_1'(\xi),\dots,B_m'(\xi)\}$$

are adjoint with respect to $A(\xi)$ iff

$$\left\langle \begin{bmatrix} B_1(\xi) \\ \vdots \\ B_m(\xi) \end{bmatrix}, \begin{bmatrix} B_1'(\xi) \\ \vdots \\ B_m'(\xi) \end{bmatrix} \right\rangle_{A(\xi)} = I.$$

Proof

Recall the equality

$$\frac{A(\xi) - A(\xi')}{\xi - \xi'} = [\xi'^{m-1},\dots,1] \mathscr{A} \begin{bmatrix} \xi^{m-1} \\ \vdots \\ 1 \end{bmatrix}.$$

Regarding both sides of the equality as polynomials in ξ and applying the operator

$$\left\langle \, , \begin{bmatrix} \xi^{m-1} \\ \vdots \\ 1 \end{bmatrix} \right\rangle_{A(\xi)}$$

we have, for the term on the left,

$$\left\langle \frac{A(\xi) - A(\xi')}{\xi - \xi'}, \begin{bmatrix} \xi^{m-1} \\ \vdots \\ 1 \end{bmatrix} \right\rangle_{A(\xi)}$$

$$= \frac{1}{2\pi i} \oint \frac{A(\xi) - A(\xi')}{\xi - \xi'} \frac{1}{A(\xi)} (\xi^{m-1},\dots,1)\, d\xi$$

$$= \frac{1}{2\pi i} \oint \frac{1}{\xi - \xi'} (\xi^{m-1},\dots,1)\, d\xi$$

$$= [\xi'^{m-1},\dots,1],$$

and for the term on the right

$$\left\langle [\xi'^{m-1},\ldots,1]\mathscr{A}\begin{bmatrix}\xi^{m-1}\\ \vdots \\ 1\end{bmatrix}, \begin{bmatrix}\xi^{m-1}\\ \vdots \\ 1\end{bmatrix}\right\rangle_{A(\xi)}$$

$$= [\xi'^{m-1},\ldots,1]\mathscr{A}\left\langle \begin{bmatrix}\xi^{m-1}\\ \vdots \\ 1\end{bmatrix}, \begin{bmatrix}\xi^{m-1}\\ \vdots \\ 1\end{bmatrix}\right\rangle_{A(\xi)}.$$

Therefore we see that

$$\left\langle \begin{bmatrix}\xi^{m-1}\\ \vdots \\ 1\end{bmatrix}, \begin{bmatrix}\xi^{m-1}\\ \vdots \\ 1\end{bmatrix}\right\rangle_{A(\xi)} = \mathscr{A}^{-1}.$$

On the other hand consider

$$\begin{bmatrix}B_1(\xi)\\ \vdots \\ B_m(\xi)\end{bmatrix} = \mathscr{B}\begin{bmatrix}\xi^{m-1}\\ \vdots \\ 1\end{bmatrix}, \quad \begin{bmatrix}B_1'(\xi)\\ \vdots \\ B_m'(\xi)\end{bmatrix} = \mathscr{B}'\begin{bmatrix}\xi^{m-1}\\ \vdots \\ 1\end{bmatrix}.$$

Then their product is

$$\left\langle \begin{bmatrix}B_1(\xi)\\ \vdots \\ B_m(\xi)\end{bmatrix}, \begin{bmatrix}B_1'(\xi)\\ \vdots \\ B_m'(\xi)\end{bmatrix}\right\rangle_A = \mathscr{B}\left\langle \begin{bmatrix}\xi^{m-1}\\ \vdots \\ 1\end{bmatrix}, \begin{bmatrix}\xi^{m-1}\\ \vdots \\ 1\end{bmatrix}\right\rangle_A {}^{t}\mathscr{B}'.$$

$$= \mathscr{B}\mathscr{A}^{-1t}\mathscr{B}'$$

Hence

$$\{B_1(\xi),\ldots,B_m(\xi)\}, \quad \{B_1'(\xi),\ldots,B_m'(\xi)\}$$

are adjoint with respect to $A(\xi)$ that is,

$${}^{t}\mathscr{B}'\mathscr{B} = \mathscr{A}$$

iff

$$\left\langle \begin{bmatrix}B_1(\xi)\\ \vdots \\ B_m(\xi)\end{bmatrix}, \begin{bmatrix}B_1'(\xi)\\ \vdots \\ B_m'(\xi)\end{bmatrix}\right\rangle_A = I. \quad \blacksquare$$

Now, suppose that $\{Q_1(\xi),\ldots,Q_m(\xi)\}$, $\{Q_1'(\xi),\ldots,Q_m'(\xi)\}$ are adjoint with respect to $A(\xi)$, and consider an arbitrary polynomial $P(\xi)$ of degree $m-1$, such that

$$P(\xi) = [c_1,\ldots,c_m]\begin{bmatrix}Q_1(\xi)\\ \vdots \\ Q_m(\xi)\end{bmatrix}.$$

Then, by application of the operator

$$\left\langle \;,\begin{bmatrix} Q_1' \\ \vdots \\ Q_m' \end{bmatrix} \right\rangle_A$$

and using Lemma 5.3 we have

$$\left\langle P, \begin{bmatrix} Q_1' \\ \vdots \\ Q_m' \end{bmatrix} \right\rangle = [c_1, \dots, c_m] \left\langle \begin{bmatrix} Q_1 \\ \vdots \\ Q_m \end{bmatrix}, \begin{bmatrix} Q_1' \\ \vdots \\ Q_m' \end{bmatrix} \right\rangle_A = [c_1, \dots, c_m].$$

Therefore

$$P(\xi) = \left\langle P, \begin{bmatrix} Q_1' \\ \vdots \\ Q_m' \end{bmatrix} \right\rangle_A \begin{bmatrix} Q_1(\xi) \\ \vdots \\ Q_m(\xi) \end{bmatrix}.$$

From this we prove

Lemma 5.3′

Let $\{Q_1(\xi), \dots, Q_m(\xi)\}$, $\{Q_1'(\xi), \dots, Q_m'(\xi)\}$ be adjoint with respect to $A(\xi)$. Then, the necessary and sufficient condition for

$$\{B_1(\xi), \dots, B_m(\xi)\}, \quad \{B_1'(\xi), \dots, B_m'(\xi)\}$$

to be adjoint with respect to $A(\xi)$ is

$$\left\langle \begin{bmatrix} B_1 \\ \vdots \\ B_m \end{bmatrix}, \begin{bmatrix} Q_1' \\ \vdots \\ Q_m' \end{bmatrix} \right\rangle_A {}^t\!\left(\left\langle \begin{bmatrix} B_1' \\ \vdots \\ B_m' \end{bmatrix}, \begin{bmatrix} Q_1 \\ \vdots \\ Q_m \end{bmatrix} \right\rangle_A \right) = I.$$

Proof

First, note that

$$\begin{bmatrix} B_1'(\xi) \\ \vdots \\ B_m'(\xi) \end{bmatrix} = \left\langle \begin{bmatrix} B_1' \\ \vdots \\ B_m' \end{bmatrix}, \begin{bmatrix} Q_1 \\ \vdots \\ Q_m \end{bmatrix} \right\rangle_A \begin{bmatrix} Q_1'(\xi) \\ \vdots \\ Q_m'(\xi) \end{bmatrix}.$$

From this we see that

$$\left\langle \begin{bmatrix} B_1 \\ \vdots \\ B_m \end{bmatrix}, \begin{bmatrix} B_1' \\ \vdots \\ B_m' \end{bmatrix} \right\rangle_A = \left\langle \begin{bmatrix} B_1 \\ \vdots \\ B_m \end{bmatrix}, \left\langle \begin{bmatrix} B_1' \\ \vdots \\ B_m' \end{bmatrix}, \begin{bmatrix} Q_1 \\ \vdots \\ Q_m \end{bmatrix} \right\rangle_A \begin{bmatrix} Q_1' \\ \vdots \\ Q_m' \end{bmatrix} \right\rangle_A$$

$$= \left\langle \begin{bmatrix} B_1 \\ \vdots \\ B_m \end{bmatrix}, \begin{bmatrix} Q_1' \\ \vdots \\ Q_m' \end{bmatrix} \right\rangle_A {}^t\!\left(\left\langle \begin{bmatrix} B_1' \\ \vdots \\ B_m' \end{bmatrix}, \begin{bmatrix} Q_1 \\ \vdots \\ Q_m \end{bmatrix} \right\rangle_A \right). \quad \blacksquare$$

5.3

With this preliminary information we now return to the original subject, the behaviour of the Lopatinski determinant in relation to the adjoint problem. We establish the following remarkable fact:

Lemma 5.4

For $(\tau, \eta) \in \mathbb{R}^n - \mathrm{i}(\gamma_0(1,0) + \dot{\Gamma})$,

$$R'(\tau, \eta) = \pm R(\tau, \eta),$$

where $+$ or $-$ applies, depending on the case.

Proof

Note that for $(\tau, \eta) \in \mathbb{R}^n - \mathrm{i}(\gamma_0(1,0) + \dot{\Gamma})$ the decomposition $A(\tau, \xi, \eta) = A_+(\tau, \xi, \eta) A_-(\tau, \xi, \eta)$ is the same one that we gave in Lemma 5.2. From this we see that

$\{A_+^{(0)}(\tau, \xi, \eta), \ldots, A_+^{(m-1)}(\tau, \xi, \eta)\}$ and $\{A_-^{(m-1)}(\tau, \xi, \eta), \ldots, A_-^{(0)}(\tau, \xi, \eta)\}$ are adjoint with respect to $A(\tau, \xi, \eta)$. This implies that if we write

$$\begin{bmatrix} B_1(\tau, \xi, \eta) \\ \vdots \\ B_m(\tau, \xi, \eta) \end{bmatrix} = C(\tau, \eta) \begin{bmatrix} A_+^{(0)}(\tau, \xi, \eta) \\ \vdots \\ A_+^{(m-1)}(\tau, \xi, \eta) \end{bmatrix},$$

$$\begin{bmatrix} B_1'(\tau, \xi, \eta) \\ \vdots \\ B_m'(\tau, \xi, \eta) \end{bmatrix} = C'(\tau, \eta) \begin{bmatrix} A_-^{(m-1)}(\tau, \xi, \eta) \\ \vdots \\ A_-^{(0)}(\tau, \xi, \eta) \end{bmatrix},$$

we have

$$C(\tau, \eta) = \left\langle \begin{bmatrix} B_1 \\ \vdots \\ B_m \end{bmatrix}, \begin{bmatrix} A_-^{(m-1)} \\ \vdots \\ A_-^{(0)} \end{bmatrix} \right\rangle_A$$

and

$$C'(\tau, \eta) = \left\langle \begin{bmatrix} B_1' \\ \vdots \\ B_m' \end{bmatrix}, \begin{bmatrix} A_+^{(0)} \\ \vdots \\ A_+^{(m-1)} \end{bmatrix} \right\rangle_A .$$

These $\{B_1(\tau, \xi, \eta), \ldots, B_m(\tau, \xi, \eta)\}$ and $\{B_1'(\tau, \xi, \eta), \ldots, B_m'(\tau, \xi, \eta)\}$ are, of course, adjoint with respect to $A(\tau, \xi, \eta)$. From Lemma 5.3′ we then see that

$$C(\tau, \eta) \cdot^{t} C'(\tau, \eta) = I.$$

Now, we wish to know the details of $C(\tau, \eta), C'(\tau, \eta)$. To this end we write

$$C = \begin{bmatrix} \overset{m_+}{\widehat{C_{11}}} & \overset{m_-}{\widehat{C_{12}}} \\ C_{21} & C_{22} \end{bmatrix} \begin{matrix})m_+ \\ )m_- \end{matrix}, \quad C' = \begin{bmatrix} \overset{m_+}{\widehat{C_{11}'}} & \overset{m_-}{\widehat{C_{12}'}} \\ C_{21}' & C_{22}' \end{bmatrix} \begin{matrix})m_+ \\ )m_- \end{matrix}.$$

We have

$C_{11}(\tau,\eta) =$

$$\begin{bmatrix} \frac{1}{2\pi i}\oint \frac{B_1(\tau,\xi,\eta)A_-^{(m-1)}(\tau,\xi,\eta)}{A(\tau,\xi,\eta)}d\xi \cdots & \frac{1}{2\pi i}\oint \frac{B_1(\tau,\xi,\eta)A_-^{(m_-)}(\tau,\xi,\eta)}{A(\tau,\xi,\eta)}d\xi \\ \vdots & \vdots \\ \frac{1}{2\pi i}\oint \frac{B_{m_+}(\tau,\xi,\eta)A_-^{(m-1)}(\tau,\xi,\eta)}{A(\tau,\xi,\eta)}d\xi \cdots & \frac{1}{2\pi i}\oint \frac{B_{m_+}(\tau,\xi,\eta)A_-^{(m_-)}(\tau,\xi,\eta)}{A(\tau,\xi,\eta)}d\xi \end{bmatrix}$$

$$= \begin{bmatrix} \frac{1}{2\pi i}\oint \frac{B_1(\tau,\xi,\eta)\xi^{m_+-1}}{A_+(\tau,\xi,\eta)}d\xi \cdots & \frac{1}{2\pi i}\oint \frac{B_1(\tau,\xi,\eta)}{A_+(\tau,\xi,\eta)}d\xi \\ \vdots & \vdots \\ \frac{1}{2\pi i}\oint \frac{B_{m_+}(\tau,\xi,\eta)\xi^{m_+-1}}{A_+(\tau,\xi,\eta)}d\xi \cdots & \frac{1}{2\pi i}\oint \frac{B_{m_+}(\tau,\xi,\eta)}{A_+(\tau,\xi,\eta)}d\xi \end{bmatrix}.$$

From this we see that

$$\det(C_{11}(\tau,\eta)) = \pm R(\tau,\eta).$$

For C' we have

$C_{22}'(\tau,\eta) =$

$$\begin{bmatrix} \frac{1}{2\pi i}\oint \frac{B_{m_++1}'(\tau,\xi,\eta)A_+^{(m_+)}(\tau,\xi,\eta)}{A(\tau,\xi,\eta)}d\xi \ldots & \frac{1}{2\pi i}\oint \frac{B_{m_++1}'(\tau,\xi,\eta)A_+^{(m-1)}(\tau,\xi,\eta)}{A(\tau,\xi,\eta)}d\xi \\ \vdots & \vdots \\ \frac{1}{2\pi i}\oint \frac{B_m'(\tau,\xi,\eta)A_+^{(m_+)}(\tau,\xi,\eta)}{A(\tau,\xi,\eta)}d\xi \ldots & \frac{1}{2\pi i}\oint \frac{B_m'(\tau,\xi,\eta)A_+^{(m-1)}(\tau,\xi,\eta)}{A(\tau,\xi,\eta)}d\xi \end{bmatrix}$$

$$= \begin{bmatrix} \frac{1}{2\pi i}\oint \frac{B_{m_++1}'(\tau,\xi,\eta)}{A_-(\tau,\xi,\eta)}d\xi \cdots & \frac{1}{2\pi i}\oint \frac{B_{m_++1}'(\tau,\xi,\eta)\xi^{m_--1}}{A_-(\tau,\xi,\eta)}d\xi \\ \vdots & \vdots \\ \frac{1}{2\pi i}\oint \frac{B_m'(\tau,\xi,\eta)}{A_-(\tau,\xi,\eta)}d\xi \cdots & \frac{1}{2\pi i}\oint \frac{B_m'(\tau,\xi,\eta)\xi^{m_--1}}{A_-(\tau,\xi,\eta)}d\xi \end{bmatrix}.$$

From this we see

$$\det(C_{22}'(\tau,\eta)) = R'(\tau,\eta).$$

We already know $C\,{}^tC' = I$. Therefore

$$\begin{bmatrix} C_{11} & C_{12} \\ C_{21} & C_{22} \end{bmatrix}\begin{bmatrix} I & {}^tC_{21}' \\ 0 & {}^tC_{22}' \end{bmatrix} = \begin{bmatrix} C_{11} & 0 \\ C_{21} & I \end{bmatrix}.$$

Taking the determinants of the matrices on both sides of this equality, and considering $\det(C) = \pm 1$ we have

$$\pm \det(C_{22}') = \det(C_{11}).$$

That is,

$$R'(\tau,\eta) = \pm R(\tau,\eta). \qquad \blacksquare$$

From this lemma, we see that for $(\tau,\eta)\in\mathbb{R}^n + \mathrm{i}\{\gamma_0(1,0) + \dot{\Gamma}\}$ we have $R^*(\tau,\eta) = \pm\overline{R(\bar{\tau},\bar{\eta})}$. Hence $R^*(\tau,\eta)$ is regular there, and the principal part of $R^*(\tau,\eta)$ is $\pm\overline{R_0(\bar{\tau},\bar{\eta})}$ satisfying

$$R^*(\tau,\eta) = \pm\overline{R(\bar{\tau},\bar{\eta})} \neq 0, \qquad (\tau,\eta)\in\mathbb{R}^n + \mathrm{i}(\gamma_1(1,0) + \dot{\Sigma}),$$

$$R_0{}^*(\tau,\eta) = \pm\overline{R_0(\tau,\eta)} \neq 0, \quad (\tau,\eta)\in\dot{\Sigma}.$$

That is, $R^*(\tau,\eta)$ is hyperbolic in the direction of $-\dot{\Sigma}$.

For this reason, it is possible to prove the existence theorem for (P*) by the same method as we employed to prove the existence theorem for (P). This implies that for

$$W_k{}^* = [(K - \Gamma')\cap\{x \geqq 0, (t,y)\in\mathbb{R}^n\}]\cup[(K-\Gamma')\cap\{x = 0, (t,y)\in\mathbb{R}^n\} - \Sigma_-{}'],$$

$$\text{where } \Sigma_- = \{(\tau,\xi,\eta)\in\mathbb{R}^{n+1} \mid (\tau,\eta)\in\dot{\Sigma}, \xi < \xi_{\max}(\tau,\eta)\}.$$

We can prove

Proposition 5.5

Let $R(\tau,\eta)$ be hyperbolic in the direction of $(1,0)$. If

$$\varphi(t,x,y)\in\mathscr{E}(\overline{\mathbb{R}_+{}^{n+1}}), \quad \mathrm{supp}[\varphi]\subset\{t\leqq T, (x,y)\in\overline{\mathbb{R}^n_+}\},$$

$$\psi_j(t,y)\in\mathscr{E}(\mathbb{R}^n), \quad \mathrm{supp}[\psi_j]\subset\{t\leqq T, y\in\mathbb{R}^{n-1}\}, \quad (j = 1,\dots,m_-)$$

then there exists

$$v(t,x,y)\in\mathscr{E}(\overline{\mathbb{R}_+{}^{n+1}}), \quad \mathrm{supp}[v]\subset\{t\leqq T, (x,y)\in\mathbb{R}^n\}$$

satisfying

$$A^*(\mathrm{D}_t,\mathrm{D}_x,\mathrm{D}_y)v = \varphi, \quad t<T, \quad x>0, \quad y\in\mathbb{R}^{n-1},$$

$$B'^*_{m_+ + j}(\mathrm{D}_t,\mathrm{D}_x,\mathrm{D}_y)v|_{x=+0} = \psi_j \quad (j=1,\dots,m_-), \quad t<T \quad y\in\mathbb{R}^{n-1}$$

as well as

$$\mathrm{supp}[v]\subset W^*_{\mathrm{supp}[\varphi]}\cup\left[\bigcup_{j=1}^{m_-}\mathrm{supp}[\psi_j] - \Sigma_-{}'\right].$$

5.4

Finally, returning to the main goal of this section, we shall prove the uniqueness of a solution for (P). We have

Theorem 5

Let $u \in C^\infty([0, \infty); \mathscr{D}'(\mathbb{R}^n))$ satisfy

$$A(\mathrm{D}_t, \mathrm{D}_x, \mathrm{D}_y)u = 0, \quad (t, x, y) \in W^*_{(t_0, x_0, y_0)},$$

$$B_j(\mathrm{D}_t, \mathrm{D}_x, \mathrm{D}_y)u|_{x=+0} = 0 \, (j = 1, \ldots, m_+), (t, y) \in W^*_{(t_0, x_0, y_0)} \cap \{x = 0, (t, y) \in \mathbb{R}^n\},$$

$$u = 0, \quad (t, x, y) \in W^*_{(t_0, x_0, y_0)} \cap \{t < 0, (x, y) \in \mathbb{R}^n\},$$

where

$$t_0 > 0, \quad x_0 \geqq 0, \quad y_0 \in \mathbb{R}^{n-1}.$$

Then

$$u(t, x, y) = 0, \quad (t, x, y) \in W^*_{(t_0, x_0, y_0)}.$$

Proof

Since

$$W^*_{(t_0, x_0, y_0)} \cap \{t \geqq 0, (x, y) \in \mathbb{R}^n\}$$

is compact, if we multiply u by an element of $\mathscr{D}$ which is identically 1 over the same compact set, we see that the support of u can be regarded as compact. Take $\rho_\varepsilon(t, y) \in \mathscr{D}(\mathbb{R}^n)$ whose support lies in an ε-neighbourhood of the origin. Setting

$$u_\varepsilon(t, x, y) = \rho_\varepsilon(t, y) \underset{(t,y)}{*} u(t, x, y) \in \mathscr{D}(\overline{\mathbb{R}_+{}^{n+1}}),$$

we have

$$A(\mathrm{D}_t, \mathrm{D}_x, \mathrm{D}_y)u_\varepsilon = \rho_\varepsilon \underset{(t,y)}{*} A(\mathrm{D}_t, \mathrm{D}_x, \mathrm{D}_y)u$$

$$B_j(\mathrm{D}_t, \mathrm{D}_x, \mathrm{D}_y)u_\varepsilon|_{x=+0} = \rho_\varepsilon \underset{(t,y)}{*} \{B_j(\mathrm{D}_t, \mathrm{D}_x, \mathrm{D}_y)u\}_{x=+0}.$$

Therefore, by taking an arbitrary open set $U \subset W^*{}_{(t_0, x_0, y_0)}$ for sufficiently small ε, we have

$$\begin{cases} Au_\varepsilon = 0, & (t, x, y) \in W_U{}^*, \\ B_j u_\varepsilon|_{x=+0} = 0 \quad (j = 1, \ldots, m_+), & (t, y) \in W_U{}^* \cap \{x = 0, (t, y) \in \mathbb{R}^n\}, \end{cases}$$

for some $u_\varepsilon \in U$.

On the other hand, if we pick an arbitrary

$$\alpha(t, x, y) \in \mathscr{D}$$

with supp $[\alpha]$ contained in U, then we see from Proposition 5.5 that there exists $v(t,x,y)\in\mathscr{E}(\overline{\mathbb{R}_+{}^{n+1}})$ such that

$$A^*(D_t, D_x, D_y)v = \alpha,$$
$$B'^*_{m_+ + j}(D_t, D_x, D_y)v|_{x=+0} = 0 \quad (j = 1,\ldots,m_-)$$

and

$$\operatorname{supp}[v] \subset W_U{}^*.$$

Then we apply Lemma 5.1 to u_ε and v to obtain

$$(u_\varepsilon, \alpha) = 0.$$

Since α is arbitrary, $u_\varepsilon = 0$ on U. Letting $\varepsilon \to +0$, we see that $u = 0$ on U. ■

Theorems 4 and 5 say that if $R(\tau,\eta)$ is hyperbolic in the direction of $(1,0)$, then the initial value problem (P) is $\mathscr{E}$-well-posed.

In other words, given a set of data $\{f, g_j, h_j\}$ for

$$\text{(P)} \quad \begin{cases} Au = f, \\ B_j u|_{x=+0} = g_j \quad (j = 1,\ldots,m_+), \\ D_t{}^j u|_{t=+0} = h_j \quad (j = 0,\ldots,m-1), \end{cases}$$

which satisfies the compatibility conditions, if each element of the data is infinitely differentiable, then, there exists a unique solution u of (P) which is infinitely differentiable.

Furthermore, by Theorems 2 and 3, we see that the converse of this statement is true. That is, if (P) is $\mathscr{E}$-well-posed, then $R(\tau,\eta)$ is hyperbolic in the direction of $(1,0)$. Therefore, we conclude that

> The necessary and sufficient condition for (P) being $\mathscr{E}$-well-posed is that $R(\tau,\eta)$ is hyperbolic in the direction of $(1,0)$.

3

Hyperbolic Boundary Value Problems with Variable Coefficients

In the previous chapter we considered boundary problems of hyperbolic equations with constant coefficients in a half-space, and looked at the possibility of constructing solutions by using Fourier–Laplace transforms. However, in the case of variable, as opposed to constant coefficients, this kind of direct construction method is not effective. In this chapter, therefore, we shall demonstrate that by recovering a given problem within the framework of an energy inequality, we can indirectly establish the existence and the uniqueness of a solution for the problem. Note, however, that in the case of variable coefficients we must narrow down our argument by confining ourselves to hyperbolic boundary value problems belonging to a special category of problem.

1. Sobolev spaces and singular integral operators

In this section we begin by defining the 'Sobolev space', an underlying space over which an energy inequality can be defined, and then define the singular integral operator operating on the space. We shall see that this operator is a conceptual extension of the Fourier–Laplace transform, and plays the same crucial rôle in the case of variable coefficients as the Fourier–Laplace transform did in the case of constant coefficients.

1.1

First, we explain the notion of 'Sobolev space'. Let Ω be an open set in $\mathbb{R}^n$, and m be a non-negative integer. Then the *Sobolev space*, denoted by $H^m(\Omega)$, is the vector space of functions $f(x), x\in\Omega$ whose partial derivatives of order less than m (inclusive) are all square integrable.† We define

† Translator's note: Notice that this implies $f\in L^2(\Omega)$ as well as $D^\nu f\in L^2(\Omega)$, $1\leqq\nu\leqq m$. This Sobolev space is often denoted by $H^m_{(2)}(\Omega)$. Note that, in general, for an integer $p\geqq 1$ and $L^p(\Omega)$, a Sobolev space $H^m_{(p)}(\Omega)$ can be defined. In fact, it is a Hilbert space. This is why we call it H.

$$(f,g)_{m,H(\Omega)} = \sum_{|\nu|\leqq m} \int_\Omega D^\nu f(x)\cdot\overline{D^\nu g(x)}\,dx,$$

$$\|f\|^2_{m,H(\Omega)} = \sum_{|\nu|\leqq m} \int_\Omega |D^\nu f(x)|^2\,dx,$$

$$|f|^2_{m,H(\Omega)} = \sum_{|\nu|= m} \int_\Omega |D^\nu f(x)|^2\,dx.$$

Next, we define the *weighted Sobolev space.* $\mathscr{H}_\eta{}^m(\Omega)$ as a space satisfying the following relation:

$$\mathscr{H}_\eta{}^m(\Omega) = e^{-\eta\cdot x}H^m(\Omega), \quad \eta\in\mathbb{R}^n,$$

i.e. $\mathscr{H}_\eta{}^m(\Omega)$ is a vector space of functions each of which becomes an element of $H^m(\Omega)$ after multiplication by $e^{\eta\cdot x}$. For $\mathscr{H}_\eta{}^m(\Omega)$ we define

$$(f,g)_{m\mathscr{H}_\eta(\Omega)} = \sum_{|\nu|\leqq m} \int_\Omega e^{2\eta\cdot x} D^\nu f(x)\overline{D^\nu g(x)}\,dx,$$

$$\|f\|^2_{m,\mathscr{H}_\eta(\Omega)} = \sum_{|\nu|\leqq m} \int_\Omega e^{2\eta\cdot x}|D^\nu f(x)|^2\,dx,$$

$$|f|^2_{m,\mathscr{H}_\eta(\Omega)} = \sum_{|\nu|= m} \int_\Omega e^{2\eta\cdot x}|D^\nu f(x)|^2\,dx.$$

If $m=0$, then we omit the 0 and write simply

$$\mathscr{H}_\eta{}^0(\Omega) = \mathscr{H}_\eta(\Omega), \quad \|f\|_{0,\mathscr{H}_\eta(\Omega)} = \|f\|_{\mathscr{H}_\eta(\Omega)}.$$

In particular, if $\Omega=\mathbb{R}^n$ we see that

$$\mathscr{H}_\eta{}^m(\mathbb{R}^n) \subset \mathscr{S}_\eta{}'(\mathbb{R}^n)$$

and for $|\nu|\leqq m$ we have

$$\mathscr{L}[D_x{}^\nu f(x)] = (\xi+i\eta)^\nu(\mathscr{L}f)(\xi+i\eta)\in L^2(\mathbb{R}_\xi{}^n).$$

By Plancherel's theorem we derive

$$(f,g)_{m,\mathscr{H}_\eta(\mathbb{R}^n)} = \frac{1}{(2\pi)^n}\int_{\mathbb{R}^n} \sum_{|\nu|\leqq m} |(\xi+i\eta)^\nu|^2(\mathscr{L}f)(\xi+i\eta)\overline{(\mathscr{L}g)(\xi+i\eta)}\,d\xi,$$

$$\|f\|^2_{m,\mathscr{H}_\eta(\mathbb{R}^n)} = \frac{1}{(2\pi)^n}\int \sum_{|\nu|\leqq m} |(\xi+i\eta)^\nu|^2|(\mathscr{L}f)(\xi+i\eta)|^2\,d\xi,$$

$$|f|^2_{m,\mathscr{H}_\eta(\mathbb{R}^n)} = \frac{1}{(2\pi)^n}\int \sum_{|\nu|= m} |(\xi+i\eta)^\nu|^2|(\mathscr{L}f)(\xi+i\eta)|^2\,d\xi.$$

On the other hand we have

$$\sum_{|\nu|=m} |(\xi+i\eta)^\nu|^2 = \sum_{|\nu|=m} (\xi_1{}^2+\eta_1{}^2)^{\nu_1}\dots(\xi_n{}^2+\eta_n{}^2)^{\nu_n}$$
$$\geqq (\xi_1{}^2+\eta_1{}^2)^m+\dots+(\xi_n{}^2+\eta_n{}^2)^m$$
$$\geqq c(|\xi|^2+|\eta|^2)^m.$$

This shows that for $\eta \neq 0$ $\|f\|_{m,\mathscr{H}_\eta(\mathbb{R}^n)}$ and $|f|_{m,\mathscr{H}_\eta(\mathbb{R}^n)}$ are equivalent norms. Writing

$$\|f\|^2_{m,\mathscr{H}_\eta(\Omega),\eta} = |\eta|^{2m}|f|^2_{0,\mathscr{H}_\eta(\Omega)} + |\eta|^{2(m-1)}|f|^2_{1,\mathscr{H}_\eta(\Omega)} + \dots + |f|^2_{m,\mathscr{H}_\eta(\Omega)},$$

we see that there exist positive numbers c_1, c_2 independent of η such that

$$c_1\|f\|_{m,\mathscr{H}_\eta(\mathbb{R}^n),\eta} \leqq |f|_{m,\mathscr{H}_\eta(\mathbb{R}^n)} \leqq c_2\|f\|_{m,\mathscr{H}_\eta(\mathbb{R}^n),\eta}.$$

Now, we reconsider the $\mathscr{E}$-well-posed problem (P) which we looked at in Chapter 2, this time in a Sobolev space:

$$Au=f, \quad 0<t<+\infty, \quad (x,y)\in\mathbb{R}_+{}^n,$$
$$B_ju|_{x=0}=g_j \quad (j=1,\dots,m_+), \quad 0<t<+\infty, \quad y\in\mathbb{R}^{n-1},$$
$$D_t{}^ju|_{t=0}=h_j \quad (j=0,\dots,m-1), \quad (x,y)\in\mathbb{R}_+{}^n.$$

For any

$$(\tau,\xi,\eta)\in\gamma_1(1,0,0)+\Gamma\cap\Sigma,$$

there exist positive constants N and C_k such that there exists a solution

$$u\in\mathscr{H}^k{}_{-(\tau,\xi,\eta)}(\mathbb{R}_+{}^1\times\mathbb{R}_+{}^n)$$

and

$$(*) \qquad \|u\|_{k,\mathscr{H}_{(\tau,\xi,\eta)}(\mathbb{R}_+{}^1\times\mathbb{R}_+{}^n)} \leqq C_k\Big\{\|f\|_{N+k,\mathscr{H}_{-(\tau,\xi,\eta)}(\mathbb{R}_+{}^1\times\mathbb{R}_+{}^n)}$$
$$+\sum_{j=1}^{m_+}\|g_j\|_{N+k,\mathscr{H}_{-(\tau,\eta)}(\mathbb{R}_+{}^1\times\mathbb{R}^{n-1})} + \sum_{j=0}^{m-1}\|h_j\|_{N+k,\mathscr{H}_{-(\xi,\eta)}(\mathbb{R}_+{}^n)}\Big\}^\dagger$$

if $f\in\mathscr{H}^{N+k}_{-(\tau,\xi,\eta)}(\mathbb{R}_+{}^1\times\mathbb{R}_+{}^n)$, $g_j\in\mathscr{H}^{N+k}_{-(\tau,\eta)}(\mathbb{R}_+{}^1\times\mathbb{R}^{n-1})$, $h_j\in\mathscr{H}^{N+k}_{-(\xi,\eta)}(\mathbb{R}_+{}^n)$.

Moreover, for an arbitrary compact set

$$K\subset\gamma_1(1,0,0)+\Gamma\cap\Sigma,$$

the above-mentioned N and C_k can be taken as constants independent of $(\tau,\xi,\eta)\in\tilde{K}$ where

$$\tilde{K}=\bigcup_{\lambda\geqq 1}(\lambda K).$$

† If this inequality is valid, we say the problem is *well-posed in the Sobolev space in the weak sense*.

1.2

Next, we shall define the singular integral operator over $\mathscr{S}_\eta'(\mathbb{R}^n)$ via a Fourier–Laplace transform defined on the same $\mathscr{S}_\eta'(\mathbb{R}^n)$. We say that $a(x, \xi, \eta)$ is a *symbol of degree m* if it satisfies the following conditions.

(i) For $(\xi,\eta)\in\mathbb{R}^n \times \mathbb{R}^n$, a is homogeneous of degree m.

(ii) For $(x,\xi,\eta)\in\mathbb{R}_x{}^n \times (\mathbb{R}_\xi{}^n \times \mathbb{R}_\eta{}^n - \{(0,0)\})$, a is infinitely differentiable, and satisfies

$$\sup_{(x,\xi,\eta)\in\mathbb{R}_x^n \times S_{\xi,\eta}{}^{2n-1}} |D_x{}^\alpha D_\xi{}^\mu D_\eta{}^\nu a(x,\xi,\eta)| < +\infty.$$

Consider a correspondence which associates

$$u(x)\in\mathscr{S}_\eta'(\mathbb{R}^n)$$

to

$$\begin{aligned} a(x, D_x', \eta)u(x) &= \mathscr{L}^{-1}[a(x,\xi,\eta)(\mathscr{L}u)(\xi+i\eta)] \\ &= e^{-x\cdot\eta}\mathscr{F}_{\xi\to x}[a(x,\xi,\eta)(\mathscr{L}u)(\xi+i\eta)]\in\mathscr{S}_\eta'(\mathbb{R}^n). \end{aligned}$$

Then, this correspondence is an operator, denoted by $a(x, D_x', \eta)$, and called the *singular integral operator* for the symbol $a(x,\xi,\eta)$.

In particular, if

$$\Lambda(\xi,\eta) = (|\xi|^2 + |\eta|^2)^{1/2}$$

then we write

$$\mathscr{H}_\eta{}^s = \{u; \|u\|_{s,\mathscr{H}_\eta} = \|\Lambda(D_x',\eta)^s u\|_{\mathscr{H}_\eta} < +\infty\}.$$

We can see that if we restrict this singular integral operator to $\mathscr{H}_\eta{}^s$, the degree of the symbol associated to the operator η is clearly reflected in the operator itself. The following lemma describes this effect.

Lemma 1.1

(i) If $a(x,\xi,\eta)$ is a symbol of degree m, then

$$\|a(x, D_x', \eta)u\|_{s,\mathscr{H}_\eta} \leqq C\|u\|_{s+m,\mathscr{H}_\eta}.$$

(ii) If $a_i(x,\xi,\eta)$ is a symbol of degree m_i for $i = 1,2$, then the singular integral operator (written $(a_1\circ a_2)(x, D_x', \eta)$) for the symbol of the product $a_1(x,\xi,\eta)\cdot a_2(x,\xi,\eta)$ satisfies

$$\|\{a_1(x,D_x',\eta)a_2(x,D_x',\eta) - (a_1\circ a_2)(x,D_x',\eta)\}u\|_{s,\mathscr{H}_\eta} \leqq C\|u\|_{s+m_1+m_2-1,\mathscr{H}_\eta}.$$

(iii) If $a(x,\xi,\eta)$ is a symbol of degree m, then the singular integral operator (written $\bar{a}(x,D_x',\eta)$) for the symbol $\overline{a(x,\xi,\eta)}$ satisfies

$$\|\{a^{(*)}(x,D_x',\eta) - \bar{a}(x,D_x',\eta)\}u\|_{s,\mathscr{H}_\eta} \leqq C\|u\|_{s+m-1,\mathscr{H}_\eta},$$

where $a^{(*)}(x,D_x',\eta)$ is a formal adjoint operator of $a(x,D_x',\eta)$ with respect to $(\,,)_{\mathscr{H}_\eta}$.

(iv) Let $A(x,\xi,\eta) = (a_{ij}(x,\xi,\eta))_{i,j=1,\ldots,N}$ be an $N \times N$ Hermitian matrix of such that

$$A(x,\xi,\eta) \geqq \delta I \quad (\delta > 0).$$

In this case if

$$A(x,\mathrm{D}_x{}',\eta) = (a_{ij}(x,\mathrm{D}_x{}',\eta))_{i,j=1,\ldots,N}, \quad u = \begin{bmatrix} u_1 \\ \vdots \\ u_N \end{bmatrix},$$

then

$$\operatorname{Re}(A(x,\mathrm{D}_x{}',\eta)u,u)_{s,\mathscr{H}_\eta} \geqq \tfrac{1}{2}\delta\|u\|^2_{s,\mathscr{H}_\eta} - C\|u\|^2_{s-1,\mathscr{H}_\eta}.$$

Note. The positive constant C is independent of u as well as η.

Although this lemma contains a weight η, these are well-known properties of a singular integral operator, so we omit the proof of the lemma.†

Next, we explain the notion of 'local symbol'. Since $a(x,\xi,\eta)$ is homogeneous with respect to (ξ,η), it is sufficient to confine ourselves to

$$(x,\xi,\eta) \in \mathbb{R}^n \times S^{2n-1}.$$

Take

$$(x_0,\xi_0,\eta_0) \in \mathbb{R}^n \times S^{2n-1},$$

and let U be its neighbourhood in $\mathbb{R}^n \times S^{2n-1}$. If the support of a symbol $\varphi(x,\xi,\eta)$ of degree m in $\mathbb{R}^n \times S^{2n-1}$ is contained in U, then $\varphi(x,\mathrm{D}_x{}',\eta)$ is called the *localising operator for U.*

If there exist two symbols of degree m,

$$a(x,\xi,\eta), \quad \tilde{a}(x,\xi,\eta)$$

which coincide in U, then with a localising operator φ, the statement

$$\|\{a(x,\mathrm{D}_x{}',\eta) - \tilde{a}(x,\mathrm{D}_x{}',\eta)\}\varphi(x,\mathrm{D}_x{}',\eta)u\|_{s,\mathscr{H}_\eta} \leqq C\|u\|_{s+m-1,\mathscr{H}_\eta}$$

is true. This is easy to prove from Lemma 1.1.

On the other hand, when the symbol $a(x,\xi,\eta)$ is defined *only* over U, we can consider a similar symbol $\varphi(x,\xi,\eta)$ of degree zero such that $\varphi = 1$ for some $U_0(\Subset U)$.‡ If we write

$$\tilde{a}(x,\xi,\eta) = \varphi(x,\xi,\eta)a(x,\xi,\eta) + \{1 - \varphi(x,\xi,\eta)\} \times a(x_0,\xi_0,\eta_0)\Lambda^m(\xi,\eta), (x_0,\xi_0,\eta_0) \in U_0,$$

† See A. P. Calderon & A. Zygmund, Singular integral operators and differential equations. *Am. J. Math.* vol. 79 (1957).

‡ Translator's note: $U_0 \Subset U$ means $\bar{U}_0$ is contained in the interior of U.

then $\tilde{a}$ is an extension of $\{a(x,\xi,\eta)|(x,\xi,\eta)\in U_0\}$. This is called a *standard extension of the symbol* $a(x,\xi,\eta)$.

1.3

In this subsection we observe a particular symbol P of degree m which becomes a polynomial of degree m in the direction of ξ_1. This means

$$P(x,\xi,\eta) = p_0(x,\xi',\eta)\xi_1{}^m + p_1(x,\xi',\eta)\xi_1{}^{m-1} + \ldots + p_m(x,\xi',\eta),$$
$$\xi' = (\xi_2,\ldots,\xi_n),$$

for some $p_0,\ldots,p_m$. The corresponding singular integral operator is

$$\begin{aligned} P(x,\mathrm{D}_x{}',\eta) &= p_0(x,\mathrm{D}_{x'}{}',\eta)\mathrm{D}_{x_1}{}'^m + p_1(x,\mathrm{D}_{x'}{}',\eta)\mathrm{D}_{x_1}{}'^{m-1} \\ &\quad + \ldots + p_m(x,\mathrm{D}_{x'}{}',\eta) \\ &= p_0(x,\mathrm{D}_{x'}{}',\eta)(\mathrm{D}_{x_1} - i\eta_1)^m \\ &\quad + p_1(x,\mathrm{D}_{x'}{}',\eta)(\mathrm{D}_{x_1} - i\eta_1)^{m-1} \\ &\quad + \ldots + p_m(x,\mathrm{D}_{x'}{}',\eta) \end{aligned}$$

which is a differential operator with respect to x_1.Therefore, if we put

$$\mathbb{R}_+{}^n = \{x_1 > 0, x' = (x_2,\ldots,x_n)\in\mathbb{R}^{n-1}\},$$
$$\mathscr{H}_\eta{}^m(\mathbb{R}_+{}^n) = \{f|\mathrm{e}^{\eta\cdot x}f\in H^m(\mathbb{R}_+{}^n)\},$$

we can see that such $P(x,\mathrm{D}_x{}',\eta)$ gives a continuous map

$$\mathscr{H}_\eta{}^m(\mathbb{R}_+{}^n)\to\mathscr{H}_\eta{}^0(\mathbb{R}_+{}^n).$$

Consider two symbols of this kind whose degrees are m and $m-1$. Write

$$P(x,\xi,\eta) = p_0(x,\xi',\eta)\xi_1{}^m + p_1(x,\xi',\eta)\xi_1{}^{m-1} + \ldots + p_m(x,\xi',\eta),$$
$$Q(x,\xi,\eta) = q_0(x,\xi',\eta)\xi_1{}^{m-1} + q_1(x,\xi',\eta)\xi_1{}^{m-2} + \ldots + q_{m-1}(x,\xi',\eta).$$

Regarding them as polynomials in ξ_1, we consider a quadratic form (*Bézout form*)

$$\begin{aligned} &\frac{P(\xi_1)Q(\bar{\xi}_1) - P(\bar{\xi}_1)Q(\xi_1)}{\xi_1 - \bar{\xi}_1} \\ &= [\bar{\xi}_1{}^{m-1},\bar{\xi}_1{}^{m-2},\ldots,1]\begin{bmatrix} C_{11} \cdots C_{1m} \\ \vdots \qquad \vdots \\ C_{m1} \cdots C_{mm}\end{bmatrix}\begin{bmatrix}\xi_1{}^{m-1} \\ \xi_1{}^{m-2} \\ \vdots \\ 1\end{bmatrix} \\ &= [\xi_1{}^{m-1},\ldots,1]C\begin{bmatrix}\xi_1{}^{m-1} \\ \vdots \\ 1\end{bmatrix}, \end{aligned}$$

for some C_{ij} (where $\bar{\xi}$ is the complex conjugate of ξ).

Then we see

$$C = p_0 Q_0 + p_1 Q_1 + \ldots + p_m Q_m,$$

where

$$Q_0 = \begin{bmatrix} q_0 & q_1 \cdots q_{m-1} \\ q_1 & \iddots \\ \vdots & \\ q_{m-1} & 0 \end{bmatrix}, \quad Q_1 = \left[\begin{array}{c|ccc} 0 & & 0 & \\ \hline & q_1 & q_2 \cdots q_{m-1} & \\ 0 & q_2 & & \\ & \vdots & \iddots & \\ & q_{m-1} & & 0 \end{array}\right],$$

$$Q_2 = \left[\begin{array}{cc|ccc} 0 & 0 & & & \\ 0 & q_0 & & 0 & \\ \hline & & q_2 & \cdots q_{m-1} & \\ & 0 & \vdots & \iddots & \\ & & q_{m-1} & & 0 \end{array}\right], \ldots,$$

$$Q_{m-1} = \left[\begin{array}{cccc|c} 0 & & & 0 & \\ & & \iddots & -q_0 & \\ & \iddots & & \vdots & 0 \\ 0 & -q_0 & \cdots & -q_{m-3} & \\ \hline & & 0 & & q_{m-1} \end{array}\right], \quad Q_m = \begin{bmatrix} 0 & & & & 0 \\ & & & \iddots & -q_0 \\ & & \iddots & & -q_1 \\ & \iddots & & & \vdots \\ 0 & -q_0 & -q_1 & \cdots & -q_{m-2} \end{bmatrix}.$$

To simplify things, we adjust the degrees of polynomials p_j, q_j to zero by multiplying by

$$\Lambda(\xi', \eta) = \sqrt{(|\xi'|^2 + |\eta|^2)}$$

and write $\tilde{p}_j, \tilde{q}_j$ for the new polynomials thus obtained. Since

$$P = \tilde{p}_0 \xi^m + \tilde{p}_1 \Lambda \xi^{m-1} + \ldots + \tilde{p}_m \Lambda^m,$$
$$Q = \tilde{q}_0 \xi^{m-1} + \tilde{q}_1 \Lambda \xi^{m-2} + \ldots + \tilde{q}_{m-1} \Lambda^{m-1},$$

we have

$$\frac{P(\xi_1)Q(\bar{\xi}_1) - P(\bar{\xi}_1)Q(\xi_1)}{\xi_1 - \bar{\xi}_1} = [\bar{\xi}_1{}^{m-1}, \Lambda \bar{\xi}_1{}^{m-2}, \ldots, \Lambda^{m-1}] \tilde{C} \begin{bmatrix} \xi_1{}^{m-1} \\ \Lambda \xi_1{}^{m-2} \\ \vdots \\ \Lambda^{m-1} \end{bmatrix},$$

where

$$\tilde{C} = \tilde{p}_0 \tilde{Q}_0 + \tilde{p}_1 \tilde{Q}_1 + \ldots + \tilde{p}_m \tilde{Q}_m.$$

For such $\tilde{C}$ we have the following property (Green's formula) which plays a crucial rôle when we consider energy inequalities later in this chapter. For $p^0 = 1$ we have

Theorem 1 *(generalised Green's formula for symbols P, Q)*
If $u, v \in \mathscr{H}_\eta^{\ m}(\mathbb{R}_+^{\ n})$ then

$$(P(x, D_x', \eta)u, Q^{(*)}(x, D_x', \eta)v)_{\mathscr{H}_\eta(\mathbb{R}_+^n)} - (Q(x, D_x', \eta)u, P^{(*)}(x, D_x', \eta)v)_{\mathscr{H}_\eta(\mathbb{R}_+^n)}$$

$$\cong \mathrm{i}\left\langle \tilde{C}(x, D_{x'}', \eta)\begin{bmatrix} D_{x_1}'^{\,m-1} \\ \Lambda D_{x_1}'^{\,m-2} \\ \vdots \\ \Lambda^{m-1}\end{bmatrix} u,\ \begin{bmatrix} D_{x_1}'^{\,m-1} \\ \Lambda D_{x_1}'^{\,m-2} \\ \vdots \\ \Lambda^{m-1}\end{bmatrix} v\right\rangle_{\mathscr{H}_{\eta}'(\mathbb{R}^{n-1})}$$

where $\langle , \rangle_{\mathscr{H}_\eta(\mathbb{R}^{n-1})}$ is the inner product of $\mathscr{H}_{\eta'}(\mathbb{R}^{n-1})$, and $\cong$ means that the difference between the right and left hand side terms is bounded by

$$C\left\{ \|u\|_{m-1, \mathscr{H}_\eta(\mathbb{R}_+^n)} \cdot \|v\|_{m-1, \mathscr{H}_\eta'(\mathbb{R}^{n-1})} + \sum_{j=0}^{m-1} \langle D_{x_1}'^{\,j} u\rangle_{m-1-j-1/2, \mathscr{H}_\eta'(\mathbb{R}^{n-1})} \times \sum_{j=0}^{m-1} \langle D_{x_1}'^{\,j} v\rangle_{m-1-j-1/2, \mathscr{H}_\eta'(\mathbb{R}^{n-1})}\right\}.$$

Note. The positive C is independent of u, v and η.

Proof
We see that

$$(Pu, Q^{(*)}v)_{\mathscr{H}_\eta(\mathbb{R}_+^n)} - (Qu, P^{(*)}v)_{\mathscr{H}_\eta(\mathbb{R}_+^n)}$$

$$= \sum_{i=0}^{m}\sum_{j=0}^{m-1} \{(p_{m-i} D_{x_1}'^{\,i} u, D_{x_1}'^{\,j} q_{m-1-j}^{(*)} v)_{\mathscr{H}_\eta(\mathbb{R}_+^n)} - (q_{m-1-j} D_{x_1}'^{\,j} u, D_{x_1}'^{\,i} p_{m-i}^{(*)} v)_{\mathscr{H}_\eta(\mathbb{R}_+^n)}\}$$

$$= \sum_{i=0}^{m}\sum_{j=0}^{m-1} I_{ij},$$

where

$$I_{ij} = (p_{m-i} D_{x_1}^{\ i}(e^{\eta_1 x_1} u), D_{x_1}^{\ j} q_{m-1-j}^{(*)}(e^{\eta_1 x_1} v))_{\mathscr{H}_\eta'(\mathbb{R}_+^n)} - (q_{m-1-j} D_{x_1}^{\ j}(e^{\eta_1 x_1} u), D_{x_1}^{\ i} p_{m-i}^{(*)}(e^{\eta_1 x_1} v))_{\mathscr{H}_\eta'(\mathbb{R}_+^n)}$$

$$= (\tilde{p}_{m-i} \Lambda^{m-i} D_{x_1}^{\ i}(e^{\eta_1 x_1} u), D_{x_1}^{\ j} \Lambda^{m-1-j} \tilde{q}_{m-1-j}^{(*)}(e^{\eta_1 x_1} v))_{\mathscr{H}_\eta'(\mathbb{R}_+^n)} - (\tilde{q}_{m-1-j} \Lambda^{m-1-j} D_{x_1}^{\ j}(e^{\eta_1 x_1} u), D_{x_1}^{\ i} \Lambda^{m-i} \tilde{p}_{m-i}^{(*)}(e^{\eta_1 x_1} v))_{\mathscr{H}_\eta'(\mathbb{R}_+^n)}.$$

Writing

$$\tilde{p}_{m-i} = p, \quad \tilde{q}_{m-1-j} = q, \quad e^{\eta_1 x_1} u = U, \quad e^{\eta_1 x_1} v = V,$$

we have

$$I_{ij} = (p\Lambda^{m-i} D_{x_1}^{\ i} U, D_{x_1}^{\ j} \Lambda^{m-1-j} q^{(*)} V)_{\mathscr{H}_\eta'(\mathbb{R}_+^n)} - (q\Lambda^{m-1-j} D_{x_1}^{\ j} U, D_{x_1}^{\ i} \Lambda^{m-i} p^{(*)} V)_{\mathscr{H}_\eta'(\mathbb{R}_+^n)}.$$

Let us consider the case $i-j=k\geqq 0$. Putting

$$\Lambda^{m-1-j}\mathrm{D}_{x_1}{}^{j}U=\varphi,\quad \Lambda^{m-1-j}\mathrm{D}_{x_1}{}^{j}V=\psi$$

we see that

$$I_{ij}\cong(\Lambda^{1-k}\mathrm{D}_{x_1}{}^{k}(p\varphi),q^{(*)}\psi)_{\mathscr{H}_{\eta}'(\mathbb{R}_+{}^n)}-(q\varphi,\Lambda^{1-k}\mathrm{D}_{x_1}{}^{k}(p^{(*)}\psi))_{\mathscr{H}_{\eta}'(\mathbb{R}_+{}^n)}=I_{ij}'.$$

In particular, if $k=0$ obviously $I_{ii}'\cong 0$. We look at the case $k>0$. Now, by integration by parts $k-1$ times we have

$$\begin{aligned}I_{ij}'=\mathrm{i}\{&\langle\Lambda^{1-k}\mathrm{D}_{x_1}{}^{k-1}(p\varphi),(q^{(*)}\psi)\rangle_{\mathscr{H}_{\eta}'(\mathbb{R}^{n-1})}\\&+\langle\Lambda^{2-k}\mathrm{D}_{x_1}{}^{k-2}(p\varphi),\Lambda^{-1}\mathrm{D}_{x_1}(q^{(*)}\psi)\rangle_{\mathscr{H}_{\eta}'(\mathbb{R}^{n-1})}\\&+\ldots+\langle(p\varphi),\Lambda^{1-k}\mathrm{D}_{x_1}{}^{k-1}(q^{(*)}\psi)\rangle_{\mathscr{H}_{\eta}'(\mathbb{R}^{n-1})}\}\\&+\{(p\varphi,\Lambda^{1-k}\mathrm{D}_{x_1}{}^{k}(q^{(*)}\psi))_{\mathscr{H}_{\eta}'(\mathbb{R}_+{}^n)}\\&\quad-(q\varphi,\Lambda^{1-k}\mathrm{D}_{x_1}{}^{k}(p^{(*)}\psi))_{\mathscr{H}_{\eta}'(\mathbb{R}_+{}^n)}\}.\end{aligned}$$

Therefore we have

$$\begin{aligned}I_{ij}'\cong\mathrm{i}\{&\langle p\circ q\Lambda^{1-k}\mathrm{D}_{x_1}{}^{k-1}\varphi,\psi\rangle_{\mathscr{H}_{\eta}'(\mathbb{R}^{n-1})}\\&+\langle p\circ q\Lambda^{2-k}\mathrm{D}_{x_1}{}^{k-2}\varphi,\Lambda^{-1}\mathrm{D}_{x_1}\psi\rangle_{\mathscr{H}_{\eta}'(\mathbb{R}^{n-1})}\\&+\ldots+\langle p\circ q\varphi,\Lambda^{1-k}\mathrm{D}_{x_1}{}^{k-1}\psi\rangle_{\mathscr{H}_{\eta}'(\mathbb{R}^{n-1})}\}.\end{aligned}$$

Hence,

$$\begin{aligned}I_{ij}\cong\mathrm{i}\{&\langle p\circ q\Lambda^{m-i}\mathrm{D}'_{x_1}{}^{i-1}u,\Lambda^{m-1-j}\mathrm{D}'_{x_1}{}^{j}v\rangle_{\mathscr{H}_{\eta}'(\mathbb{R}^{n-1})}\\&+\langle p\circ q\Lambda^{m-i+1}\mathrm{D}'_{x_1}{}^{i-2}u,\Lambda^{m-2-j}\mathrm{D}'_{x_1}{}^{j+1}v\rangle_{\mathscr{H}_{\eta}'(\mathbb{R}^{n-1})}\\&+\ldots+\langle p\circ q\Lambda^{m-1-j}\mathrm{D}'_{x_1}{}^{j}u,\Lambda^{m-i}\mathrm{D}'_{x_1}{}^{i-1}v\rangle_{\mathscr{H}_{\eta}'(\mathbb{R}^{n-1})}\}.\end{aligned}$$

We now observe the case $i<j$. By interchanging i and j we see that

$$\begin{aligned}I_{ij}\cong-\mathrm{i}\{&\langle p\circ q\Lambda^{m-1-i}\mathrm{D}'_{x_1}{}^{i}u,\Lambda^{m-j}D'_{x_1}{}^{j-1}v\rangle_{\mathscr{H}_{\eta}'(\mathbb{R}^{n-1})}\\&+\langle p\circ q\Lambda^{m-2-i}\mathrm{D}'_{x_1}{}^{i+1}u,\Lambda^{m-j+1}\mathrm{D}'_{x_1}{}^{j-2}v\rangle_{\mathscr{H}_{\eta}'(\mathbb{R}^{n-1})}\\&+\ldots+\langle p\circ q\Lambda^{m-j}\mathrm{D}'_{x_1}{}^{j-1}u,\Lambda^{m-1-i}\mathrm{D}'_{x_1}{}^{i}v\rangle_{\mathscr{H}_{\eta}'(\mathbb{R}^{n-1})}\}.\end{aligned}\qquad\blacksquare$$

If P and Q are both real symbols, then we can take P,Q themselves instead of $P^{(*)}$, $Q^{(*)}$ in the statement of Theorem 1. We shall come across this in §3. If this is the case, letting $u=v$ we have:

Corollary to Theorem 1

Under the same conditions of Theorem 1, if $P(x,\xi,\eta)$, $Q(x,\xi,\eta)$ are real symbols, then

$$2\,\mathrm{Im}\,(P(x,\mathrm{D}_x',\eta)u,Q(x,\mathrm{D}_x',\eta)u)_{\mathscr{H}_{\eta}(\mathbb{R}_+{}^n)}$$

$$\cong\left\langle\tilde{C}(x,\mathrm{D}_{x'}',\eta)\begin{bmatrix}\mathrm{D}'_{x_1}{}^{m-1}\\ \vdots\\ \Lambda^{m-1}\end{bmatrix}u,\begin{bmatrix}\mathrm{D}'_{x_1}{}^{m-1}\\ \vdots\\ \Lambda^{m-1}\end{bmatrix}u\right\rangle_{\mathscr{H}_{\eta}'(\mathbb{R}^{n-1})},$$

where the difference between the right and left hand side terms is bounded by

$$C\left\{ \|u\|^2_{m-1,\mathscr{H}_\eta(\mathbb{R}_+{}^n)} + \sum_{j=0}^{m-1} \langle \mathrm{D}'_{x_1}{}^j u\rangle^2_{m-1-j-1/2,\mathscr{H}_\eta'(\mathbb{R}^{n-1})} \right\}$$

and the positive constant C is independent of u and η.

2. The uniform Lopatinski condition

In this section, we observe various energy inequalities of Dirichlet type, and demonstrate that the uniform Lopatinski condition is, in fact, the necessary and sufficient condition for the energy inequality of Dirichlet type ($\mathscr{H}$) being valid. Here we shall prove only the 'necessary' part leaving the 'sufficient' part to the next section.

2.1

In the subsequent sections of this chapter, we shall assume roughly the same problem setting as in Chapter 2. This means that we are concerned with the well-posedness of an initial value problem of a hyperbolic equation in a half-space. Remember, however, that in this chapter we are treating the case in which a differential operator has variable coefficients.

Note that the assumption 'in a half-space' is purely a convention: a domâin with a smooth boundary can be transformed locally to a half-space by a transformation of variables.

As we expect, since we are treating variable coefficients, we must begin to work under stronger restrictions compared with the case of constant coefficients. Consider

$$A = A(t,x,y;\mathrm{D}_t,\mathrm{D}_x,\mathrm{D}_y) = \sum_{i+k+|\nu|\leq m} a_{ik\nu}(t,x,y)\mathrm{D}_t{}^i\mathrm{D}_x{}^k\mathrm{D}_y{}^\nu,$$

$$B_j = B_j(t,y;\mathrm{D}_t,\mathrm{D}_x,\mathrm{D}_y) = \sum_{i+k+|\nu|\leq v_j} b_{ik\nu}{}^{(j)}(t,y)\mathrm{D}_t{}^i\mathrm{D}_x{}^k\mathrm{D}_y{}^\nu,$$

where all the coefficients are infinitely differentiable and become constants outside a certain ball. In place of Hypothesis (A) in Chapter 2, we set up the following stronger hypothesis.

Hypothesis (A′)

(i) A_0 is strongly hyperbolic in the direction of the t-axis, i.e. $A_0(t,x,y;1,0,0)\neq 0$ and if $(\xi,\eta)\in\mathbb{R}^n\backslash\{0\}$ then the roots of $A_0(t,x,y;\tau,\xi,\eta)=0$ with respect to τ are distinct real numbers.

(ii) $A_0(t,0,y;0,1,0) \neq 0$,

(iii) $B_{j0}(t,y;0,1,0) \neq 0, \quad 0 \leqq r_j \leqq m-1, \quad r_i \neq r_j \quad (i \neq j)$,

where A_0, B_{j0} are the principal parts of A, B_j.

Given $\{A, B_j\}$ which satisfy the above conditions, we consider the problem of seeking a solution x of the equations

$$
\text{(P)} \qquad \begin{cases} Au = f, & 0 < t < +\infty, \qquad (x,y) \in \Omega (= \mathbb{R}_+{}^n), \\ B_j u = g_j \quad (j = 1, \ldots, m_+), & 0 < t < +\infty, \quad (x,y) \in \partial\Omega, \\ \mathrm{D}_t{}^j u = h_j \quad (j = 0, \ldots, m-1), & t = 0, \quad (x,y) \in \Omega \end{cases}
$$

for the data $\{f, g, h_j\}$. This is called the *initial boundary value problem for the half-space* $\mathbb{R}_+{}^n$.

In Chapter 2, we observed the $\mathscr{E}$-well-posedness of (P) and in §1 of this chapter we remarked that the above $\mathscr{E}$-well-posedness induces well-posedness in Sobolev spaces in the weak sense (see the inequality (*) immediately before §1.2).In the following sections we shall focus our attention on well-posedness in Sobolev spaces in the strong sense (see §2.2), which will assure the $\mathscr{E}$-well-posedness after the consideration of finiteness of the dependence domain. We shall only consider Sobolev spaces that are weighted in the direction of the t-axis, therefore we employ the following convention:

$$
\mathscr{H}_{-\gamma(1,0,0)}{}^k = \mathscr{H}_{-\gamma}{}^k, \quad \mathscr{H}_{\gamma(1,0,0)}{}^k = \mathscr{H}_{\gamma}{}^k.
$$

2.2

For $\{A, B_j\}$ we say *the energy inequality of Dirichlet type* (*H*) holds iff for arbitrary real numbers $S, T (S < T)$

$$
\begin{aligned}
\text{(H)} \quad & \|u\|^2_{m-1,\mathscr{H}_{-\gamma_0}((S,T)\times\Omega)} + \sum_{j=0}^{m-1} \|\mathrm{D}_x{}^j u\|^2_{m-1-j,\mathscr{H}_{-\gamma_0}((S,T)\times\partial\Omega)} \\
& + \sum_{j=0}^{m-1} \|\mathrm{D}_t{}^j u(T)\|^2_{m-1-j,\mathscr{H}_{-\gamma_0}(\Omega)} \\
& \leqq C \Bigg\{ \|Au\|^2_{\mathscr{H}_{-\gamma_0}((S,T)\times\Omega)} + \sum_{j=1}^{m_+} \|B_j u\|^2_{m-1-r_j,\mathscr{H}_{-\gamma_0}((S,T)\times\partial\Omega)} \\
& + \sum_{j=0}^{m-1} \|\mathrm{D}_t{}^j u(S)\|^2_{m-1-j,\mathscr{H}_{-\gamma_0}(\Omega)} \Bigg\}
\end{aligned}
$$

is true,† where $u(T) = u|_{t=T}$ and γ_0, C are positive constants independent of u as well as S, T. Assume that (H) is true for the principal part $\{A_0, B_{j0}\}$

† Sometimes this is called '*the condition of well-posedness in the Sobolev space in the strong sense*' (the existence of a solution is presupposed).

of $\{A, B_j\}$. Write

$$\tilde{u}(t, x, y) = u(\lambda t, \lambda x, \lambda y), \quad \lambda = \gamma_0/\gamma > 0,$$
$$\tilde{T} = (1/\lambda)T, \quad \tilde{S} = (1/\lambda)S$$

and apply (H) with $\{A_0, B_{j0}\}$. Then, we see that the following inequality holds:

$$(\mathrm{H})_\gamma \qquad \gamma\|u\|^2_{m-1,\mathscr{H}_{-\gamma}((S,T)\times\Omega),\gamma} + \sum_{j=0}^{m-1} \|\mathrm{D}_x{}^j u\|^2_{m-1-j,\mathscr{H}_{-\gamma}((S,j)\times\partial\Omega),\gamma}$$
$$+ \sum_{j=0}^{m-1} \|\mathrm{D}_t{}^j u(T)\|^2_{m-1-j,\mathscr{H}_{-\gamma}(\Omega),\gamma}$$
$$\leqq C\left\{\frac{1}{\gamma}\|A_0 u\|^2_{\mathscr{H}_{-\gamma}((S,T)\times\Omega)} + \sum_{j=1}^{m_+} \|B_{j0}u\|^2_{m-1-r_j,\mathscr{H}_{-\gamma}((S,T)\times\partial\Omega),\gamma}\right.$$
$$\left. + \sum_{j=0}^{m-1} \|\mathrm{D}_t{}^j u(S)\|^2_{m-1-j,\mathscr{H}_{-\gamma}(\Omega),\gamma}\right\},$$

where

$$\|u\|^2_{k,\mathscr{H}_{-\gamma},\gamma} = |u|^2_{k,\mathscr{H}_{-\gamma}} + \gamma^2|u|^2_{k-1,\mathscr{H}_{-\gamma}} + \dots + \gamma^{2k}|u|^2_{\mathscr{H}_{-\gamma}}.$$

Now, we take a sufficiently large γ_0, and consider $(\mathrm{H})_\gamma$ for $\gamma \geqq \gamma_0$. In this case, we can replace $\{A_0, B_{j0}\}$ with $\{A, B_j\}$ and vice versa. This implies that if (H) is true for $\{A_0, B_{j0}\}$, then (H) is also true for $\{A, B_j\}$.

With the above observation, we define the well-posedness of (P) as follows:

We say that (P) is *H-well-posed* iff the following conditions hold.

(i) For the principal part $\{A_0, B_{j0}\}$ the energy inequality of Dirichlet type (H) holds. (Therefore, it also holds for $\{A, B_j\}$)

(ii) For the arbitrary data with compatibility condition

$$\{f \in H^0((0,T)\times\Omega), g_j \in H^{m-1-r_j}((0,T)\times\partial\Omega), h_j \in H^{m-1-j}(\Omega)\},$$

there exists a unique solution $u \in H^{m-1}((0,T)\times\Omega)$ satisfying (P).

Let us consider the operator $\{A_0{}^{(0)}, B_{j0}{}^{(0)}\}$ with constant coefficients which is obtained from $\{A_0, B_{j0}\}$ by fixing its coefficients at (t_0, x_0, y_0) $(t_0 \geqq 0, x_0 = 0)$. Using this notation we prove

Lemma 2.1

If $\{A_0, B_{j0}\}$ satisfies (H), then $\{A_0{}^{(0)}, B_{j0}{}^{(0)}\}$ also satisfies (H).

Proof

Letting

$$(\tilde{t}, \tilde{x}, \tilde{y}) = (t_0 + \varepsilon^{-1}(t - t_0), x_0 + \varepsilon^{-1}(x - x_0), y_0 + \varepsilon^{-1}(y - y_0)),$$
$$(\tilde{\tilde{t}}, \tilde{\tilde{x}}, \tilde{\tilde{y}}) = (t_0 + \varepsilon(t - t_0), x_0 + \varepsilon(x - x_0), y_0 + \varepsilon(y - y_0)),$$

and

$$A_0^{(\varepsilon)}(t,x,y;D_t,D_x,D_y) = A_0(\tilde{\tilde{t}},\tilde{\tilde{x}},\tilde{\tilde{y}};D_t,D_x,D_y),$$
$$B_{j0}^{(\varepsilon)}(t,y;D_t,D_x,D_y) = B_{j0}(\tilde{\tilde{t}},\tilde{\tilde{y}};D_t,D_x,D_y),$$

we have

$$(A_0\tilde{u})(t,x,y) = (1/\varepsilon^m)(A_0^{(\varepsilon)}u)(\tilde{t},\tilde{x},\tilde{y}),$$
$$(B_{j0}\tilde{u})(t,x,y) = (1/\varepsilon^{r_j})(B_{j0}^{(\varepsilon)}u)(\tilde{t},\tilde{x},\tilde{y})$$

for $\tilde{u}(t,x,y) = u(\tilde{t},\tilde{x},\tilde{y})$. Now, we set

$$\tilde{u}(t,x,y) = u(\tilde{t},\tilde{x},\tilde{y}), \quad \tilde{\gamma} = \varepsilon^{-1}\gamma,$$
$$\tilde{\tilde{T}} = t_0 + \varepsilon(T-t_0), \quad \tilde{\tilde{S}} = t_0 + \varepsilon(S-t_0)$$

and insert them in $(\mathrm{H})_\gamma$ for $\{A_0, B_{j0}\}$.
Then we have the term on the left

$$\begin{aligned} &= \tilde{\gamma}\,\|\tilde{u}\|^2_{m-1,\mathscr{H}_{-\tilde{\gamma}}((\tilde{\tilde{S}},\tilde{\tilde{T}})\times\Omega),\tilde{\gamma}} \\ &\quad + \sum_{j=0}^{m-1} \|D_x^{\,j}\tilde{u}\|^2_{m-1-j,\mathscr{H}_{-\tilde{\gamma}}((\tilde{\tilde{S}},\tilde{\tilde{T}})\times\partial\Omega),\tilde{\gamma}} \\ &\quad + \sum_{j=0}^{m-1} \|D_t^{\,j}\tilde{u}(\tilde{\tilde{T}})\|^2_{m-1-j,\mathscr{H}_{-\tilde{\gamma}}(\Omega),\tilde{\gamma}} \\ &= \varepsilon^{n-2(m-1)}e^{-2\gamma(1/\varepsilon-1)t_0}\Big\{\gamma\|u\|^2_{m-1,\mathscr{H}_{-\tilde{\gamma}}((S,T)\times\Omega),\gamma} \\ &\qquad + \sum_{j=0}^{m-1} \|D_x^{\,j}u\|^2_{m-1-j,\mathscr{H}_{-\gamma}((S,T)\times\partial\Omega),\gamma} \\ &\qquad + \sum_{j=0}^{m-1} \|D_t^{\,j}u(T)\|^2_{m-1-j,\mathscr{H}_{-\gamma}(\Omega),\gamma}\Big\}, \end{aligned}$$

the term on the right

$$\begin{aligned} &= C\varepsilon^{n-2(m-1)}e^{-2\gamma(1/\varepsilon-1)t_0}\Big\{\frac{1}{\gamma}\|A_0^{(\varepsilon)}u\|^2_{\mathscr{H}_{-\gamma}((S,T)\times\Omega)} \\ &\qquad + \sum_{j=1}^{m_+} \|B_{j0}^{(\varepsilon)}u\|^2_{m-1-r_j,\mathscr{H}_{-\gamma}((S,T)\times\partial\Omega),\gamma} \\ &\qquad + \sum_{j=0}^{m-1} \|D_t^{\,j}u(S)\|^2_{m-1-j,\mathscr{H}_{-\gamma}(\Omega),\gamma}\Big\}. \end{aligned}$$

Dividing both sides by $\varepsilon^{n-2(m-1)}e^{-2\gamma(1/\varepsilon-1)t_0}$ and letting $\varepsilon \to +0$, we obtain $(\mathrm{H})_\gamma$ for $\{A_0^{(0)}, B_{j0}^{(0)}\}$. ■

2.3

Note that the initial boundary value problem (P) can be regarded as a simple boundary value problem

$$(\mathrm{P}'')\qquad \begin{cases} Au=f, & -\infty<t<+\infty,\quad (x,y)\in\Omega,\\ B_{ju}=g_j\quad (j=1,\dots,m_+), & -\infty<t<+\infty,\quad (x,y)\in\partial\Omega^{\cdot}\end{cases}$$

We assume that for $\{A,B_j\}$, the inequality $(\mathrm{H})_\gamma(\gamma\geqq\gamma_0)$ holds. Then for $\gamma\geqq\gamma_0$ we have

$$(\mathscr{H})_\gamma\qquad \gamma\|u\|^2_{m-1,\mathscr{H}_{-\gamma}(\mathbb{R}^1\times\Omega)}+\sum_{j=0}^{m-1}\|\mathrm{D}_x{}^ju\|^2_{m-1-j,\mathscr{H}_{-\gamma}(\mathbb{R}^1\times\partial\Omega)}$$

$$\leqq C\left\{\frac{1}{\gamma}\|Au\|^2_{\mathscr{H}_{-\gamma}(\mathbb{R}^1\times\Omega)}+\sum_{j=1}^{m_+}\|B_ju\|^2_{m-1-r_j,\mathscr{H}_{-\gamma}(\mathbb{R}^1\times\partial\Omega)}\right\}.$$

The inequality $(\mathscr{H}_\gamma)$ shows the properties of the principal part $\{A_0,B_{jo}\}$ of $\{A,B_j\}$ (we have already seen this in the case of $(\mathrm{H})_\gamma$).

For $\{A,B_j\}$ we say *the energy inequality of Dirichlet type* $(\mathscr{H})$ holds iff

$$(\mathscr{H})\qquad \|u\|^2_{m-1,\mathscr{H}_{-\gamma_0}(\mathbb{R}^1\times\Omega)}+\sum_{j=0}^{m-1}\|\mathrm{D}_x{}^ju\|^2_{m-1-j,\mathscr{H}_{-\gamma_0}(\mathbb{R}^1\times\partial\Omega)}$$

$$\leqq C\left\{\|Au\|^2_{\mathscr{H}_{-\gamma_0}(\mathbb{R}^1\times\Omega)}+\sum_{j=1}^{m_+}\|B_ju\|^2_{m-1-r_j,\mathscr{H}_{-\gamma_0}(\mathbb{R}^1\times\partial\Omega)}\right\}.$$

In this case we see that $(\mathscr{H})$ is true for $\{A_0,B_{jo}\}$ iff $(\mathscr{H})_\gamma$ is true for $\{A_0,A_{jo}\}$.

We now define well-posedness for (P″) as follows:

We say that (P″) is $\mathscr{H}$*-well-posed* iff the following conditions hold.

(i) $(\mathscr{H})$ is true for $\{A_0,B_{jo}\}$.

(ii) For the arbitrary data

$$\{f\in\mathscr{H}_{-\gamma}(\mathbb{R}^1\times\Omega),g_j\in\mathscr{H}_{-\gamma}{}^{m-1-r_j}(\mathbb{R}^1\times\partial\Omega)\},\quad \gamma\geqq\gamma_0,$$

there exists a unique solution

$$u\in\mathscr{H}_{-\gamma}{}^{m-1}(\mathbb{R}^1\times\Omega)$$

of (P″).

It is obvious that $(\mathscr{H})$ has a similar property to that claimed for (H) (see Lemma 2.1).

Lemma 2.1′

If $(\mathscr{H})$ is true for $\{A_0,B_{jo}\}$ then it is also true for $\{A_0{}^{(0)},B_{jo}{}^{(0)}\}$.

Since $\{A_0{}^{(0)}, B_{j0}{}^{(0)}\}$ has constant coefficients, if $u \in \mathscr{H}_{-\gamma}{}^m(\mathbb{R}^1 \times \Omega)$, then setting

$$\hat{u}(x;\tau,\eta) = \mathscr{L}_{(t,y)\to(\tau,\eta)}[u(t,x,y)],$$

for $\operatorname{Im}\tau = -\gamma, \quad \operatorname{Im}\eta = 0$, we have

$$\mathscr{L}[A_0{}^{(0)}u] = A_0{}^{(0)}(\tau, D_x, \eta)\hat{u}(x;\tau,\eta),$$
$$\mathscr{L}[B_{j0}{}^{(0)}u|_{x=0}] = B_{j0}{}^{(0)}(\tau, D_x, \eta)\hat{u}(0,\tau,\eta).$$

In this case we can establish

Lemma 2.2
If $(\mathscr{H})$ is true for $\{A_0{}^{(0)}, B_{j0}{}^{(0)}\}$, then for $\operatorname{Im}\tau < 0, \eta \in \mathbb{R}^{n-1}$, the following inequality holds.

$$|\operatorname{Im}\tau| \sum_{j=0}^{m-1} (|\tau| + |\eta|)^{2(m-1-j)} \|D_x{}^j v\|^2_{H(\mathbb{R}_+{}^1)}$$
$$+ \sum_{j=0}^{m-1} (|\tau| + |\eta|)^{2(m-1-j)} |D_x{}^j v(0)|^2$$
$$\leqq C\left\{\frac{1}{|\operatorname{Im}\tau|}\|A_0{}^{(0)}(\tau, D_x, \eta)v\|^2_{H(\mathbb{R}_+{}^1)}\right.$$
$$\left. + \sum_{j=1}^{m_+} (|\tau| + |\eta|)^{2(m-1-r_j)} |B_{j0}{}^{(0)}(\tau, D_x, \eta)v(0)|^2\right\}.$$

Proof
First, note that $(\mathscr{H})_\gamma$ is true for $\{A_0{}^{(0)}, B_{j0}{}^{(0)}\}$ iff for $\gamma > 0$

$$\gamma|u|^2_{m-1,\mathscr{H}_{-\gamma}(\mathbb{R}^1\times\Omega)} + \sum_{j=0}^{m-1} D_x{}^j u|^2_{m-1-j,\mathscr{H}_{-\gamma}(\mathbb{R}^1\times\partial\Omega)}$$
$$\leqq C\left\{\frac{1}{\gamma}|A_0{}^{(0)}u|^2_{\mathscr{H}_{-\gamma}(\mathbb{R}^1\times\Omega)} + \sum_{j=1}^{m_+} |B_{j0}{}^{(0)}u|^2_{m-1-r_j,\mathscr{H}_{-\gamma}(\mathbb{R}^1\times\partial\Omega)}\right\}.$$

By Plancherel's theorem we have

$$\int_{\substack{\operatorname{Im}\tau=-\gamma\\ \eta\in\mathbb{R}^{n-1}}} \left\{\gamma \sum_{j=0}^{m-1} (|\tau| + |\eta|)^{2(m-1-j)} \|D_x{}^j \hat{u}(x;\tau,\eta)\|^2_{S(\mathbb{R}_+{}^1)}\right.$$
$$\left. + \sum_{j=1}^{m-1} (|\tau| + |\eta|)^{2(m-1-j)} |D_x{}^j \hat{u}(0;\tau,\eta)|^2\right\} d\tau\, d\eta$$
$$\leqq C \int_{\substack{\operatorname{Im}\tau=-\gamma\\ \eta\in\mathbb{R}^{n-1}}} \left\{\frac{1}{\gamma}\|A_0{}^{(0)}(\tau, D_x, \eta)\hat{u}(x;\tau,\eta)\|^2_{H(\mathbb{R}_+{}^1)}\right.$$
$$\left. + \sum_{j=1}^{m_+} (|\tau| + |\eta|)^{2(m-1-r_j)} |B_{j0}{}^{(0)}(\tau, D_x, \eta)\hat{u}(0;\tau,\eta)|^2\right\} d\tau\, d\eta.$$

Therefore, in particular, if we write

$$\hat{u}(\tau,\eta;x) = \varphi_\varepsilon(\sigma,\eta)v(x), \quad v(x)\in H^m(\mathbb{R}_+{}^1)$$

where $\sigma = \operatorname{Re}\tau$ and φ_ε satisfies

$$\varphi_\varepsilon(\sigma,\eta) = \frac{1}{\varepsilon^{n/2}}\varphi\left(\frac{\sigma-\sigma_0}{\varepsilon}, \frac{\eta-\eta_0}{\varepsilon}\right),$$

$$\varphi(\sigma,\eta)\in\mathscr{D}, \quad \int\varphi^2(\sigma,\eta)\,\mathrm{d}\sigma\,\mathrm{d}\eta = 1,$$

then, by substitution, from the previous inequality we have

$$\gamma\sum_{j=0}^{m-1}\int(|\tau|+|\eta|)^{2(m-1-j)}\varphi_\varepsilon(\sigma,\eta)^2\,\mathrm{d}\sigma\,\mathrm{d}\eta\,\|\mathrm{D}_x{}^j v\|^2_{H(\mathbb{R}_+{}^1)}$$

$$+\sum_{j=0}^{m-1}\int(|\tau|+|\eta|)^{2(m-1-j)}\varphi_\varepsilon(\sigma,\eta)^2\,\mathrm{d}\sigma\,\mathrm{d}\eta|\mathrm{D}_x{}^j v(0)|^2$$

$$\leqq C\left\{\frac{1}{\gamma}\int\varphi_\varepsilon(\sigma,\eta)^2\|A_0{}^{(0)}(\tau,\mathrm{D}_x,\eta)v\|^2_{H(\mathbb{R}_+{}^1)}\,\mathrm{d}\sigma\,\mathrm{d}\eta\right.$$

$$\left.+\sum_{j=1}^{m_+}\int(|\tau|+|\eta|)^{2(m-1-r_j)}\varphi_\varepsilon(\sigma,\eta)^2|B_{j0}{}^{(0)}(\tau,\mathrm{D}_x,\eta)v(0)|^2\,\mathrm{d}\sigma\,\mathrm{d}\eta\right\}.$$

Letting $\varepsilon\to+0$ we see that

$$\gamma\sum_{j=0}^{m-1}(|\tau_0|+|\eta_0|)^{2(m-1-j)}\|\mathrm{D}_x{}^j v\|^2_{H(\mathbb{R}_+{}^1)}$$

$$+\sum_{j=0}^{m-1}(|\tau_0|+|\eta_0|)^{2(m-1-j)}|\mathrm{D}_x{}^j v(0)|^2$$

$$\leqq C\left\{\frac{1}{\gamma}\|A_0{}^{(0)}(\tau_0,\mathrm{D}_x,\eta_0)v\|^2_{H(\mathbb{R}_+{}^1)}\right.$$

$$\left.+\sum_{j=1}^{m_+}(|\tau_0|+|\eta_0|)^{2(m-1-r_j)}|B_{j0}{}^{(0)}(\tau_0,\mathrm{D}_x,\eta_0)v(0)|^2\right\},$$

where $\tau_0 = \sigma_0 - \mathrm{i}\gamma$. ■

2.4

Let us write $R_0(t,y;\tau,\eta)$ for the Lopatinski determinant for the principal part $\{A_0, B_{j0}\}$ of $\{A, B_j\}$. This implies that for $\operatorname{Im}\tau<0, \eta\in\mathbb{R}^{n-1}$ if

$$A_0(t,x,y;\tau,\xi,\eta) = \prod_{j=1}^{m_+}(\xi-\xi_{j0}{}^+(t,x,y;\tau,\eta))\prod_{j=1}^{m_-}(\xi-\xi_{j0}{}^-(t,x,y;\tau,\eta)),$$

$$\operatorname{Im}\xi_{j0}{}^\pm(t,x,y;\tau,\eta)\gtrless 0,$$

then R_0 is given by

$$R_0(t,y;\tau,\eta)=\det\left(\frac{1}{2\pi i}\oint\frac{B_{j0}(t,y;\tau,\xi,\eta)\xi^{k-1}}{\prod_{i=1}^{m_+}(\xi-\xi_{i0}{}^{+}(t,0,y;\tau,\eta))}d\xi\right)_{j,k=1,\ldots,m_+}$$

Note that R_0 can be continuously extended to $\operatorname{Im}\tau\leqq 0$, $\eta\in\mathbb{R}^{n-1}$, $(\tau,\eta)\neq(0,0)$. If we fix $(t,y)\in\mathbb{R}^n$ we see that $R_0(t,y;\tau,\eta)$ is hyperbolic in the direction of $(1,0)$ iff

$$R_0(t,y;\tau,\eta)\neq 0,\quad \operatorname{Im}\tau<0,\quad \eta\in\mathbb{R}^{n-1}.$$

We now consider a stronger condition than this, i.e.

$$R_0(t,y;\tau,\eta)\neq 0,\quad \operatorname{Im}\tau\leqq 0,\quad \eta\in\mathbb{R}^{n-1},\quad (\tau,\eta)\neq(0,0).$$

We call this condition the *uniform Lopatinski condition*. For this condition we have

Theorem 2
If $(\mathscr{H})$ is true for $\{A_0,B_{j0}\}$, then the uniform Lopatinski condition holds for $\{A_0,B_{j0}\}$.

Proof
On this assumption, from Lemma 2.1′ we see that $(\mathscr{H})$ is true for $\{A_0{}^{(0)},B_{j0}{}^{(0)}\}$. Then, by Lemma 2.2, for $v(x;\tau,\eta)\in H^m(\mathbb{R}_+{}^1)$ satisfying

$$A_0{}^{(0)}(\tau,D_x,\eta)v(x;\tau,\eta)=0$$

we have

$$\sum_{j=0}^{m-1}|D_x{}^jv(0;\tau,\eta)|^2\leqq C\sum_{j=1}^{m_+}|B_{j0}{}^{(0)}(\tau,D_x,\eta)v(0;\tau,\eta)|^2,$$

$$\operatorname{Im}\tau<0,\quad \eta\in\mathbb{R}^{n-1},\quad |\tau|^2+|\eta|^2=1.$$

In particular, if $\{D_x{}^jv(0;\tau,\eta)\}_{j=0,\ldots,m-1}$ is continuous at $\operatorname{Im}\tau\leqq 0$, $\eta\in\mathbb{R}^{n-1}$, $(\tau,\eta)\neq(0,0)$, then, obviously the following inequality holds

$$\sum_{j=0}^{m-1}|D_x{}^jv(0;\tau,\eta)|^2\leqq C\sum_{j=1}^{m_+}|B_{j0}{}^{(0)}(\tau,D_x,\eta)v(0;\tau,\eta)|^2$$

$$\operatorname{Im}\tau\leqq 0,\quad \eta\in\mathbb{R}^{n-1},\quad |\tau|^2+|\eta|^2=1.$$

Now we assume that $R_0(t_0,y_0;\tau_0,\eta_0)=0$ for $(t_0,y_0)\in\mathbb{R}^n$, $\operatorname{Im}\tau_0\leqq 0$, $\eta_0\in\mathbb{R}^{n-1}$, $|\tau_0|^2+|\eta_0|^2=1$. (Translator's note: From this we shall derive a contradiction.) First we see there exists $(c_1,\ldots,c_{m_+})(\sum_{j=1}^{m_+}|c_j|^2=1)$ such that

$$\sum_{k=1}^{m_+}c_k\frac{1}{2\pi i}\oint\frac{B_{j0}(t_0,y_0;\tau_0,\xi,\eta_0)\xi^{k-1}}{\prod_{i=1}^{m_+}(\xi-\xi_{i0}{}^{+}(t_0,0,y_0;\tau_0,\eta_0))}d\xi=0\quad (j=0,\ldots,m_+)$$

For $\operatorname{Im}\tau < 0$, $\eta \in \mathbb{R}^{n-1}$ we set

$$v(x;\tau,\eta) = \sum_{k=1}^{m+} c_k \frac{1}{2\pi i} \oint \frac{e^{ix\xi}\xi^{k-1}}{\prod_{i=1}^{m+}(\xi - \xi_{i0}{}^{+}(t_0, 0, y_0; \tau, \eta))} d\xi.$$

Then we have

$$v(x;\tau,\eta) \in H^m(\mathbb{R}_+{}^1), \quad A_0(t_0, 0, y_0, \tau, D_x, \eta)v(x;\tau,\eta) = 0.$$

Since

$$D_x{}^j v(0;\tau,\eta) = \sum_{k=1}^{m+} c_k \frac{1}{2\pi i} \oint \frac{\xi^{j+k-1}}{\prod_{i=1}^{m+}(\xi - \xi_{i0}{}^{+}(t_0, 0, y_0; \tau, \eta))} d\xi,$$

it is continuous for $\operatorname{Im}\tau \leqq 0$. $\eta \in \mathbb{R}^{n-1}$, $(\tau,\eta) \neq (0,0)$, therefore for $\operatorname{Im}\tau \leqq 0$, $\eta \in \mathbb{R}^{n-1}$, $|\tau|^2 + |\eta|^2 = 1$, we have

$$\sum_{j=0}^{m-1} |D_x{}^j v(0;\tau,\eta)|^2 \leqq C \sum_{j=1}^{m+} |B_{j0}(t_0, y_0; \tau, D_x, \eta)v(0;\tau,\eta)|^2$$

Hence, setting $(\tau,\eta) = (\tau_0, \eta_0)$ we see that

$$\sum_{j=0}^{m-1} |D_x{}^j v(0;\tau_0,\eta_0)|^2 \leqq C \sum_{j=1}^{m+} |B_{j0}(t_0, y_0; \tau_0, D_x, \eta_0)v(0;\tau_0,\eta_0)|^2,$$

where clearly the right hand term is zero (see the definition of v). Also we know that

$$\det\left(\frac{1}{2\pi i} \oint \frac{\xi^{j+k-1}}{\prod_{i=1}^{m+}(\xi - \xi_{i0}{}^{+}(t_0, 0, y_0; \tau_0, \eta_0))} d\xi\right)_{j,k=1,\dots,m+} \neq 0$$

and $(c_1, \dots, c_{m+}) \neq (0, \dots, 0)$, so that

$$(v(0;\tau_0,\eta_0), D_x v(0;\tau_0,\eta_0), \dots, D_x{}^{m+-1} v(0;\tau_0,\eta_0)) \neq (0, \dots, 0).$$

Hence the left hand term is greater than zero; a contradiction. ■

3. The energy inequalities of Dirichlet type $(\mathscr{H})$

In the previous section we obtained the uniform Lopatinski condition as a necessary condition for the energy inequality of Dirichlet type $(\mathscr{H})$ to hold for $\{A_0, B_{j0}\}$. In this section, we shall prove the converse of this, that if the uniform Lopatinski condition is satisfied, then for $\{A_0, B_{j0}\}$ the energy inequality of Dirichlet type $(\mathscr{H})$ is valid. Note, however, that in this section we consider only $\{A_0, B_{j0}\}$, assuming that $\{A, B_j\}$ consists only of its principal part.

3.1

Let us assume that, $\{A(t, x, y; \tau, \xi, \eta), B_j(t, y; \tau, \xi, \eta)\}$ are homogeneous polynomials of degree m, r_j, respectively, and satisfy Hypothesis (A′). If we introduce the parameter

$$X = (t, x, y; \tau, \eta) \in \mathbb{R}_{t,x,y}^{n+1} \times \mathbb{C}_\tau^{1} \times \mathbb{R}_\eta^{n-1},$$

then A, B_j become polynomials in ξ, and we write $\{A(X, \xi), B_j(X, \xi)\}$. We wish to know the behaviour of these polynomials in a neighbourhood of $X = X_0$.

First, we look at $A(X, \xi)$. By Hypothesis (A′), we see that the coefficients $a_j(X)$ of $A(X, \xi) = \xi^m + a_1(X)\xi^{m-1} + \ldots + a_m(X)$ are real when X is real, where we consider the normalised A as $A(t, x, y; 0, 1, 0) = 1$. For a fixed $X = X_0$, we write $\{\xi_1{}^0, \ldots, \xi_h{}^0\}$ for the real roots of the equation $A(X_0, \xi) = 0$ where $\xi_i{}^0 \neq \xi_j{}^0 (i \neq j)$ and each $\xi_j{}^0$ is an m_j-tuple root. If X_0 is not real, there are no real roots of A. Therefore, the following result applies only when X_0 is real. Let X move around within a sufficiently small neighbourhood of $X = X_0$. Since a root ξ of $A(X, \xi) = 0$ is a continuous function of X, it follows that if we write

$$\{\xi_{jk}(X)\}_{k=1,\ldots,m_j}$$

for the set of roots with the property $\xi = \xi_j{}^0$ at $X = X_0$, then we may suppose that the $\xi_{jk}(X)$ are contained in C_j and all the other roots are outside C_j, where C_j is a small ball centred on $\xi = \xi_j{}^0$. Let

$$\begin{aligned} H_j(X, \xi) &= \prod_{k=1}^{m_j} (\xi - \xi_{jk}(X)) \\ &= (\xi - \xi_j{}^0)^{m_j} + h_{j1}(X)(\xi - \xi_j{}^0)^{m_j - 1} + \ldots + h_{jm_j}(X). \end{aligned}$$

Then the $h_{jk}(X)$ are infinitely differentiable in a neighbourhood of $X = X_0$ and satisfy $h_{jk}(X_0) = 0$.

Let us also consider the roots ξ of $A(X, \xi) = 0$ that are complex at $X = X_0$ and satisfy $\operatorname{Im} \xi > 0$, and call them $\{\xi_1{}^+(X), \ldots, \xi_{m_0{}^+}{}^+(X)\}$, not necessarily all distinct. Similarly for $\operatorname{Im} \xi < 0$ we have $\{\xi_1{}^-(X), \ldots, \xi_{m_0{}^-}{}^-(X)\}$.

Setting

$$E_\pm(X, \xi) = \prod_{j=1}^{m_0{}^\pm} (\xi - \xi_j{}^\pm(X)),$$

$$E(X, \xi) = E_+(X, \xi) E_-(X, \xi),$$

and considering $E_\pm(X, \xi)$ as polynomials in ξ, we see that their coefficients are infinitely differentiable in a neighbourhood of $X = X_0$. Therefore $A(X, \xi)$ decomposes into smooth functions as

$$A(X, \xi) = \prod_{j=1}^{h} H_j(X, \xi) E(X, \xi)$$

in a neighbourhood of $X = X_0$.

Lemma 3.1

(i) The coefficients of the polynomials $H_j(X,\xi), E(X,\xi)$ in ξ are real if X is real.

(ii) If X_0 is real and $(\tau_0, \eta_0) \neq (0,0)$, then $\dfrac{\partial h_{jm_j}}{\partial \tau}(X) \neq 0$.

Convention: In the following, we write $-\operatorname{sgn}\dfrac{\partial h_{jm_j}}{\partial \tau}(X_0) = \varepsilon_j$.

Proof

(i) If X is real, then $A(X,\xi)$ is a polynomial in ξ with real coefficients, therefore, the zeros of the polynomial are real, or complex conjugate pairs. We take a complex conjugate pair of roots of $A(X,\xi)=0$.If one of them is equal to a real root $\xi_j{}^0$ at $X = X_0$, then they are zeros of $H_j(X,\xi)$. Otherwise, they are zeros of $E(X,\xi)$. Therefore, $H_j(X,\xi), E(X,\xi)$ are polynomials of ξ with real coefficients.

(ii) From the facts $A(X_0, \xi_j{}^0) = 0, (\tau_0, \eta_0) \neq (0,0)$, we see that $(\eta_0, \xi_j{}^0) \neq (0,0)$. Since A is strongly hyperbolic, we know that $(\partial A/\partial \tau)(X_0, \xi_j{}^0) \neq 0$. On the other hand

$$\frac{\partial A}{\partial \tau}(X_0, \xi_j{}^0) = \frac{\partial H_j}{\partial \tau}(X_0, \xi_j{}^0) \prod_{k \neq j} H_k(X_0, \xi_j{}^0) E(X_0, \xi_j{}^0),$$

$$\frac{\partial H_j}{\partial \tau}(X_0, \xi_j{}^0) = \frac{\partial h_{jm_j}}{\partial \tau}(X_0),$$

therefore

$$\frac{\partial h_{jm_j}}{\partial \tau}(X_0) \neq 0. \quad ■$$

We shall observe the behaviour of $H_j(X,\xi)$ in a neighbourhood of $X = X_0$ in $\{\tau \in \mathbb{C}^1, (t,x,y,\eta) \in \mathbb{R}^{2n}\}$ when X_0 is real. If $X \in U_-(X_0) = U(X_0) \cap \{\operatorname{Im}\tau < 0, (t,x,y,\eta) \in \mathbb{R}^{2n}\}$, then $A(X,\xi) = 0$ has no real roots, therefore $H_j(X,\xi) = 0$ also has no real roots in $U_-(X_0)$. We look at the roots of $H_j(X,\xi) = 0$ in $U_-(X_0)$. They include some roots with positive imaginary parts and some with negative imaginary parts. In this case we see that the number $m_j{}^+$ of the former type of roots and $m_j{}^-$ of the latter are independent of $X \in U_-(X_0)$ if we count each multiple root as the number of roots of its multiplicity.

Clearly $m_j{}^+ + m_j{}^- = m_j$. In particular, when X becomes

$$X_\gamma = (t_0, x_0, y_0; \tau_0 - i\gamma, \eta_0), \text{ for some small } \gamma > 0,$$

we wish to know the behaviour of the roots of $H_j(X_\gamma, \xi) = 0$. Applying a

Taylor expansion to $h_{jk}(X_\gamma)$ at X_0, we see that $h_{jk}(X_0)=0$, therefore

$$h_{jk}(X_\gamma) = -\mathrm{i}\gamma\frac{\partial h_{jk}}{\partial \tau}(X_0) + O(\gamma^2).$$

If we consider the roots $(\xi_{j1}{}^1,\ldots,\xi_{jm_j}{}^1)$ of

$$(\xi-\xi_j{}^0)^{m_j} - \mathrm{i}\gamma\frac{\partial h_{jm_j}}{\partial\tau}(X_0) = 0$$

instead of the roots of $H_j(X_\gamma,\xi)=0$, we can write

$$\xi_{jk}{}^1 = \xi_j{}^0 + \mathrm{e}^{\mathrm{i}(\varepsilon_j\pi/2-\pi+2k\pi)\alpha_j}a_j\gamma^{\alpha_j} \text{ where } a_j = \left|\frac{\partial h_{jm_j}}{\partial\tau}(X_0)\right|^{1/m_j}, \alpha_j = \frac{1}{m_j}.$$

This implies that the $\xi_{jk}{}^1$ are lying *on* the circle whose centre and radius are $\xi_j{}^0$ and $a_j\gamma^{\alpha_j}$ respectively; and, moreover, they are at equal distances from each other. Suppose that $|\xi_{jk}{}^1-\xi_{jl}{}^1| \geqq 2b_j\gamma^{\alpha_j}$ $(k \neq l)$. We write D_j for the interior of the circle with centre and radius $\xi_j{}^0$ and $(a_j+b_j)\gamma^{\alpha_j}$, respectively (see Fig. 9). Then, since for $\xi\in D_j$ we have

$$\begin{aligned} H_j(X_\gamma,\xi) &= (\xi-\xi_j{}^0)^{m_j} + h_{j1}(X_\gamma)(\xi-\xi_j{}^0)^{m_j-1} + \cdots + h_{jm_j}(X_\gamma) \\ &= (\xi-\xi_j{}^0)^{m_j} - \mathrm{i}\gamma\frac{\partial h_{jm_j}}{\partial\tau}(X_0) + O(\gamma^{1+\alpha_j}), \end{aligned}$$

we see that

$$\left|H_j(X_\gamma,\xi) - \prod_{k=1}^{m_j}(\xi-\xi_{jk}{}^1)\right| \leqq C\gamma^{1+\alpha_j}.$$

Let us choose an arbitrary k_0 $(1 \neq k_0 \neq m_j)$. Consider a small circle $C_{jk_0}(\subset D_j)$ whose centre and radius are $\xi_{jk_0}{}^1$ and $R\gamma^{2\alpha_j}$ respectively. Then, we see that if $\xi\in C_{jk_0}$ then

$$\left|\prod_{k\neq k_0}(\xi-\xi_{jk}{}^1)\right| \geqq c\gamma^{(m_j-1)\alpha_j}, \quad |\xi-\xi_{jk_0}{}^1| = R\gamma^{2\alpha_j},$$

therefore

$$\left|\prod_{k=1}^{m+}(\xi-\xi_{jk}{}^1)\right| \geqq cR\gamma^{1+\alpha_j}$$

Choosing R such that $R > C/c$ we see that for a sufficiently small γ we have $C_{jk_0}\subset D_j$.Therefore, by Rouché's theorem, we have only one root of the equation $H_j(X_\gamma,\xi)=0$ in every C_{jk}, which we write as $\xi_{jk}(X_\gamma)$:

$$\xi_{jk}(X_\gamma) = \xi_{jk}{}^1 + O(\gamma^{2\alpha_j}).$$

Therefore, we see that

(a) if m_j is even, $m_j{}^+ = m_j{}^- = \frac{1}{2}m_j$,
(b) if m_j is odd, $m_j{}^+ = \frac{1}{2}(m_j-\varepsilon_j)$, $m_j{}^- = \frac{1}{2}(m_j+\varepsilon_j)$,

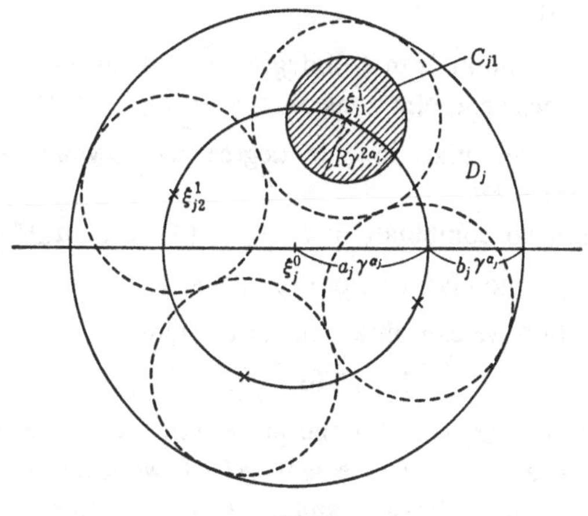

Fig. 9

and, at the same time, we have

$$\operatorname{Im}\xi_{jk}(X_\gamma)>0 \quad (k=1,\ldots,m_j^+),$$
$$\operatorname{Im}\xi_{jk}(X_\gamma)<0 \quad (k=m_j^+ + 1,\ldots,m_j).$$

Hence, we see that if $\operatorname{Im}\tau<0$ in a neighbourhood of $X=X_0$ then

$$A_+(X,\xi)=\prod_{j=1}^{h}\prod_{k=1}^{m_j^+}(\xi-\xi_{jk}(X))E_+(X,\xi),$$

$$A_-(X,\xi)=\prod_{j=1}^{h}\prod_{k=m_j^+ +1}^{m_j}(\xi-\xi_{jk}(X))E_-(X,\xi)$$

is true for $X\in U_-(X_0)$. We now observe the behaviour of the Lopatinski determinant of $\{A(X,\xi),B_j(X,\xi)\}$. In general, if we put

$$A_\pm(X,\xi)=\xi^{m_\pm}+a_1{}^\pm(X)\xi^{m_\pm-1}+\cdots+a_m{}^\pm(X)$$

then $a_j{}^\pm(X)$ is continuous for $\{\operatorname{Im}\tau\neq 0,(t,x,y,\eta)\in\mathbb{R}^{2n}\}$. In this case, as we did in Chapter 2, §5, we set

$$A_\pm{}^{(j)}(X,\xi)=\begin{cases}\xi^j+a_1{}^\pm(X)\xi^{j-1}+\ldots+a_j{}^\pm(X) & (j\leqq m_\pm-1),\\ \xi^{j-m_\pm}A_\pm(\xi) & (j\geqq m_\pm)\end{cases}.$$

Then, as the polynomials of $\xi,\{A_+{}^{(0)}(X,\xi),\ldots,A_+{}^{(m-1)}(X,\xi)\}$, $\{A_-{}^{(m-1)}(X,\xi),\ldots,A_-{}^{(0)}(X,\xi)\}$ are adjoint with respect to $A(X,\xi)$, this means that

$$\sum_{j=0}^{m-1}A_+{}^{(j)}(X,\xi)A_-{}^{(m-1-j)}(X,\xi')=\frac{A(X,\xi)-A(X,\xi')}{\xi-\xi'}.$$

Hence we can write

$$L^+ = \{\text{the polynomials in } \xi \text{ of degree } m_+ - 1 \text{ (or less)}\}$$
$$= \{\text{the linear combinations of } A_+^{(0)}(X,\xi), \ldots, A_+^{(m_+ - 1)}(X,\xi)\}.$$

$$L_A^-(X) = \{\text{the polynomials in } \xi \text{ of degree } m-1, \text{ which are divisible by } A_+(X,\xi)\}$$
$$= \{\text{the linear combinations of } A_+^{(m_+)}(X,\xi), \ldots, A_+^{(m-1)}(X,\xi)\}.$$

$$L = \{\text{the polynomials of } \xi \text{ of degree } m-1\}.$$

Using this notation we can obtain the direct sum

$$L = L^+ \oplus L_A^-(X).$$

Let us write $P^+(X)$, $P_A^-(X)$ for the projections of L onto L^+, $L_A^-(X)$, respectively. For an arbitrary element $Q(\xi) \in L$ we shall see how $Q(\xi)$ is decomposed in terms of $P^+(X)Q$ and $P_A^-(X)Q$ as follows:

$$Q(\xi) = \left\langle Q(\xi), \begin{bmatrix} A_-^{(m-1)}(X,\xi) \\ \vdots \\ A_-^{(0)}(X,\xi) \end{bmatrix} \right\rangle_{A(X,\xi)} \begin{bmatrix} A_+^{(0)}(X,\xi) \\ \vdots \\ A_+^{(m-1)}(X,\xi) \end{bmatrix}$$

$$= \left[\frac{1}{2\pi i} \oint \frac{Q(\xi) A_-^{(m-1)}(X,\xi)}{A(X,\xi)} \mathrm{d}\xi, \ldots, \frac{1}{2\pi i} \oint \frac{Q(\xi) A_-^{(0)}(X,\xi)}{A(X,\xi)} \mathrm{d}\xi \right]$$

$$\times \begin{bmatrix} A_+^{(0)}(X,\xi) \\ \vdots \\ A_+^{(m-1)}(X,\xi) \end{bmatrix}$$

$$= \left[\frac{1}{2\pi i} \oint \frac{Q(\xi)\xi^{m_+ - 1}}{A_+(X,\xi)} \mathrm{d}\xi, \ldots, \frac{1}{2\pi i} \oint \frac{Q(\xi)}{A_+(X,\xi)} \mathrm{d}\xi \right] \begin{bmatrix} A_+^{(0)}(X,\xi) \\ \vdots \\ A_+^{(m_+ - 1)}(X,\xi) \end{bmatrix}$$

$$+ \left[\frac{1}{2\pi i} \oint \frac{Q(\xi) A_-^{(m_- - 1)}(X,\xi)}{A(X,\xi)} \mathrm{d}\xi, \ldots, \frac{1}{2\pi i} \oint \frac{Q(\xi) A_-^{(0)}(X,\xi)}{A(X,\xi)} \mathrm{d}\xi \right]$$

$$\times \begin{bmatrix} A_+^{(m_+)}(X,\xi) \\ \vdots \\ A_+^{(m-1)}(X,\xi) \end{bmatrix}$$

$$= P^+(X)[Q(\xi)] + P_A^-(X)[Q(\xi)].$$

We now take $B_1(X,\xi_1), \ldots, B_{m_+}(X,\xi) \in L$ and write

$$L_B(X) = \{\text{linear combinations of } B_1(X,\xi), \ldots, B_{m_+}(X,\xi)\}.$$

Since

$$R(X) = \pm \det \left\langle \begin{bmatrix} B_1(X,\xi) \\ \vdots \\ B_{m_+}(X,\xi) \end{bmatrix}, \begin{bmatrix} A_-^{(m-1)}(X,\xi) \\ \vdots \\ A_-^{(m_-)}(X,\xi) \end{bmatrix} \right\rangle_{A(X,\xi)}$$

Therefore we have the following:

Lemma 3.2

Let $X \in \{\operatorname{Im}\tau \leqq 0, (t,x,y,\eta) \in \mathbb{R}^{2n}\}$. Then the statement $R(X) \neq 0$ is equivalent to

$$L = L_B(X) \oplus L_A^-(X).$$

Again we look at the neighbourhood $U(X_0)$. Consider the following polynomials of degree $m-1$, which possess smooth coefficients in $U(X_0)$:

$$P_{jk}(X,\xi) = (\xi - \xi_j^0)^k \prod_{l \neq j} H_l(X,\xi) E(X,\xi) \quad \begin{pmatrix} j = 1, \dots, h \\ k = 0, \dots m_j - 1 \end{pmatrix},$$

$$Q_k^+(X,\xi) = \xi^k \prod_l H_l(X,\xi) E_-(X,\xi) \quad (k = 0, \dots, m_0^+ - 1),$$

$$Q_k^-(X,\xi) = \xi^k \prod_l H_l(X,\xi) E_+(X,\xi) \quad (k = 0, \dots, m_0^- - 1).$$

These polynomials form a basis of L. However, at $X = X_0$,

$$\{P_{jk}(X_0,\xi)\}_{\substack{j=1,\dots,h \\ k = m_j^+, \dots, m_j - 1}}, \quad \{Q_k^-(X_0,\xi)\}_{k=0,\dots,m_0^- - 1}$$

are divisible by $A_+(X_0,\xi)$. This implies that they form a basis of $L_A^-(X_0)$. Let us write $\tilde{L}_A^-(X)$ for the subspace of L generated by

$$\{P_{jk}(X,\xi)\}_{\substack{j=1,\dots,h \\ k = m_j^+, \dots, m_j - 1}}, \quad \{Q_k^-(X,\xi)\}_{k=0,\dots,m_0^- - 1}.$$

Obviously we have $\tilde{L}_A^-(X_0) = L_A^-(X_0)$. If $R(X_0) \neq 0$ then by Lemma 3.2 we see that

$$L = L_B(X_0) \oplus \tilde{L}_A^-(X_0).$$

Since these $L_B(X)$, $\tilde{L}_A^-(X)$ are continuous in $U(X_0) \cap \{\operatorname{Im}\tau \leqq 0, (t,x,y,\eta) \in \mathbb{R}^{2n}\}$, we have

Corollary to Lemma 3.2

Let $X_0 \in \{\operatorname{Im}\tau \leqq 0, (t,x,y,\eta) \in \mathbb{R}^{2n}\}$. If $R(X_0) \neq 0$ then

$$L = L_B(X) \oplus \tilde{L}_A^-(X)$$

in $U(X_0) \cap \{\operatorname{Im}\tau \leqq 0, (t,x,y,\eta) \in \mathbb{R}^{2n}\}$.

3.2

Consider the decomposition of $A(X,\xi)$ in $U(X_0)$. Then, each term on the right hand side of

$$A(X,\xi) = \prod_{j=1}^{h} H_j(X,\xi)E_+(X,\xi)E_-(X,\xi)$$

is a local symbol (see p. 123). We shall extend these local symbols by standard extension (see §1) and observe the singular integral operators for extended symbols.

First, we observe the symbol of degree m_j

$$H_j(X,\xi) = (\xi - \xi_j{}^0\Lambda)^{m_j} + h_{j1}(X)(\xi - \xi_j{}^0\Lambda)^{m_j-1} + \cdots + h_{jm_j}(X).$$

Since $h_{jk}(X)$ is a symbol of degree k, from Lemma 3.1 we have

$$h_{jk}(X) = h_{jk}'(X) - \mathrm{i}\gamma h_{jk}''(X), \quad \tau = \sigma - \mathrm{i}\gamma,$$

where $h_{jk}'(X)$ is a real symbol of degree k, and $h_{jk}''(X)$ is a real symbol of degree $k-1$. Setting

$$\begin{aligned} H_j'(X,\xi) &= (\xi - \xi_j{}^0\Lambda)^{m_j} + h_{j1}'(X)(\xi - \xi_j{}^0\Lambda)^{m_j-1} + \cdots + h_{jm_j}'(X) \\ &= (\xi - \xi_j{}^0\Lambda)^{m_j} + \tilde{h}_{j1}'(X)\Lambda(\xi - \xi_j{}^0\Lambda)^{m_j-1} + \ldots + \tilde{h}_{jm_j}'(X)\Lambda^{m_j}, \\ H_j''(X,\xi) &= h_{j1}''(X)(\xi - \xi_j{}^0\Lambda)^{m_j-1} + h_{j2}''(X)(\xi - \xi_j{}^0\Lambda)^{m_j-2} \\ &\quad + \cdots + h_{jm_j}''(X) \\ &= \tilde{h}_{j1}''(X)(\xi - \xi_j{}^0\Lambda)^{m_j-1} + \tilde{h}_{j2}''(X)\Lambda(\xi - \xi_j{}^0\Lambda)^{m_j-2} \\ &\quad + \ldots + \tilde{h}_{jm_j}''(X)\Lambda^{m_j-1}, \end{aligned}$$

we see that

$$H_j(X,\xi) = H_j'(X,\xi) - \mathrm{i}\gamma H_j''(X,\xi).$$

Take an arbitrary real symbol Q of degree $m_j - 1$, such that

$$Q(X,\xi) = q_0(X)(\xi - \xi_j{}^0\Lambda)^{m_j-1} + q_1(X)(\xi - \xi_j{}^0\Lambda)^{m_j-2} + \ldots + q_{m_j-1}(X).$$

We now consider

$$\begin{aligned} &(H_ju, Qu)_{\mathscr{H}_{-\gamma}} - (Qu, H_ju)_{\mathscr{H}_{-\gamma}} \\ &= \{(H_j'u, Qu)_{\mathscr{H}_{-\gamma}} - (Qu, H_j'u)_{\mathscr{H}_{-\gamma}}\} - \mathrm{i}\gamma\{(H_j''u, Qu)_{\mathscr{H}_{-\gamma}} + (Qu, H_j''u)_{\mathscr{H}_{-\gamma}}\} \\ &= \mathrm{i}I_1 + \mathrm{i}\gamma I_2. \end{aligned}$$

For I_1, by Theorem 1 we see that

$$I_1 = (1/\mathrm{i})\{(H_j'u, Qu)_{\mathscr{H}_{-\gamma}} - (Qu, H_j'u)_{\mathscr{H}_{-\gamma}}\}$$

$$\cong \left\langle \tilde{\mathscr{H}}_j' \begin{bmatrix} (\mathrm{D}_x - \xi_j{}^0\Lambda)^{m_j-1} \\ \Lambda(\mathrm{D}_x - \xi_j{}^0\Lambda)^{m_j-2} \\ \vdots \\ \Lambda^{m_j-1} \end{bmatrix} u, \begin{bmatrix} (\mathrm{D}_x - \xi_j{}^0\Lambda)^{m_j-1} \\ \Lambda(\mathrm{D}_x - \xi_j{}^0\Lambda)^{m_j-2} \\ \vdots \\ \Lambda^{m_j-1} \end{bmatrix} u \right\rangle_{\mathscr{H}_{-\gamma}},$$

where

$$\tilde{\mathscr{H}}_j' = \tilde{Q}_0 + \bar{h}_{j1}'\tilde{Q}_1 + \ldots + \bar{h}_{jm_j'}\tilde{Q}_{m_j}$$

and $\cong$ means that the difference is bounded by a constant times

$$\|u\|^2_{m_j-1,\mathscr{H}_{-\gamma}} + \sum_{k=0}^{m_j-1} \langle D_x{}^k u\rangle^2_{m_j-1-k-1/2,\mathscr{H}_{-\gamma}}.$$

Let us assume that $\tilde{Q}_k(k=0,1,\ldots,m_j)$ are constant matrices independent of X. Then writing

$$\mathscr{H}_j'(X_0) = \tilde{Q}_0 = \begin{bmatrix} \tilde{q}_0 & \tilde{q}_1 \cdots \tilde{q}_{m_j-1} \\ \tilde{q}_1 & \\ \vdots & \cdot^{\cdot^{\cdot}} \\ \tilde{q}_{m_j-1} & 0 \end{bmatrix},$$

we have

Lemma 3.3

We can find $\tilde{Q}_0$ as follows. There exist $c_1 > 0, c_2 > 0$ such that

$$[\bar{u}_{m_j-1},\ldots,\bar{u}_0]\tilde{Q}_0 \begin{bmatrix} u_{m_j-1} \\ \vdots \\ u_0 \end{bmatrix} \geqq c_1 \sum_{k=m_j^+}^{m_j-1} |u_k|^2 - c_2 \sum_{k=0}^{m_j^+-1} |u_k|^2,$$

and

$$\operatorname{sgn} \tilde{q}_{m_j-1} = \varepsilon_j.$$

Proof

To prove the first part consider the decomposition

$$\tilde{Q}_0 = \begin{array}{c} \begin{matrix} m_j^- & m_j^+ \end{matrix} \\ \begin{bmatrix} \widehat{Q}_{11} & \widehat{Q}_{12} \\ Q_{21} & Q_{22} \end{bmatrix} \end{array} \begin{matrix})m_j^- \\ )m_j^+ \end{matrix}.$$

Then, we have

$$Q_{11} = \begin{bmatrix} \tilde{q}_0 & \tilde{q}_1 \cdots & \tilde{q}_{m_j^--1} \\ \tilde{q}_1 & & \vdots \\ \vdots & \cdot^{\cdot^{\cdot}} & \\ \tilde{q}_{m_j^--1} & & \cdots \tilde{q}_{2(m_j^--1)} \end{bmatrix}$$

where

(i) If m_j is even, then $m_j^- = m_j/2$, therefore $2(m_j^- - 1) = m_j - 2$,
(ii) If m_j is odd, and $\varepsilon_j = 1$, then $m_j^- = (m_j+1)/2$, therefore $2(m_j^- - 1) = m_j - 1$,

(iii) If m_j is odd, and $\varepsilon_j = -1$, then $m_j^- = (m_j - 1)/2$, therefore $2(m_j^- - 1) = m_j - 3$.

Now letting

$$\tilde{q}_k = a^{2k} \quad (a > 0), \quad k = 0, 1, \ldots, 2(m_j^- - 1)$$

we have

$$\tilde{q}_k \leqq a\tilde{q}_{k-l}\tilde{q}_{k+l} \quad (l \geqq 1).$$

Therefore, for a sufficiently small we see that

$$Q_{11} > 0.$$

Hence, for any undetermined $\tilde{q}_k (k > 2(m_j^- - 1))$, the inequality of the lemma holds. To see that $\operatorname{sgn} \tilde{q}_{m_j - 1} = \varepsilon_j$ notice that, when $\tilde{q}_{m_j - 1}$ is determined this is simply case (ii) above, and the equality is obviously true. When $\tilde{q}_{m_j - 1}$ is not determined, it is easy to make it satisfy the equality.

■

Corollary to Lemma 3.3

Let μ be a parameter such that $0 < \mu < 1$. There exists $\tilde{Q}_0^{(\mu)}$ for μ such that

$$[\bar{u}_{m_j - 1}, \ldots, \bar{u}_0] \tilde{Q}_0^{(\mu)} \begin{bmatrix} u_{m_j - 1} \\ \vdots \\ u_0 \end{bmatrix} \geqq c_1 \sum_{k = m_j^+}^{m_j - 1} |u_k|^2 - c_2 \mu^2 \sum_{k=0}^{m_j^+ - 1} |u_k|^2,$$

$$\operatorname{sgn} \tilde{q}_{m_j - 1} = \varepsilon_j,$$

where c_1, c_2 are positive constants independent of μ.

Proof

Let

$$M = \left[\begin{array}{ccc|ccc} \mu^{-(m_j^- - 1)} & & 0 & & & \\ & \ddots & & & \mathbf{0} & \\ 0 & & \mu^{-1} & & & \\ & & 1 & & & \\ \hline & & & \mu & & \\ & \mathbf{0} & & & \ddots & 0 \\ & & & 0 & & \mu^{m_j^+} \end{array}\right].$$

By setting

$$\tilde{Q}_0^{(\mu)} = M\tilde{Q}_0 M, \quad U = \begin{bmatrix} u_{m_j - 1} \\ \vdots \\ u_0 \end{bmatrix}$$

we see that

$$ {}^t\bar{U}\tilde{Q}_0^{(\mu)}U = {}^t\bar{U}M\tilde{Q}_0MU = {}^t\,\overline{(MU)}\tilde{Q}_0(MU). $$

By Lemma 3.3, we have

$$ {}^t\bar{U}\tilde{Q}_0^{(\mu)}U \geqq c_1 \sum_{k=m_j^+}^{m_j-1} |\mu^{m_j^+-k}u_k|^2 - c_2 \sum_{k=0}^{m_j^+-1} |\mu^{m_j^+-k}u_k|^2 $$

$$ \geqq c_1 \sum_{k=m_j^+}^{m_j-1} |u_k|^2 - c_2\mu^2 \sum_{k=0}^{m_j^+-1} |u_k|^2. \quad \blacksquare $$

From now on we take $Q = Q^{(\mu)}$ satisfying the above corollary. We intend to make μ sufficiently small in the argument in the following §3.4. For the moment, we suppose that μ is a positive parameter. Then we see that the property of $\tilde{Q}_0$ mentioned in Lemma 3.3 in simply that of $\tilde{\mathscr{H}}_j'(X_0)$. Let us regard the standard extension of H_j as an extension starting from a $U^{(\mu)}(X_0)$, a sufficiently small neighbourhood of $X = X_0$. This implies that the property $\tilde{\mathscr{H}}_j'(X_0)$ is just that of $\tilde{\mathscr{H}}_j'(X)$. That is,

$$ [\tilde{u}_{m_j-1},\ldots,\tilde{u}_0]\tilde{\mathscr{H}}_j'(X)\begin{bmatrix} u_{m_j-1} \\ \vdots \\ u_0 \end{bmatrix} \geqq \sum_{k=m_j^+}^{m_j-1} |u_k|^2 - c_2\mu^2 \sum_{k=0}^{m_j^+-1} |u_k|^2, $$

where

$$ X = (t, 0, y; \tau, \eta), \quad |\tau|^2 + |\eta|^2 = 1. $$

From this we have

$$ I_1 \geqq c_1 \sum_{k=m_j^+}^{m_j-1} \langle (\mathrm{D}_x - \xi_j^0\Lambda)^k u\rangle^2_{m_j-1-k,\mathscr{H}_{-\gamma}} $$

$$ - c_2\mu^2 \sum_{k=0}^{m_j^+-1} \langle (\mathrm{D}_x - \xi_j^0\Lambda)^k u\rangle^2_{m_j-1-k,\mathscr{H}_{-\gamma}} $$

$$ - C_\mu\left\{ \|u\|^2_{m_j-1,\mathscr{H}_{-\gamma}} + \sum_{k=0}^{m_j-1} \langle \mathrm{D}_x^k u\rangle^2_{m_j-1-k-1/2,\mathscr{H}_{-\gamma}} \right\}. $$

Next, under the same conditions, we consider

$$ I_2 = -\{(H_j''u, Qu)_{\mathscr{H}_{-\gamma}} + (Qu, H_j''u)_{\mathscr{H}_{-\gamma}}\}. $$

Since we have

$$H_j''(X,\xi)=\tilde{h}_{j0}''(X)(\xi-\xi_j^0\Lambda)^{m_j-1}+\ldots+\tilde{h}_{jm_j}''(X)\Lambda^{m_j-1},$$
$$Q(X,\xi)=\tilde{q}_0(\xi-\xi_j^0\Lambda)^{m_j-1}+\ldots+\tilde{q}_{m_j-1}\Lambda^{m_j-1},$$
$$\operatorname{sgn}\tilde{q}_{m_j-1}=\varepsilon_j=-\operatorname{sgn}\tilde{h}_{jm_j}''(X_0)$$

we see clearly that there exist $c_\mu>0$, $C_\mu>0$ such that

$$I_2\geqq c_\mu\|\Lambda^{m_j-1}u\|^2_{\mathscr{H}_{-\gamma}}-C_\mu\sum_{k=1}^{m_j-1}\|\Lambda^{m_j-1-k}(\mathrm{D}_x-\xi_j^0\Lambda)^k u\|^2_{\mathscr{H}_{-\gamma}}.$$

Also, we see that by the interpolation theorem, for any $\varepsilon>0$,

$$\sum_{k=1}^{m_j-1}\|\Lambda^{m_j-1-k}(\mathrm{D}_x-\xi_j^0\Lambda)^k u\|^2_{\mathscr{H}_{-\gamma}}$$
$$\leqq\varepsilon\|\Lambda^{m_j-1}u\|^2_{\mathscr{H}_{-\gamma}}+C_\varepsilon\|\Lambda^{-1}(\mathrm{D}_x-\xi_j^0\Lambda)^{m_j}u\|^2_{\mathscr{H}_{-\gamma}},$$

so that

$$I_2\geqq c_\mu\sum_{k=0}^{m_j-1}\|\Lambda^{m_j-1-k}(\mathrm{D}_x-\xi_j^0\Lambda)^k u\|^2_{\mathscr{H}_{-\gamma}}-C_\mu\|\Lambda^{-1}(\mathrm{D}_x-\xi_j^0\Lambda)^{m_j}u\|^2_{\mathscr{H}_{-\gamma}}.$$

On the other hand we know that

$$H_j(X,\xi)=(\xi-\xi_j^0\Lambda)^{m_j}+\tilde{h}_{j1}(X)\Lambda(\xi-\xi_j^0\Lambda)^{m_j-1}+\cdots+\tilde{h}_{jm_j}(X)\Lambda^{m_j},$$
$$H_j(X_0,\xi)=(\xi-\xi_j^0)^{m_j}.$$

Therefore, as before, we take $U^{(\mu)}(X_0)$, an even smaller neighbourhood of $X=X_0$ from which H_j is extended. Then we see that

$$I_2\geqq c_\mu\|u\|^2_{m_j-1,\mathscr{H}_{-\gamma}}-C_\mu\{\|\Lambda^{-1}H_j u\|^2_{\mathscr{H}_{-\gamma}}+\|u\|^2_{m_j-1,\mathscr{H}_{-\gamma}}\}.$$

We now synthesise the estimations of I_1 and I_2 to obtain

Lemma 3.4
For an arbitrary $0<\mu<1$ there exists a neighbourhood $U^{(\mu)}$ of $X=X_0$. Let H_j be the standard extension with regard to $U^{(\mu)}$. Then

$$c_\mu\gamma\|u\|^2_{m_j-1,\mathscr{H}_{-\gamma}}+\sum_{k=m_j^+}^{m_j-1}\langle(\mathrm{D}_x-\xi_j^0\Lambda)^k u\rangle^2_{m_j-1-k,\mathscr{H}_{-\gamma}}$$

$$\leqq C\mu^2\sum_{k=0}^{m_j^+-1}\langle(\mathrm{D}_x-\xi_j^0\Lambda)^k u\rangle^2_{m_j-1-k,\mathscr{H}_{-\gamma}}$$

$$+C_\mu\left\{\frac{1}{\gamma}\|H_j u\|^2_{\mathscr{H}_{-\gamma}}+\|u\|^2_{m_j-1,\mathscr{H}_{-\gamma}}\right\},\quad \gamma>0.$$

3.3

Let us consider a singular integral operator for the symbol

$$E_{\pm}(X,\xi)=\xi^{m_0\pm}+e_1{}^{\pm}(X)\xi^{m_0\pm-1}+\ldots+e_{m_0\pm}{}^{\pm}(X).$$

Then let us separate this into the real and imaginary parts to obtain

$$E_{\pm}(X,\xi)=E_{\pm}'(X,\xi)-\mathrm{i}E_{\pm}''(X,\xi),$$
$$E_{\pm}'(X,\xi)=\xi^{m_0\pm}+e_1{}^{\pm\prime}(X)\xi^{m_0\pm-1}+\ldots+e_{m_0\pm}{}^{\pm\prime}(X),$$
$$E_{\pm}''(X,\xi)=e_1{}^{\pm\prime\prime}(X)\xi^{m_0\pm-1}+\ldots+e_{m_0\pm}{}^{\pm\prime\prime}(X).$$

We know that the zeros of $E_+(X,\xi)$ and $E_-(X,\xi)$ are in the regions $\operatorname{Im}\xi>0$ and $\operatorname{Im}\xi<0$ respectively. Therefore, we have

(i) $e_1{}^{+\prime\prime}(X)>0,\quad e_1{}^{-\prime\prime}(X)<0,$

(ii) all the zeros of $E_{\pm}'(X,\xi)$ and $E_{\pm}''(X,\xi)$ are on the real axis of the ξ-plane, and separate each other (use *Hermite's theorem*).†

We write

$$\begin{aligned}\|\Lambda^{-1/2}E_{\pm}u\|^2_{\mathscr{H}_{-\gamma}} &= \|\Lambda^{-1/2}(E_{\pm}'-\mathrm{i}E_{\pm}'')u\|^2_{\mathscr{H}_{-\gamma}}\\ &=\{\|\Lambda^{-1/2}E_{\pm}'u\|^2_{\mathscr{H}_{-\gamma}}+\|\Lambda^{-1/2}E_{\pm}''u\|^2_{\mathscr{H}_{-\gamma}}\}\\ &\quad+\mathrm{i}\{(E_{\pm}'u,\Lambda^{-1}E_{\pm}''u)_{\mathscr{H}_{-\gamma}}\\ &\quad-(\Lambda^{-1}E_{\pm}''u,E_{\pm}'u)_{\mathscr{H}_{-\gamma}}\}\\ &=I_1{}^{\pm}+I_2{}^{\pm}.\end{aligned}$$

For $I_1{}^{\pm}$ we have

$$I_1{}^{\pm}\geqq c\|\Lambda^{-1/2}u\|^2_{m_0\pm,\mathscr{H}_{-\gamma}}-C\|u\|^2_{m_0\pm-1,\mathscr{H}_{-\gamma}}.$$

And for $I_2{}^{\pm}$, by Theorem 1, we see that

$$I_2{}^{\pm}\cong-\left\langle\tilde{\mathscr{E}}_{\pm}\begin{bmatrix}D_x{}^{m_0\pm-1}\\ \Lambda D_x{}^{m_0\pm-2}\\ \vdots\\ \Lambda^{m_0\pm-1}\end{bmatrix}u,\begin{bmatrix}D_x{}^{m_0\pm-1}\\ \Lambda D_x{}^{m_0\pm-2}\\ \vdots\\ \Lambda^{m_0\pm-1}\end{bmatrix}u\right\rangle_{\mathscr{H}_{-\gamma}},$$

where

$$[\bar{\xi}^{m_0\pm-1},\ldots,\Lambda^{m_0\pm-1}]\tilde{\mathscr{E}}_{\pm}(X)\begin{bmatrix}\xi^{m_0\pm-1}\\ \vdots\\ \Lambda^{m_0\pm-1}\end{bmatrix}$$
$$=\frac{E_{\pm}'(X,\xi)\Lambda^{-1}E_{\pm}''(X,\bar{\xi})-E_{\pm}'(X,\bar{\xi})\Lambda^{-1}E_{\pm}''(X,\xi)}{\xi-\bar{\xi}}$$

† Translator's note: (Hermite's theorem) Given a polynomial $f(z)=u+\mathrm{i}v$; $u,v\in\mathbb{R}$, if the signs of the imaginary parts of the roots of the equation $f(z)=0$ are the same, then u, v have real roots only that appear alternately on the real axis.

and $\cong$ is an estimation bounded by a constant times

$$\|u\|^2_{m_0\pm-1,\mathscr{H}_{-\gamma}}+\sum_{k=0}^{m_0\pm-1}\langle D_x{}^k u\rangle^2_{m_0\pm-1-k-1/2,\mathscr{H}_{-\gamma}}.$$

In this connection we have

Lemma 3.5
If $X=(t,x,y;\tau,\eta)$, $|\tau|^2+|\eta|^2=1$ then

$$\pm[\bar{u}_{m_0\pm-1},\ldots,\bar{u}_0]\tilde{\mathscr{E}}_\pm(X)\begin{bmatrix}u_{m_0\pm}{}^{-1}\\ \vdots\\ u_0\end{bmatrix}\geqq c\sum_{k=0}^{m_0\pm-1}|u_k|^2\quad(c>0).$$

Proof
Let us take $E_\pm'(X,\xi), E_\pm''(X,\xi)$ as polynomials of ξ, and apply the Euclidean algorithm. Let $E_\pm{}^{(0)}=E_\pm', E_\pm{}^{(1)}=E_\pm''$ and

$$E_\pm{}^{(0)}(X,\xi)=K_\pm{}^{(1)}(X,\xi)E_\pm{}^{(1)}(X,\xi)-E_\pm{}^{(2)}(X,\xi).$$

Then we see that the roots of $E_\pm{}^{(0)}(X,\xi)=0$, and those of $E_\pm{}^{(1)}(X,\xi)=0$ are distinct reals which separate each other. Therefore, the roots of $E_\pm{}^{(2)}(X,\xi)=0$ are also distinct reals which separate the roots of $E_\pm{}^{(1)}(X,\xi)=0$. At the same time we have

$$e_{+0}{}^{(2)}(X)>0,\quad e_{-0}{}^{(2)}(X)>0$$

in

$$E_\pm{}^{(2)}(X,\xi)=e_{\pm0}{}^{(2)}(X)\xi^{m_0\pm-2}+e_{\pm1}{}^{(2)}(X)\xi^{m_0\pm-3}+\ldots+e_{\pm m_0\pm-2}{}^{(2)}(X).$$

Repeating this process we have

$$\begin{aligned}&E_\pm{}^{(1)}(X,\xi)=K_\pm{}^{(2)}(X,\xi)E_\pm{}^{(2)}(X,\xi)-E_\pm{}^{(3)}(X,\xi),\\ &\quad\vdots\\ &E_\pm{}^{(m_0\pm-2)}(X,\xi)=K_\pm{}^{(m_0\pm-1)}(X,\xi)E_\pm{}^{(m_0\pm-1)}(X,\xi)-E_\pm{}^{(m_0\pm)}(X),\\ &E_\pm{}^{(m_0\pm-1)}(X,\xi)=K_\pm{}^{(m_0\pm)}(X,\xi)E_\pm{}^{(m_0\pm)}(X)\end{aligned}$$

and so as a result we have

$$E_\pm{}^{(k)}(X,\xi)=e_{\pm0}{}^{(k)}(X)\xi^{m_0\pm-k}+\ldots+e_{\pm m_0\pm-k}{}^{(k)}(X),$$
$$e_{+0}{}^{(k)}(X)>0,\quad(-1)^k e_{-0}{}^{(k)}(X)>0.$$

Now if we write

$$K_\pm{}^{(k)}(X,\xi)=\kappa_{\pm0}{}^{(k)}(X)\xi+\kappa_{\pm1}{}^{(k)}(X),$$

we have

$$\kappa_{\pm0}{}^{(k)}(X)=e_{\pm0}{}^{(k-1)}/e_{\pm0}{}^{(k)},$$

therefore

$$\kappa_{+0}{}^{(k)}(X) > 0, \quad \kappa_{-0}{}^{(k)}(X) < 0.$$

Then we consider the Bézout form of $E_\pm{}^{(0)}(X,\xi), E_\pm{}^{(1)}(X,\xi)$ with respect to ξ as follows:

$$\frac{E_\pm{}^{(0)}(X,\xi)E_\pm{}^{(1)}(X,\bar\xi) - E_\pm{}^{(0)}(X,\bar\xi)E_\pm{}^{(1)}(X,\xi)}{\xi - \bar\xi}$$

$$= \frac{\{K_\pm{}^{(1)}(X,\xi)E_\pm{}^{(1)}(X,\xi) - E_\pm{}^{(2)}(X,\xi)\}E_\pm{}^{(1)}(X,\bar\xi)}{\xi - \bar\xi}$$

$$- \frac{\{K_\pm{}^{(1)}(X,\bar\xi)E_\pm{}^{(1)}(X,\bar\xi) - E_\pm{}^{(2)}(X,\bar\xi)\}E_\pm{}^{(1)}(X,\xi)}{\xi - \bar\xi}$$

$$= \kappa_{\pm 0}{}^{(1)}(X)E_\pm{}^{(1)}(X,\xi)E_\pm{}^{(1)}(X,\bar\xi)$$

$$+ \frac{E_\pm{}^{(1)}(X,\xi)E_\pm{}^{(2)}(X,\bar\xi) - E_\pm{}^{(1)}(X,\bar\xi)E_\pm{}^{(2)}(X,\xi)}{\xi - \bar\xi}$$

$$= \sum_{j=1}^{m_0\pm} \kappa_{\pm 0}{}^{(j)}(X)E_\pm{}^{(j)}(X,\xi)E_\pm{}^{(j)}(X,\bar\xi).$$

Note that

$$\{E_\pm{}^{(1)}(X,\xi), \ldots, E_\pm{}^{(m_0\pm)}(X,\xi)\}$$

is a basis for the space of polynomials of ξ of degree $m_0{}^\pm - 1$. ■

Now, we return to $I_2{}^\pm$. By Lemma 3.5, we obtain the following estimations:

$$I_2{}^+ \geqq -C\left\{\sum_{k=0}^{m_0{}^+-1} \langle D_x{}^k u\rangle^2_{m_0{}^+-1-k,\mathscr{H}_{-\gamma}} + \|u\|^2_{m_0{}^+-1,\mathscr{H}_{-\gamma}}\right\},$$

$$I_2{}^- \geqq c\sum_{k=0}^{m_0{}^--1} \langle D_x{}^k u\rangle^2_{m_0{}^--1-k,\mathscr{H}_{-\gamma}}$$

$$-C\left\{\|u\|^2_{m_0{}^--1,\mathscr{H}_{-\gamma}} + \sum_{k=0}^{m_0{}^--1} \langle D_x{}^k u\rangle^2_{m_0{}^--1-k-1/2,\mathscr{H}_{-\gamma}}\right\}.$$

Therefore we establish

Lemma 3.6

For $\gamma > 0$ the following inequalities hold

$$\gamma\|u\|^2_{m_0{}^+-1,\mathscr{H}_{-\gamma}} \leqq C\left\{\sum_{k=0}^{m_0{}^+-1} \langle D_x{}^k u\rangle^2_{m_0{}^+-1-k,\mathscr{H}_{-\gamma}} + \frac{1}{\gamma}\|E_+u\|^2_{\mathscr{H}_{-\gamma}}\right.$$

$$\left. + \|u\|^2_{m_0{}^+-1,\mathscr{H}_{-\gamma}}\right\},$$

$$\gamma\|u\|^2_{m_0^- - 1,\mathscr{H}_{-\gamma}} + \sum_{k=0}^{m_0^- - 1} \langle D_x{}^k u\rangle_{m_0^- - 1 - k,\mathscr{H}_{-\gamma}}$$
$$\leqq C\left\{\frac{1}{\gamma}\|E_- u\|^2_{\mathscr{H}_{-\gamma}} + \|u\|^2_{m_0^- - 1,\mathscr{H}_{-\gamma}}\right\}.$$

3.4

Up to now we have considered energy inequalities for H_j and $E_\pm$. Let us show how from these observations we can obtain the energy inequality for A itself. To this end we consider the symbol A in a neighbourhood of $X = X_0$ as

$$\begin{aligned} A(X,\xi) &= \prod_{j=1}^{h} H_j(X,\xi)E_+(X,\xi)E_-(X,\xi) \\ &= H_1(X,\xi)P_{10}(X,\xi) = \cdots = H_h(X,\xi)P_{h0}(X,\xi) \\ &= E_+(X,\xi)Q_0{}^+(X,\xi) = E_-(X,\xi)Q_0{}^-(X,\xi), \end{aligned}$$

and the corresponding singular integral operator. Then we have

$$\begin{aligned} \|A\varphi u\|_{\mathscr{H}_{-\gamma}} &\cong \|H_1P_{10}\varphi u\|_{\mathscr{H}_{-\gamma}} \cong \ldots \cong \|H_hP_{h0}\varphi u\|_{\mathscr{H}_{-\gamma}} \\ &\cong \|E_+Q_0{}^+\varphi u\|_{\mathscr{H}_{-\gamma}} \cong \|E_-Q_0{}^-\varphi u\|_{\mathscr{H}_{-\gamma}} \end{aligned}$$

where φ is a localising operator for a neighbourhood of $X = X_0$ and the difference in the relation (resulting from the estimation $\cong$) is bounded by a constant times $\|u\|_{m-1,\mathscr{H}_{-\gamma}}$.

We now apply Lemma 3.4 to $P_{j0}\varphi u$ instead of u, and obtain

(A) $$c_\mu\gamma\|P_{j0}\varphi u\|^2_{m_j-1,\mathscr{H}_{-\gamma}} + \sum_{k=m_j^+}^{m_j-1} \langle P_{jk}\varphi u\rangle^2_{m_j-1-k,\mathscr{H}_{-\gamma}}$$
$$\leqq C\mu^2\sum_{k=0}^{m_j^+-1} \langle P_{jk}\varphi u\rangle^2_{m_j-1-k,\mathscr{H}_{-\gamma}} + C_\mu\left\{\frac{1}{\gamma}\|Au\|^2_{\mathscr{H}_{-\gamma}} + \|u\|^2_{m-1,\mathscr{H}_{-\gamma}}\right\}$$

for $\gamma > 0$ and an arbitrary $0 < \mu < 1$. Next, we use the inequality of Lemma 3.6 for $Q_0{}^\pm\varphi u$ instead of u. We see that

(B) $$\gamma\|Q_0{}^+\varphi u\|^2_{m_0^+-1,\mathscr{H}_{-\gamma}} \leqq C\left\{\sum_{k=0}^{m_0^+-1} \langle Q_k{}^+\varphi u\rangle^2_{m_0^+-1-k,\mathscr{H}_{-\gamma}} + \frac{1}{\gamma}\|Au\|^2_{\mathscr{H}_{-\gamma}} + \|u\|^2_{m-1,\mathscr{H}_{-\gamma}}\right\},$$

(C) $$\gamma\|Q_0{}^-\varphi u\|^2_{m_0^- - 1,\mathscr{H}_{-\gamma}} + \sum_{k=0}^{m_0^- - 1} \langle Q_k{}^-\varphi u\rangle^2_{m_0^- - 1-k,\mathscr{H}_{-\gamma}}$$
$$\leqq C\left\{\frac{1}{\gamma}\|Au\|^2_{\mathscr{H}_{-\gamma}} + \|u\|^2_{m-1,\mathscr{H}_{-\gamma}}\right\}.$$

Then, we add the inequalities (A) and (C) term by term to obtain

$$\sum_{j=1}^{h}\sum_{k=m_j^+}^{m_j-1}\langle P_{jk}\varphi u\rangle^2_{m_j-1-k,\mathscr{H}_{-\gamma}}+\sum_{k=0}^{m_0^--1}\langle Q_k^-\varphi u\rangle^2_{m_0^--1-k,\mathscr{H}_{-\gamma}}$$

$$\leqq C\mu^2\sum_{j=1}^{h}\sum_{k=0}^{m_j^+-1}\langle P_{jk}\varphi u\rangle^2_{m_j-1-k,\mathscr{H}_{-\gamma}}+C_\mu\left\{\frac{1}{\gamma}\|Au\|^2_{\mathscr{H}_{-\gamma}}+\|u\|^2_{m-1,\mathscr{H}_{-\gamma}}\right\}.$$

When $R(X_0)\neq 0$, by the corollary to Lemma 3.2, we have

$$\sum_{k=0}^{m-1}\langle D_x^k\varphi u\rangle^2_{m-1-k,\mathscr{H}_{-\gamma}}\leqq C\Bigg\{\sum_{j=1}^{h}\sum_{k=m_j^+}^{m_j-1}\langle P_{jk}\varphi u\rangle^2_{m_j-1-k,\mathscr{H}_{-\gamma}}$$

$$+\sum_{k=0}^{m_0^--1}\langle Q_k^-\varphi u\rangle^2_{m_0^--1-k,\mathscr{H}_{-\gamma}}$$

$$+\sum_{j=1}^{m+}\langle B_j\varphi u\rangle^2_{m-1-r_j,\mathscr{H}_{-\gamma}}$$

$$+\sum_{k=0}^{m-1}\langle D_x^k u\rangle^2_{m-1-k-1/2,\mathscr{H}_{-\gamma}}\Bigg\},$$

therefore

$$\sum_{k=0}^{m-1}\langle D_x^k\varphi u\rangle^2_{m-1-k,\mathscr{H}_{-\gamma}}\leqq C\mu^2\sum_{j=1}^{h}\sum_{k=0}^{m_j^+-1}\langle P_{jk}\varphi u\rangle^2_{m_j-1-k,\mathscr{H}_{-\gamma}}$$

$$+C_\mu\Bigg\{\frac{1}{\gamma}\|Au\|^2_{\mathscr{H}_{-\gamma}}+\sum_{j=1}^{m+}\langle B_ju\rangle^2_{m-1-r_j,\mathscr{H}_{-\gamma}}$$

$$+\|u\|^2_{m-1,\mathscr{H}_{-\gamma}}\Bigg\}.$$

If we choose μ sufficiently small to make the left hand side of the above inequality absorb the first term of the right hand side, then, we have

$$\sum_{k=0}^{m-1}\langle D_x^k\varphi u\rangle^2_{m-1-k,\mathscr{H}_{-\gamma}}\leqq C\Bigg\{\frac{1}{\gamma}\|Au\|^2_{\mathscr{H}_{-\gamma}}+\sum_{j=1}^{m+}\langle B_ju\rangle^2_{m-1-r_j,\mathscr{H}_{-\gamma}}$$

$$+\|u\|^2_{m-1,\mathscr{H}_{-\gamma}}\Bigg\}.$$

We again add the inequalities (A), (B) and (C) term by term to obtain

$$(*)_\varphi\qquad \gamma\|\varphi u\|^2_{m-1,\mathscr{H}_{-\gamma}}+\sum_{k=0}^{m-1}\langle D_x^k\varphi u\rangle^2_{m-1-k,\mathscr{H}_{-\gamma}}$$

$$\leqq C\left\{\frac{1}{\gamma}\|Au\|^2_{\mathscr{H}_{-\gamma}}+\sum_{j=1}^{m_+}\langle B_ju\rangle^2_{m-1-r_j,\mathscr{H}_{-\gamma}}+\|u\|^2_{m-1,\mathscr{H}_{-\gamma}}\right\}.$$

Notice that in the foregoing argument we fixed

$$X_0 = (t_0, x_0, y_0; \tau_0, \eta_0), \quad (t_0, x_0, y_0) \in \mathbb{R}^{n+1}, \quad (\tau_0, \eta_0) \in \mathscr{K}_-,$$

where

$$\mathscr{K}_- = \{(\tau, \eta) \in \mathbb{C}^1 \times \mathbb{R}^{n-1} | \operatorname{Im} \tau \leqq 0, |\tau|^2 + |\eta|^2 = 1\}.$$

Recall that, according to our hypothesis, the coefficients of $\{A, B_j\}$ are constants outside the compact set K of $\mathbb{R}^{n+1}$. Therefore, for every point X of $K \times \mathscr{K}_-$ there is a small neighbourhood $U(X)$ such that $(*)_\varphi$ holds and a finite number of these $\{U(X_l)\}$ covers $K \times \mathscr{K}_-$, i.e.

$$K \times \mathscr{K}_- \subset \bigcup_{l=1}^{N} U(X_l) = \bigcup_{l=1}^{N} U_l.$$

We choose a localising operator φ_l for each U_l. Then we see that in the inequality $(*)_\varphi$ we can replace φ with φ_l. Considering all such φ_l, and adding the corresponding inequalities term by term, we have

$$(*)_\alpha \quad \gamma \|\alpha u\|^2_{m-1,\mathscr{H}-\gamma} + \sum_{k=0}^{m-1} \langle D_x{}^j \alpha u \rangle^2_{m-1-k,\mathscr{H}-\gamma}$$

$$\leqq C \left\{ \frac{1}{\gamma} \| A u \|^2_{\mathscr{H}-\gamma} + \sum_{j=1}^{m+} \langle B_j u \rangle^2_{m-1-r_j,\mathscr{H}-\gamma} + \|u\|^2_{m-1,\mathscr{H}-\gamma} \right\},$$

where $\alpha = \alpha(t, x, y)$ is a fixed infinitely differentiable function with supp $[\alpha]$ in a neighbourhood of K, and $\alpha = 1$ in K. Note that outside K $\{A, B_j\}$ becomes an operator $\{A^{(0)}, B_j{}^{(0)}\}$ with constant coefficients. For $\{A^{(0)}, B_j{}^{(0)}\}$ we can perform a localisation similar to the one above, this time confining ourselves to (τ, η)-space. For this reason, the compactness of $\mathscr{K}_-$ enables us to use a finite number of localisation operators so that

$$\gamma \|u\|^2_{m-1,\mathscr{H}-\gamma} + \sum_{k=0}^{m-1} \langle D_x{}^k u \rangle^2_{m-1-k,\mathscr{H}-\gamma}$$

$$\leqq C \left\{ \frac{1}{\gamma} \|A^{(0)} u\|^2_{\mathscr{H}-\gamma} + \sum_{j=1}^{m+} \langle B_j{}^{(0)} u \rangle^2_{m-1-r_j,\mathscr{H}-\gamma} \right\}.$$

Substituting $(1-\alpha)u$ for u we have

$(*)_{1-\alpha}$

$$\gamma \|(1-\alpha)u\|^2_{m-1,\mathscr{H}-\gamma} + \sum_{k=0}^{m-1} \langle D_x{}^k((1-\alpha)u) \rangle^2_{m-1-k,\mathscr{H}-\gamma}$$

$$\leqq C \left\{ \frac{1}{\gamma} \|A(1-\alpha)u)\|^2_{\mathscr{H}-\gamma} + \sum_{j=1}^{m+} \langle B_j((1-\alpha)u) \rangle^2_{m-1-r_j,\mathscr{H}-\gamma} \right\}$$

$$\leqq C' \left\{ \frac{1}{\gamma} \|Au\|^2_{\mathscr{H}-\gamma} + \sum_{j=1}^{m+} \langle B_j u \rangle^2_{m-1-r_j,\mathscr{H}-\gamma} + \|u\|^2_{m-1,\mathscr{H}-\gamma} \right\}.$$

Therefore, from $(*)_\alpha$ and $(*)_{1-\alpha}$ we obtain

$$(*)\quad \gamma\|u\|^2_{m-1,\mathscr{H}_{-\gamma}} + \sum_{k=0}^{m-1} \langle D_x{}^k u\rangle^2_{m-1-k,\mathscr{H}_{-\gamma}} \leqq C\left\{\frac{1}{\gamma}\|Au\|^2_{\mathscr{H}_{-\gamma}} + \sum_{j=1}^{m_+} \langle B_j u\rangle^2_{m-1-r_j,\mathscr{H}_{-\gamma}} + \|u\|^2_{m-1,\mathscr{H}_{-\gamma}}\right\}.$$

In the foregoing argument we assumed that $\gamma > 0$ was arbitrary. However, if we choose γ_0 sufficiently large and make $\gamma \geqq \gamma_0$, then the left hand side of the inequality absorbs the last term of the right hand side. This implies that for $\gamma \geqq \gamma_0$ we have

$$\gamma\|u\|^2_{m-1,\mathscr{H}_{-\gamma}} + \sum_{k=0}^{m-1} \langle D_x{}^k u\rangle^2_{m-1-k,\mathscr{H}_{-\gamma}} \leqq C\left\{\frac{1}{\gamma}\|Au\|^2_{\mathscr{H}_{-\gamma}} + \sum_{j=1}^{m_+} \langle B_j u\rangle^2_{m-1-r_j,\mathscr{H}_{-\gamma}}\right\}.$$

Summing up we have proved

Theorem 3 *(the energy inequality of Dirichlet type ($\mathscr{H}$))*
If $\{A, B_j\}$ satisfies Hypothesis (A′) and the uniform Lopatinski condition, then the energy inequality of Dirichlet type $(\mathscr{H})_\gamma$ is valid. ■

3.5

We take this opportunity to derive some results from the energy inequality of Dirichlet type $(\mathscr{H})_\gamma$. First of all, for energy inequality of Dirichlet type of higher order, we establish the following.

Lemma 3.7 *(the energy inequality of Dirichlet type ($\mathscr{H}$) for a higher order)*
If the energy inequality of Dirichlet type $(\mathscr{H})_\gamma$ holds for $\{A, B_j\}$ then for an arbitrary real number s,

$$\gamma\|\Lambda^s u\|^2_{m-1,\mathscr{H}_{-\gamma}} + \sum_{k=0}^{m-1} \langle D_x{}^k u\rangle^2_{m-1-k+s,\mathscr{H}_{-\gamma}} \leqq C_s\left\{\frac{1}{\gamma}\|\Lambda^s Au\|^2_{\mathscr{H}_{-\gamma}} + \sum_{j=1}^{m_+} \langle B_j u\rangle^2_{m-1-r_j+s,\mathscr{H}_{-\gamma}}\right\},\quad \gamma \geqq \gamma_s.$$

In particular, if s is a positive integer, then

$$\gamma\|u\|^2_{m-1+s,\mathscr{H}_{-\gamma}} + \sum_{k=0}^{m-1+s} \langle D_x{}^k u\rangle^2_{m-1-k+s,\mathscr{H}_{-\gamma}} \leqq C_s\left\{\frac{1}{\gamma}\|Au\|^2_{s,\mathscr{H}_{-\gamma}} + \sum_{j=1}^{m_+} \langle B_j u\rangle^2_{m-1-r_j+s,\mathscr{H}_{-\gamma}}\right\},\quad \gamma \geqq \gamma_s.$$

Proof
Replacing u with $\Lambda^s u$ in Theorem 3, we have

$$\gamma\|\Lambda^s u\|^2_{m-1,\mathscr{H}_{-\gamma}} + \sum_{k=0}^{m-1} \langle D_x{}^k u\rangle^2_{m-1-k+s,\mathscr{H}_{-\gamma}}$$

$$\leqq C\left\{\frac{1}{\gamma}\|A\Lambda^s u\|^2_{\mathscr{H}_{-\gamma}} + \sum_{j=1}^{m_+} \langle B_j\Lambda^s u\rangle^2_{m-1-r_j,\mathscr{H}_{-\gamma}}\right\}$$

$$\leqq C'\left\{\frac{1}{\gamma}\|\Lambda^s A u\|^2_{\mathscr{H}_{-\gamma}} + \sum_{j=1}^{m_+} \langle B_j u\rangle^2_{m-1-r_j+s,\mathscr{H}_{-\gamma}}\right.$$

$$\left. + \frac{1}{\gamma}\|\Lambda^s u\|^2_{m-1,\mathscr{H}_{-\gamma}} + \sum_{k=0}^{m-1} \langle D_x{}^k u\rangle^2_{m-1-k+s-1,\mathscr{H}_{-\gamma}}\right\}.$$

From this, we see that if we take γ_0 as a sufficiently large γ_s, then for $\gamma \geqq \gamma_s$ the first inequality holds. If s is a positive integer, then from the equations, we know that the derivatives with respect to x are

$$D_x{}^m u = Au - (a_1 D_x{}^{m-1} + \ldots + a_m)u,$$
$$D_x{}^{m+1} u = D_x Au - D_x(a_1 D_x{}^{m-1} + \ldots + a_m)u,$$
$$\vdots$$

so that

$$\|\Lambda^{s-1} D_x{}^m u\|_{\mathscr{H}_{-\gamma}} \leqq \|\Lambda^{s-1} Au\|_{\mathscr{H}_{-\gamma}} + C\|\Lambda^s u\|_{m-1,\mathscr{H}_{-\gamma}},$$

$$\|\Lambda^{s-2} D_x{}^{m+1} u\|_{\mathscr{H}_{-\gamma}} \leqq \|\Lambda^{s-2} D_x Au\|_{\mathscr{H}_{-\gamma}} + C\{\|\Lambda^{s-1} D_x{}^m u\|_{\mathscr{H}_{-\gamma}} + \|\Lambda^s u\|_{m-1,\mathscr{H}_{-\gamma}}\}.$$

We can therefore estimate

$$\sum_{k=m}^{m-1+s} \|\Lambda^{m-1+s-k} D_x{}^k u\|^2_{\mathscr{H}_{-\gamma}} \leqq C\{\|Au\|^2_{s-1,\mathscr{H}_{-\gamma}} + \|\Lambda^s u\|^2_{m-1,\mathscr{H}_{-\gamma}}\}$$

$$\leqq C\left\{\frac{1}{\gamma^2}\|Au\|^2_{s,\mathscr{H}_{-\gamma}} + \|\Lambda^s u\|^2_{m-1,\mathscr{H}_{-\gamma}}\right\}.$$

The result which we obtained at the beginning of the proof can be applied to the second term on the right hand side of the above inequality. Similarly, for the boundary norm we conclude that

$$\sum_{k=m}^{m-1+s} \langle D_x{}^k u\rangle^2_{m-1+s-k,\mathscr{H}_{-\gamma}}$$

$$\leqq C\left\{\langle Au\rangle^2_{s-1,\mathscr{H}_{-\gamma}} + \sum_{k=0}^{m-1} \langle D_x{}^k u\rangle^2_{m-1+s-k,\mathscr{H}_{-\gamma}}\right\}$$

$$\leqq C\left\{\frac{1}{\gamma}\|Au\|^2_{s,\mathscr{H}_{-\gamma}} + \sum_{k=0}^{m-1} \langle D_x{}^k u\rangle^2_{m-1+s-k,\mathscr{H}_{-\gamma}}\right\}. \qquad \blacksquare$$

We can define $\mathscr{H}_{-\gamma,-\gamma'}$ which contains $\mathscr{H}_{-\gamma}, \mathscr{H}_{-\gamma'}(\gamma \leqq \gamma')$. We take an infinitely differentiable function α such that $\alpha(t) = 1$ for $t > 0$ and $\alpha(t) = 0$ for $t < -1$. Using this for synthesising the weight functions $e^{-\gamma t}$ and $e^{-\gamma' t}$, we obtain a function

$$e_{-\gamma,-\gamma'}(t) = \alpha(t)e^{-\gamma t} + \alpha(-t)e^{-\gamma' t}.$$

This function satisfies the following condition:

$$e_{-\gamma,-\gamma'}(t) = \begin{cases} e^{-\gamma t}, & t > 1, \\ e^{-\gamma' t}, & t < -1. \end{cases}$$

Considering the new function as a weight function we write

$$\mathscr{H}_{-\gamma,-\gamma'}{}^k = \{f \mid e_{-\gamma,-\gamma'}(t)f \in H^k\},$$
$$\|f\|_{k,\mathscr{H}_{\gamma-\gamma'}} = \|e_{-\gamma,-\gamma'}(t)f\|_{k,H}.$$

Then using the $\mathscr{H}_{-\gamma,-\gamma'}$-norm instead of the $\mathscr{H}_{-\gamma}$-norm, we can establish a similar energy inequality. To see this we prove

Lemma 3.8

If the energy ineuqality $(\mathscr{H})_\gamma$ holds for $\{A, B_j\}$, then there exists $\gamma_\delta > 0$ for an arbitrary $\delta > 0$ such that

$$\gamma\|u\|^2_{m-1,\mathscr{H}_{-\gamma,-\gamma}} + \sum_{k=0}^{m-1} \langle D_x{}^k u\rangle^2_{m-1-k,\mathscr{H}_{-\gamma},\mathscr{H}_{-\gamma'}}$$
$$\leqq C\left\{\frac{1}{\gamma}\|Au\|^2_{\mathscr{H}_{-\gamma,-\gamma'}} + \sum_{j=1}^{m+} \langle B_j u\rangle^2_{m-1-r_j,\mathscr{H}_{-\gamma,-\gamma'}}\right\},$$

where $\gamma_\delta \leqq \gamma \leqq \gamma' \leqq \gamma + \delta$.

Proof

We can establish that

$$e_{-\gamma,-\gamma'}(t) = e^{-\gamma t}e_{0,-\delta'}(t) \quad (\gamma' = \gamma + \delta', 0 \leqq \delta' \leqq \delta).$$

Setting

$$e(t) = e_{0,-\delta'}(t),$$

we have an equality

$$\|u\|_{k,\mathscr{H}_{-\gamma,-\gamma'}} = \|e(t)u\|_{k,\mathscr{H}_{-\gamma}}.$$

If we substitute $e(t)u$ for u in $(\mathscr{H})_\gamma$, we have

$$\gamma\|u\|^2_{m-1,\mathscr{H}_{-\gamma,-\gamma}} + \sum_{k=0}^{m-1} \langle D_x{}^k u\rangle^2_{m-1-k,\mathscr{H}_{-\gamma,-\gamma'}}$$
$$\leqq C\left\{\frac{1}{\gamma}\|A(e(t)u)\|^2_{\mathscr{H}_{-\gamma}} + \sum_{j=1}^{m+} \langle B_j(e(t)u)\rangle^2_{m-1-r_j,\mathscr{H}_{-\gamma}}\right\}.$$

On the other hand we have

$$D_t(e(t)u) = e(t)\left(D_t - i\frac{e'(t)}{e(t)}\right)u$$

therefore

$$\begin{aligned} A(t,x,y;D_t,D_x,D_y)(e(t)u) &= e(t)A\left(t,x,y;D_t - i\frac{e'(t)}{e(t)},D_x,D_y\right)u \\ &= e(t)\{A(t,x,y;D_t,D_x,D_y) \\ &\qquad + A^{(\delta')}(t,x,y;D_t,D_x,D_y)\}u \end{aligned}$$

and

$$B_j(t,y;D_t,D_x,D_y)(e(t)u) = e(t)\{B_j(t,y;D_t,D_x,D_y) + B_j^{(\delta')}(t,y;D_t,D_x,D_y)\}.$$

Hence we obtain

$$(**) \qquad \begin{aligned} &\gamma\|u\|^2_{m-1,\mathscr{H}_{-\gamma,-\gamma'}} + \sum_{k=0}^{m-1}\langle D_x{}^k u\rangle^2_{m-1-k,\mathscr{H}_{-\gamma,-\gamma'}} \\ &\leqq C\left\{\frac{1}{\gamma}\|Au\|^2_{\mathscr{H}_{-\gamma,-\gamma'}} + \sum_{j=1}^{m_+}\langle B_j u\rangle^2_{m-1-r_j,\mathscr{H}_{-\gamma,-\gamma'}}\right\} \\ &\quad + C\left\{\frac{1}{\gamma}\|A^{(\delta')}u\|^2_{\mathscr{H}_{-\gamma,-\gamma'}} + \sum_{j=1}^{m_+}\langle B_j{}^{(\delta')}u\rangle^2_{m-1-r_j,\mathscr{H}_{-\gamma,-\gamma'}}\right\}. \end{aligned}$$

Let us write I for the second term on the right hand side of the above inequality. We wish to obtain an estimation of I. To do this, we first observe that

$$e(t) = \begin{cases} 1, & t > 1, \\ e^{-\delta' t}, & t < -1, \end{cases}$$

so that

$$\frac{e'(t)}{e(t)} = \begin{cases} 0, & t > 1, \\ -\delta', & t < -1. \end{cases}$$

At the same time, we can establish that

$$\left|\left(\frac{d}{dt}\right)^k\left(\frac{e'(t)}{e(t)}\right)\right| \leqq C_{k,\delta}, \quad 0 \leqq \delta' \leqq \delta,$$

for $|t| \leqq 1$. (This is clear from the definition of $e(t)$. Therefore, for $0 < \delta' < \delta$ we have

$$\|A^{(\delta')}u\|_{\mathscr{H}_{-\gamma,-\gamma'}} \leqq C_\delta\|u\|_{m-1,\mathscr{H}_{-\gamma,-\gamma'}},$$

$$\langle B_j{}^{(\delta')}u\rangle_{m-1-r_j,\mathscr{H}_{-\gamma,-\gamma'}} \leqq C_\delta\sum_{k=0}^{m-2}\langle D_x{}^k u\rangle_{m-2-k,\mathscr{H}_{-\gamma,-\gamma'}}.$$

Hence we obtain the estimation

$$I \leqq C_\delta\left\{\frac{1}{\gamma}\|u\|^2_{m-1,\mathscr{H}_{-\gamma,-\gamma'}} + \sum_{k=0}^{m-2}\langle D_x{}^k u\rangle^2_{m-2-k,\mathscr{H}_{-\gamma,-\gamma'}}\right\}$$

$$\leqq \frac{C_\delta}{\gamma^2}\left\{\gamma\|u\|^2_{m-1,\mathscr{H}_{-\gamma,-\gamma'}} + \sum_{k=0}^{m-1}\langle D_x{}^k u\rangle^2_{m-1-k,\mathscr{H}_{-\gamma,-\gamma'}}\right\}.$$

From this we conclude that if we choose γ_δ sufficiently large for any δ, then for $\gamma \geq \gamma_\delta$ the left hand side of the inequality $(**)$ absorbs I. ■

4. The existence theorems

In Chapter 2 we directly showed the existence of solutions, and then proved their uniqueness by way of the existence of solutions of the adjoint problem. In this chapter, on the other hand, we have obtained an energy inequality of Dirichlet type $(\mathscr{H})$ which guarantees the uniqueness of solutions in $(\mathscr{H})$-space. In the following subsection, we assume that $\{A, B_j\}$ satisfies the uniform Lopatinski condition. Then, by demonstrating the validity of the energy inequality of Dirichlet type $(\mathscr{H})$ for the adjoint problem, we establish the existence of solutions for the original problem.

4.1

In Chapter 2, we saw that in the case of constant coefficients we can derive the adjoint problem from the original problem. In the case of variable coefficients we can derive the adjoint problem from the original one by the same procedure. But this time we can not use the commutativity of a differential operator, and consequently, instead of the conjugacy between symbols, we have a problem arising from the appearance of some lower-order terms in the differential operator.

Lemma 4.1

For $C(y,\eta) = (c_{ij}(y,\eta))_{i,j=1,\ldots,m}$ let $c_{ij}(y,\eta)$ be polynomials of degree $r_i - s_j$ with respect to η satisfying the condition

$$\sum_{i=1}^{m} r_i = \sum_{j=1}^{m} s_j.$$

Write $c_{ij}{}^{(0)}(y,\eta)$ for the principal parts of $c_{ij}(y,\eta)$ with respect to η, and $C^{(0)}(y,\eta) = (c_{ij}{}^{(0)}(y,\eta))_{ij}$. Assume $\det(C^{(0)}(y,\eta)) = c(y) \neq 0$. Then, there exists

$$C'(y,\eta) = (c_{ij}{}'(y,\eta))_{i,j=1,\ldots,m}$$

such that

$$C'(y,D_y)\ C(y,D_y) = I(\text{identity}),$$

where the $c_{ij}{}'(y,\eta)$ are polynomials of degree $s_i - r_j$ with respect to η.

Proof

We decompose $C(y,\eta)$ according to degree, and write

$$C(y,\eta) = C^{(0)}(y,\eta) + C^{(1)}(y,\eta) + \ldots,$$
$$C^{(k)}(y,\eta) = (c_{ij}^{(k)}(y,\eta))_{i,j=1,\ldots,m},$$

where the $c_{ij}^{(k)}(y,\eta)$ are homogeneous polynomials of degree $r_i - s_j - k$ with respect to η. Similarly, by decomposition of $C'(y,\eta)$ we obtain

$$C'(y,\eta) = C'^{(0)}(y,\eta) + C'^{(1)}(y,\eta) + \ldots,$$
$$C'^{(k)}(y,\eta) = (c_{ij}'^{(k)}(y,\eta))_{i,j=1,\ldots,m},$$

where the $c_{ij}'^{(k)}(y,\eta)$ are homogeneous polynomials of degree $s_i - r_j - k$ with respect to η. Using these expansions we rewrite the condition which is to be satisfied as

$$\begin{aligned} I &= C'(y,\mathrm{D}_y)C(y,\mathrm{D}_y) \\ &= \{C'^{(0)}(y,\mathrm{D}_y) + C^{(1)}(y,\mathrm{D}_y) + \ldots\}\{C^{(0)}(y,\mathrm{D}_y) + C^{(1)}(y,\mathrm{D}_y) + \ldots\} \\ &= C'^{(0)}(y,\mathrm{D}_y)C^{(0)}(y,\mathrm{D}_y) + \{C'^{(1)}(y,\mathrm{D}_y)C^{(0)}(y,\mathrm{D}_y) \\ &\quad + C'^{(0)}(y,\mathrm{D}_y)C^{(1)}(y,\mathrm{D}_y)\} + \{C'^{(2)}(y,\mathrm{D}_y)C^{(0)}(y,\mathrm{D}_y) \\ &\quad + C'^{(1)}(y,\mathrm{D}_y)C^{(1)}(y,\mathrm{D}_y) + C'^{(0)}(y,\mathrm{D}_y)C^{(2)}(y,\mathrm{D}_y)\} + \ldots. \end{aligned}$$

In addition we can see that

$$\begin{aligned} C'^{(k)}(y,\mathrm{D}_y)C^{(l)}(y,\mathrm{D}_y) &= C'^{(k)}(y,\mathrm{D}_y)\circ C^{(l)}(y,\mathrm{D}_y) \\ &\quad + \sum_{|\nu|=1} (\mathrm{D}_\eta{}^\nu C'^{(k)})(y,\mathrm{D}_y)\circ(\partial_y{}^\nu C^{(l)})(y,\mathrm{D}_y) \\ &\quad + \sum_{|\nu|=2} \frac{1}{\nu!}(\mathrm{D}_\eta{}^\nu C'^{(k)})\,(y,\mathrm{D}_y)\circ(\partial_y{}^\nu C^{(l)})(y,\mathrm{D}_y) + \ldots \end{aligned}$$

so that the condition becomes

$$C'^{(0)}(y,\mathrm{D}_y)\circ C^{(0)}(y,\mathrm{D}_y) = I,$$

$$\begin{aligned} &C'^{(1)}(y,\mathrm{D}_y)\circ C^{(0)}(y,\mathrm{D}_y) + C'^{(0)}(y,\mathrm{D}_y)\circ C^{(1)}(y,\mathrm{D}_y) \\ &\quad + \sum_{|\nu|=1} (\mathrm{D}_\eta{}^\nu C'^{(0)})(y,\mathrm{D}_y)\circ(\partial_y{}^\nu C^{(0)})(y,\mathrm{D}_y) = 0, \end{aligned}$$

$$\begin{aligned} &\{C'^{(2)}(y,\mathrm{D}_y)\circ C^{(0)}(y,\mathrm{D}_y) + C'^{(1)}(y,\mathrm{D}_y)\circ C^{(1)}(y,\mathrm{D}_y) + C'^{(0)}(y,\mathrm{D}_y)\circ C^{(2)}(y,\mathrm{D}_y)\} \\ &\quad + \sum_{|\nu|=1} \{(\mathrm{D}_\eta{}^\nu C'^{(1)}(y,\mathrm{D}_y)\circ(\partial_y{}^\nu C^{(0)})(y,\mathrm{D}_y) \\ &\quad + (\mathrm{D}_\eta{}^\nu C'^{(0)})(y,\mathrm{D}_y)\circ(\partial_y{}^\nu C^{(1)})(y,\mathrm{D}_y)\} \\ &\quad + \sum_{|\nu|=2} \frac{1}{\nu!}(\mathrm{D}_n{}^\nu C'^{(0)})(y,\mathrm{D}_y)\circ(\partial_y{}^\nu C^{(0)})(y,\mathrm{D}_y) = 0, \end{aligned}$$

Hence, we can subsequently determine $C'^{(0)}(y,\eta), C'^{(1)}(y,\eta), \ldots$ by

$$C'^{(0)}(y,\eta) = C^{(0)}(y,\eta)^{-1},$$

$$C'^{(1)}(y,\eta) = -\Big\{C'^{(0)}(y,\eta)C^{(1)}(y,\eta) + \sum_{|\nu|=1} (D_\eta{}^\nu C'^{(0)})(y,\eta)(\partial_y{}^\nu C^{(0)}(y,\eta)\Big\}C^{(0)}(y,\eta)^{-1},$$

$$\vdots$$

Then, $C'^{(0)}(y,\eta)$ which is determined by the first of the above equations, has homogeneous polynomials $c_{ij}'^{(0)}(y,\eta)$ of degree $s_i - r_j$ as its components. Therefore the components $c_{ij}'^{(k)}(y,\eta)$ of the subsequent $C'^{(k)}(y,\eta)$ are homogeneous polynomials of degree $s_i - r_j - k$. ■

Now, for $\{B_j(t,y;\tau,\xi,\eta)\}_{j=1,\ldots,m_+}$, by the annexation of

$$\{B_j(t,y;\tau,\xi,\eta) = \xi^{r_j}\}_{j=m_+ +1,\ldots,m}$$

and by using condition (iii) of Hypothesis (A′) we obtain

$$\begin{bmatrix} B_1(t,y;\tau,\xi,\eta) \\ \vdots \\ B_m(t,y;\tau,\xi,\eta) \end{bmatrix} = \begin{bmatrix} b_{11}(t,y;\tau,\eta)\ldots b_{1m}(t,y;\tau,\eta) \\ \vdots \\ b_{m1}(t,y;\tau,\eta)\ldots b_{mm}(t,y;\tau,\eta) \end{bmatrix} \begin{bmatrix} \xi^{m-1} \\ \vdots \\ 1 \end{bmatrix}$$

$$= \mathscr{B}(t,y;\tau,\eta) \begin{bmatrix} \xi^{m-1} \\ \vdots \\ 1 \end{bmatrix},$$

where the $b_{ij}(t,y;\tau,\eta)$ are polynomials of degree $r_i - (m-j)$ with respect to (τ,η) and

$$\det(\mathscr{B}(t,y;\tau,\eta)) = b(t,y) \neq 0.$$

By Lemma 4.1, we see that there exists $\mathscr{B}'(t,y;\tau,\eta)$ such that

$$\mathscr{B}'(t,y;D_t,D_y)\mathscr{B}(t,y;D_t,D_y) = I.$$

Therefore we apply the operator $\mathscr{B}'(t,y;D_t,D_y)$ on the left of

$$\begin{bmatrix} B_1(t,y;D_t,D_x,D_y) \\ \vdots \\ B_m(t,y;D_t,D_x,D_y) \end{bmatrix} = \mathscr{B}(t,y;D_t,D_y) \begin{bmatrix} D_x{}^{m-1} \\ \vdots \\ 1 \end{bmatrix}$$

to obtain

$$\begin{bmatrix} D_x{}^{m-1} \\ \vdots \\ 1 \end{bmatrix} = \mathscr{B}'(t,y;D_t,D_y) \begin{bmatrix} B_1(t,y;D_t,D_x,D_y) \\ \vdots \\ B_m(t,y;D_t,D_x,D_y) \end{bmatrix}.$$

Also, for

$$A(t,x,y;D_t,D_x,D_y) = \sum_{j+k+|\nu|\leqq m} a_{jk\nu}(t,x,y)D_t^jD_x^kD_y^\nu$$

$$= \sum_{k=0}^{m} a_{m-k}(t,x,y;D_t,D_y)D_x^k,$$

we have its formal adjoint of the L^2-inner product given by

$$A^*(t,x,y;D_t,D_x,D_y) = \sum_{j+k+|\nu|\leqq m} D_t^jD_x^kD_y^\nu\overline{a_{jk\nu}(t,x,y)}$$

$$= \sum_{k=0}^{m} D_x^k a_{m-k}{}^*(t,x,y;D_t,D_y).$$

We write (,) for the L^2-inner product of the half-space $\mathbb{R}_+{}^{n+1} = \{x>0, (t,y)\in\mathbb{R}^n\}$, and $\langle\,,\rangle$ for the L^2-inner product at the boundary $\mathbb{R}^n = \{x=0, (t,y)\in\mathbb{R}^n\}$. Using this notation, for $u, v\in\mathscr{D}(\overline{\mathbb{R}_+{}^{n+1}})$ we have

$$\begin{aligned}
&(A(t,x,y;D_t,D_x,D_y)u,v) - (u, A^*(t,x,y;D_t,D_x,D_y)v)\\
&= \sum\{(a_{m-k}D_x^k u, v) - (u, D_x^k a_{m-k}{}^*v)\}\\
&= \sum\{(D_x^k u, a_{m-k}{}^*v) - (u, D_x^k a_{m-k}{}^*v)\}\\
&= i\sum\{\langle D_x^{k-1}u\cdot a_{m-k}{}^*v\rangle + \langle D_x^{k-2}u, D_x a_{m-k}{}^*v\rangle\\
&\quad + \ldots + \langle u, D_x^{k-1}a_{m-k}{}^*v\rangle\}
\end{aligned}$$

$$= i\left\langle \begin{bmatrix} D_x^{m-1}\\ D_x^{m-2}\\ \vdots\\ 1\end{bmatrix}u, \begin{bmatrix} a_0{}^*\\ D_x a_0{}^* + a_1{}^*\\ \vdots\\ D_x^{m-1}a_0{}^* + D_x^{m-2}a_1{}^* + \ldots + a_{m-1}{}^*\end{bmatrix}v\right\rangle$$

$$= i\left\langle \begin{bmatrix} D_x^{m-1}\\ D_x^{m-2}\\ \vdots\\ 1\end{bmatrix}u, \begin{bmatrix} A^{(0)*}\\ A^{(1)*}\\ \vdots\\ A^{(m-1)*}\end{bmatrix}v\right\rangle$$

and since we know that

$$\begin{bmatrix} D_x^{m-1}\\ \vdots\\ 1\end{bmatrix} = \mathscr{B}'\begin{bmatrix} B_1\\ \vdots\\ B_m\end{bmatrix},$$

obviously

$$(Au,v) - (u,A^*v) = i\left\langle \begin{bmatrix} B_1\\ \vdots\\ B_m\end{bmatrix}u, \mathscr{B}'^*\begin{bmatrix} A^{(0)*}\\ \vdots\\ A^{(m-1)*}\end{bmatrix}v\right\rangle$$

Setting

$$(A^{(0)}(t,0,y;\mathrm{D}_t,\mathrm{D}_x,\mathrm{D}_y),\ldots,A^{(m-1)}(t,0,y;\mathrm{D}_t,\mathrm{D}_x,\mathrm{D}_y))\mathscr{B}'(t,y;\mathrm{D}_t,\mathrm{D}_y)$$
$$=(B_1'(t,y;\mathrm{D}_t,\mathrm{D}_x,\mathrm{D}_y),\ldots,B_m'(t,y;\mathrm{D}_t,\mathrm{D}_x,\mathrm{D}_y))$$

we conclude that

$$(Au,v)-(u,A^*v)=\mathrm{i}\left\langle\begin{bmatrix}B_1\\ \vdots\\ B_m\end{bmatrix}u,\begin{bmatrix}B_1'^*\\ \vdots\\ B_m'^*\end{bmatrix}v\right\rangle.$$

Summing up our argument, we have the following lemma.

Lemma 4.2 *(Green's formula)*

If $u,v\in\mathscr{D}(\overline{\mathbb{R}_+{}^{n+1}})$, then

$$(A(t,x,y;\mathrm{D}_t,\mathrm{D}_x,\mathrm{D}_y)u,v)_{L^2(\mathbb{R}_+{}^{n+1})}-(u,A^*(t,x,y;\mathrm{D}_t,\mathrm{D}_x,\mathrm{D}_y)v)_{L^2(\mathbb{R}_+{}^{n+1})}$$
$$=\mathrm{i}\sum_{j=1}^{m}\langle B_j(t,y;\mathrm{D}_t,\mathrm{D}_x,\mathrm{D}_y)u,B_j'^*(t,y;\mathrm{D}_t,\mathrm{D}_x,\mathrm{D}_y)v\rangle_{L^2(\mathbb{R}^n),x=0},$$

where $A^*, B_j'^*$ are the formal adjoints of the L^2-inner products of A, B_j' respectively.

Now, recall the definition of $B_j'(t,y;\tau,\xi,\eta)$ (see above). They are polynomials of degree $r_j'=m-1-r_j$ with respect to (τ,ξ,η). If we write the principal part of B_j' as $B_{j0}'(t,y;\tau,\xi,\eta)$, then

$$\{B_{10}(t,y;\tau,\xi,\eta),\ldots,B_{m0}(t,y;\tau,\xi,\eta)\},\{B_{10}'(t,y;\tau,\xi,\eta),\ldots,B_{m0}'(t,y;\tau,\xi,\eta)\}$$

become adjoint with respect to $A_0(t,0,y;\tau,\xi,\eta)$ when we regard them as polynomials of ξ. That is

$$\frac{A_0(t,0,y;\tau,\xi,\eta)-A_0(t,0,y;\tau,\bar{\xi},\eta)}{\xi-\bar{\xi}}=\sum_{j=1}^{m}B_{j0}(t,y;\tau,\xi,\eta)B_{j0}'(t,y;\tau,\bar{\xi},\eta).$$

With these facts in mind, for the original initial value problem

$$(\mathrm{P})\quad\begin{cases}A(t,x,y;\mathrm{D}_t,\mathrm{D}_x,\mathrm{D}_y)u=f, & (t,x,y)\in(0,T)\times\mathbb{R}_+{}^n,\\ B_j(t,y;\mathrm{D}_t,\mathrm{D}_x,\mathrm{D}_y)u|_{x=0}=g_j\quad(j=1,\ldots,m_+), & (t,y)\in(0,T)\times\mathbb{R}^{n-1},\\ D_t{}^ju|_{t=0}=u_j\quad(j=0,\ldots,m-1), & (x,y)\in\mathbb{R}_+{}^n,\end{cases}$$

we define its adjoint problem as

$$(\mathrm{P}^*)\quad\begin{cases}A^*(t,x,y;\mathrm{D}_t,\mathrm{D}_x,\mathrm{D}_y)v=\varphi,\quad(t,x,y)\in(0,T)\times\mathbb{R}_+{}^n,\\ B_j'^*(t,y;\mathrm{D}_t,\mathrm{D}_x,\mathrm{D}_y)v|_{x=0}=\psi_j\quad(j=m_++1,\ldots,m),\\ \qquad(t,y)\in(0,T)\times\mathbb{R}^{n-1},\\ \mathrm{D}_t{}^jv|_{t=T}=v_j\quad(j=0,\ldots,m-1),\quad(x,y)\in\mathbb{R}_+{}^n,\end{cases}$$

where we note that

$$A_0{}^*(t,x,y;\tau,\xi,\eta) = A_0(t,x,y;\tau,\xi,\eta),$$
$$B_{j0}{}'^*(t,y;\tau,\xi,\eta) = \overline{B_{j0}{}'(t,y;\bar{\tau},\bar{\xi},\bar{\eta})},$$
$$\frac{A_0(t,0,y;\tau,\xi,\eta) - A_0(t,0,y;\tau,\bar{\xi},\eta)}{\xi - \bar{\xi}} = \sum_{j=1}^{m} B_{j0}(t,y;\tau,\xi,\eta) B_{j0}{}'(t,y;\tau,\bar{\xi},\eta).$$

Then, since (P) satisfies Hypothesis (A′), so does (P*).

Furthermore, if we write $R_0(t,y;\tau,\eta)$ for the Lopatinski determinant with respect to $\{A_0(t,0,y;\tau,\xi,\eta),\ B_{j0}(t,y;\tau,\xi,\eta)\ \ (j=1,\dots,m_+)\}$, and $R_0{}^*(t,y;\tau,\eta)$ for the Lopatinski determinant with respect to $\{A_0{}^*(t,0,y;\tau,\xi,\eta),\ B_{j0}{}^*(t,y;\tau,\xi,\eta)\ \ (j=m_+ + 1,\dots,m)\}$, then from the result obtained in Chapter 2, we have

$$R_0{}^*(t,y;\tau,\eta) = c(t,y)\overline{R_0(t,y;\bar{\tau},\bar{\eta})}, \quad c(t,y) \neq 0.$$

Therefore, if we assume the uniform Lopatinski condition

$$R_0(t,y;\tau,\eta) \neq 0, \quad \operatorname{Im}\tau \leqq 0, \quad \eta \in \mathbb{R}^{n-1}, \quad (\tau,\eta) \neq (0,0)$$

for (P), then for (P*) we see that the uniform Lopatinski condition

$$R_0{}^*(t,y;\tau,\eta) \neq 0, \quad \operatorname{Im}\tau \geqq 0, \quad \eta \in \mathbb{R}^{n-1}, \quad (\tau,\eta) \neq (0,0)$$

holds. This shows that we can apply almost the same argument to (P*) as we did to (P) (see §3). Writing

$$\|\ \ \|_{k,\mathscr{H}_{-\gamma(1,0,0)}(\mathbb{R}_+{}^{n+1})} = \|\ \ \|_{k,\mathscr{H}_\gamma},$$
$$\|\ \ \|_{k,\mathscr{H}_{-\gamma(1,0)}(\mathbb{R}^n),x=0} = \langle\ \ \rangle_{k,\mathscr{H}_\gamma},$$

we have

Lemma 4.3
If (P) satisfies the uniform Lopatinski condition, then the energy inequality of Dirichlet type ($\mathscr{H}^*$) holds, i.e. for $\gamma \geqq \gamma_0$

$$\gamma\|v\|^2_{m-1,\mathscr{H}_\gamma} + \sum_{k=0}^{m-1} \langle \mathrm{D}_x{}^k v\rangle^2_{m-1-k,\mathscr{H}_\gamma}$$
$$\leqq C\left\{\frac{1}{\gamma}\|A^*v\|^2_{\mathscr{H}_\gamma} + \sum_{j=m_+ + 1}^{m} \langle B_j{}'^* v\rangle^2_{m-1-r_j{}',\mathscr{H}_\gamma}\right\}.$$

Recall that the singular integral operator for the symbol

$$\Lambda^s(\tau,\eta) = (|\tau|^2 + |\eta|^2)^{s/2}$$

operates over $\mathscr{S}_{-\gamma(1,0)}{}'(\mathbb{R}^n)$ and takes the form

$$\Lambda^s = \mathscr{L}^{-1}_{(\tau,\eta)\to(t,y)} \Lambda^s(\tau,\eta) \mathscr{L}_{(t,y)\to(\tau,\eta)}, \quad \operatorname{Im}\tau = -\gamma.$$

However the same symbol $\Lambda^s(\tau,\eta)$ defines another singular integral operator Λ'^s which operates over $\mathscr{S}_{\gamma(1,0)}'(\mathbb{R}^n)$, i.e.

$$\Lambda'^s = \mathscr{L}^{-1}_{(\tau,\eta)\to(t,y)}\Lambda^s(\tau,\eta)\mathscr{L}_{(t,y)\to(\tau,\eta)}, \quad \operatorname{Im}\tau=\gamma.$$

In this case we can show

Lemma 4.3′

For an arbitrary real number s, and $\gamma \geqq \gamma_s$,

$$\gamma\|\Lambda'^s v\|^2_{m-1,\mathscr{H}_\gamma} + \sum_{k=0}^{m-1} \langle D_x{}^k v\rangle^2_{m-1-k+s,\mathscr{H}_\gamma}$$

$$\leqq C_s\left\{\frac{1}{\gamma}\|\Lambda'^s A^* v\|^2_{\mathscr{H}_\gamma} + \sum_{j=m_++1}^{m} \langle B_j'^* v\rangle^2_{m-1-r_j'+s,\mathscr{H}_\gamma}\right\},$$

where

$$\langle u\rangle_{s,\mathscr{H}_\gamma} = \langle \Lambda'^s u\rangle_{\mathscr{H}_\gamma}.$$

In particular, if $s \geqq 0$ is an integer, then

$$\gamma\|v\|^2_{m-1+s,\mathscr{H}_\gamma} + \sum_{k=0}^{m-1+s} \langle D_x{}^k v\rangle^2_{m-1-k+s,\mathscr{H}_\gamma}$$

$$\leqq C_s\left\{\frac{1}{\gamma}\|A^* v\|^2_{s,\mathscr{H}_\gamma} + \sum_{j=m_++1}^{m} \langle B_j'^* v\rangle^2_{m-1-r_j'+s,\mathscr{H}_\gamma}\right\}.$$

4.2

We wish to introduce a Hilbert space V^s relating to the adjoint problem (P*). To this end we write

$$|||v|||_s{}^2 = \|\Lambda'^s A^* v\|^2_{\mathscr{H}_\gamma} + \sum_{j=m_++1}^{m} \langle B_j'^* v\rangle^2_{m-1-r_j'+s,\mathscr{H}_\gamma}.$$

By Lemma 4.3′ we can establish the inequality

$$c_{s,\gamma}\left\{\|\Lambda'^{s-1} v\|^2_{m,\mathscr{H}_\gamma} + \sum_{k=0}^{m-1} \langle D_x{}^k v\rangle^2_{m-1-k+s,\mathscr{H}_\gamma}\right\}$$

$$\leqq |||v|||_s{}^2 \leqq C_{s,\gamma}\left\{\|\Lambda'^s v\|^2_{m,\mathscr{H}_\gamma} + \sum_{k=0}^{m-1} \langle D_x{}^k v\rangle^2_{m-1-k+s,\mathscr{H}_\gamma}\right\},$$

so that if we put

$$V^s = \{v \mid \Lambda'^{s-1} v \in \mathscr{H}_\gamma{}^m(R_+{}^{n+1}),\ \ |||v|||_s < +\infty\},$$

then we obtain a Hilbert space with the inner product defined as

$$((v,w))_s = (\Lambda'^s A^* v, \Lambda'^s A^* w)_{\mathscr{H}_\gamma} + \sum_{j=m_++1}^{m} \langle B_j'^* v, B_j'^* w\rangle_{m-1-r_j'+s,\mathscr{H}_\gamma}.$$

Recall that *Riesz's theorem*† says that any continuous linear functional defined over V^s can be represented by an element of V^s. We intend to use this fact to seek the solution of (P″).

Theorem 4(a)
Let $s \geqq 0$ be an integer. Then, for $\gamma \geqq \gamma_s$, if

$$f \in \mathscr{H}_{-\gamma}{}^s(\mathbb{R}_+{}^{n+1}), \quad g_j \in \mathscr{H}_{-\gamma}{}^{m-1-r_j+s}(\mathbb{R}^n) \quad (j = 1, \ldots, m_+),$$

then there exists

$$\Lambda^{-1} u \in \mathscr{H}_{-\gamma}{}^{m+s}(\mathbb{R}_+{}^{n+1})$$

such that

$$\text{(P″)} \qquad \begin{cases} Au = f, & (t, x, y) \in \mathbb{R}_+{}^{n+1}, \\ B_j u|_{x=0} = g_j & (j = 1, \ldots, m_+), \quad (t, y) \in \mathbb{R}^n. \end{cases}$$

Proof
We divide our proof into four steps as follows.
Step 1. For $v \in V^{-m-s+1}$ we write

$$l(v) = (v, f) + i \sum_{j=1}^{m_+} \langle B_j'^* v, g_j \rangle.$$

This is a continuous linear functional defined over V^{-m-s+1}. To see this we observe that

$$\begin{aligned} |(v, f)| &= |(\Lambda'^{-s} v, \Lambda^s f)| \leqq \|\Lambda'^{-s} v\|_{\mathscr{H}_\gamma} \|\Lambda^s f\|_{\mathscr{H}_{-\gamma}} \\ &\leqq \|\Lambda'^{-s-m} v\|_{m, \mathscr{H}_\gamma} \|\Lambda^s f\|_{\mathscr{H}_{-\gamma}} \\ &\leqq C |||v|||_{-m-s+1} \|\Lambda^s f\|_{\mathscr{H}_{-\gamma}}, \\ |\langle B_j'^* v, g_j \rangle| &= |\langle \Lambda'^{-(m-1-r_j+s)} B_j'^* v, \Lambda^{m-1-r_j+s} g_j \rangle| \\ &\leqq C \sum_{k=0}^{m-1} \langle D_x{}^k v \rangle_{-k-s, \mathscr{H}_\gamma} \langle g_j \rangle_{m-1-r_j+s, \mathscr{H}_{-\gamma}} \\ &\leqq C |||v|||_{-m-s+1} \langle g_j \rangle_{m-1-r_j+s, \mathscr{H}_{-\gamma}} . \end{aligned}$$

Therefore by Riesz's theorem again, there exists $w \in V^{-m-s+1}$ such that

$$l(v) = ((v, w))_{-m-s+1}.$$

This implies that

$$\begin{aligned} (f, v) - \mathrm{i} \sum_{j=1}^{m_+} \langle g_j, B_j'^* v \rangle &= (\Lambda'^{-m-s+1} A^* w, \Lambda'^{-m-s+1} A^* v)_{\mathscr{H}_\gamma} \\ &\quad + \sum_{j=m_+ +1}^{m} \langle B_j'^* w, B_j'^* v \rangle_{-r_j'-s, \mathscr{H}_\gamma} . \end{aligned}$$

† Translator's note: See Mizohata (4), p. 81. or J. Weidmann, *Linear operators in Hilbert spaces*. Springer, Berlin (1980) p. 61.

If we write

$$u = \Lambda^{-m-s+1} e^{2\gamma t} \Lambda'^{-m-s+1} A^* w,$$

then we see that

$$\Lambda^{m+s-1} u \in \mathscr{H}_{-\gamma},$$

and for an arbitrary $v \in V^{-m-s+1}$ we have

$$(f, v) - \mathrm{i} \sum_{j=1}^{m_+} \langle g_j, B_j'^* v \rangle = (u, A^* v) + \sum_{j=m_+ +1}^{m} \langle B_j'^* w, B_j'^* v \rangle_{-r_j' - s, \mathscr{H}_\gamma}.$$

Therefore, in particular, for $v \in \mathscr{D}(\mathbb{R}_+^{n+1})$ we can show that

$$(f, v) = (u, A^* v).$$

Hence, using the language of distribution, we establish that

$$Au = f$$

in $\mathbb{R}_+^{n+1}$.

Step 2. Next, we wish to see the smoothness of u which we have just obtained. Let us set

$$A = A(t, x, y; \mathrm{D}_t, \mathrm{D}_x, \mathrm{D}_y) = \sum_{j=0}^{m} a_{m-j}(t, x, y; \mathrm{D}_t, \mathrm{D}_y) \mathrm{D}_x^{\,j}, \quad a_0 = 1.$$

Then we can write

$$\begin{aligned} Au &= A\Lambda^{-m-s+1} e^{2\gamma t} \Lambda'^{-m-s+1} A^* w \\ &= A(t, x, y; \mathrm{D}_t, \mathrm{D}_x, \mathrm{D}_y) \Lambda^{2(-m-s+1)} A^*(t, x, y; \mathrm{D}_t + 2\mathrm{i}\gamma, \mathrm{D}_x, \mathrm{D}_y)(e^{2\gamma t} w) \\ &= \sum_{j=0}^{2m} \mathrm{D}_x^{\,j} \Lambda^{-j+2(-s+1)} \alpha_{2m-j}(e^{2\gamma t} w), \quad \alpha_0 = 1, \end{aligned}$$

where α_j is a bounded operator of $\mathscr{H}_{-\gamma}^{\,k}$. We already know that

$$\Lambda'^{-m-s} w \in \mathscr{H}_\gamma^{\,m}$$

so, letting

$$w_j = \Lambda^{-s} \alpha_j(e^{2\gamma t} w),$$

we have

$$\Lambda^{-m} w_j \in \mathscr{H}_{-\gamma}^{\,m},$$

and

$$\begin{aligned} \Lambda^{s-2} Au &= \mathrm{D}_x^{\,m} \Lambda^{-m}(\mathrm{D}_x^{\,m} \Lambda^{-m} w_0 + \ldots + w_m) \\ &\quad + (\mathrm{D}_x^{\,m-1} \Lambda^{-(m-1)} w_{m-1} + \ldots + w_{2m}) \\ &= \mathrm{D}_x^{\,m} \Lambda^{-m} \varphi + \psi, \quad \varphi, \psi \in \mathscr{H}_{-\gamma}. \end{aligned}$$

That is,

$$\mathrm{D}_x^{\,m} \Lambda^{-m} \varphi = \Lambda^{s-2} f - \psi \in \mathscr{H}_{-\gamma}, \quad \varphi \in \mathscr{H}_{-\gamma}.$$

Using the interpolation theorem we now obtain

$$\Lambda^{-m}\varphi \in \mathscr{H}_{-\gamma}{}^{m}.$$

Note that

$$\varphi = D_x{}^m \Lambda^{-m} w_0 + \ldots + w_m ,$$

therefore we see that

$$D_x \Lambda^{-1}\varphi = D_x{}^{m+1}\Lambda^{-(m+1)} w_0 + \ldots + D_x \Lambda^{-1} w_m .$$

This implies

$$D_x{}^{m+1}\Lambda^{-(m+1)} w_0 = D_x \Lambda^{-1}\varphi - (D_x{}^m \Lambda^{-m} w_1 + \ldots + D_x \Lambda^{-1} w_m) \in \mathscr{H}_{-\gamma},$$

so that

$$\Lambda^{-(m+1)} w_0 \in \Lambda_{-\gamma}{}^{m+1}.$$

From this we see

$$\Lambda^{-(m+1)} w_j \in \mathscr{H}_{-\gamma}{}^{m+1}.$$

We repeat this process until we obtain

$$\Lambda^{-2m} w_j \in \mathscr{H}_{-\gamma}{}^{2m}.$$

In particular, we have

$$\Lambda^{-2m} w_0 = \Lambda^{-2m-s}(e^{2\gamma t} w) \in \mathscr{H}_{-\gamma}{}^{2m}.$$

On the other hand we know that

$$u = \Lambda^{2(-m-s+1)} A^*(t, x, y; D_t + 2i\gamma, D_x, D_y)(e^{2\gamma t} w)$$

therefore we obtain

$$\Lambda^{s-2} u \in \mathscr{H}_{-\gamma}{}^{m}.$$

Furthermore, if $s > 0$ then since $Au = f \in \mathscr{H}_{-\gamma}{}^{s}$, we have

$$D_x{}^{m+1} u = D_x f - D_x\{a_1 D_x{}^{m-1} u + \ + a_m u\} .$$

This means

$$\Lambda^{s-3} u \in \Lambda_{-\gamma}{}^{m+1}.$$

We now repeat the same argument until we obtain

$$\Lambda^{-2} u \in \mathscr{H}_{-\gamma}{}^{m+s}.$$

Step 3. Let us assume $s \geqq 2$. We wish to establish the fact that u satisfies the given boundary condition. Using Lemma 4.2, for $u \in \mathscr{H}_{-\gamma}{}^{m}, v \in \mathscr{H}_{\gamma}{}^{m}$, we obtain

$$(Au, v) - (u, A^* v) = i \sum_{j=1}^{m} \langle B_j u, B_j{}'^* v \rangle .$$

In this case, we know that $Au = f$ and

$$(f, v) - i \sum_{j=1}^{m_+} \langle g_j, B_j{}'^* v \rangle = (u, A^* v) + \sum_{j=m_+ +1}^{m} \langle B_j{}'^* w, B_j{}'^* v \rangle_{-r_j{}'-s, \mathscr{H}_\gamma} .$$

From these facts, we see that

$$\mathrm{i}\sum_{j=1}^{m_+} \langle B_j u - g_j, B_j'^* v\rangle = \sum_{j=m_++1}^{m} \{-\mathrm{i}\langle B_j u, B_j'^* v\rangle + \langle B_j'^* w, B_j'^* v\rangle_{-r_j'-s,\mathscr{H}_\gamma}\}.$$

On the other hand we can choose arbitrary

$$\{B_j'^* v|_{x=0}\}_{j=1,\ldots,m}$$

so that

$$B_j u|_{x=0} = g_j \quad (j=1,\ldots,m_+).$$

Step 4. Finally, we wish to establish that

$$\Lambda^{-1} u \in \mathscr{H}_{-\gamma}^{m+s}$$

by means of approximation. For this purpose we choose smooth data as follows. Letting

$$\|f_l - f\|_{s,\mathscr{H}_{-\gamma}} \to 0, \quad f_l \in \mathscr{D}(\overline{\mathbb{R}_+^{n+1}}),$$
$$\langle g_{jl} - g_j\rangle_{m-1-r_j+s,\mathscr{H}_{-\gamma}} \to 0, \quad g_{jl} \in \mathscr{D}(\mathbb{R}^n)$$

and considering the results obtained in steps 1–3, we see that for $u_l \in \mathscr{H}_{-\gamma}^{m+s}(\mathbb{R}_+^{n+1})$

$$Au_l = f_l, \quad x>0,$$
$$B_j u_l|_{x=0} = g_{jl} \quad (j=1,\ldots,m_+).$$

At the same time, from the energy inequality we have

$$\|\Lambda^{-1}(u_l - u_k)\|_{m+s,\mathscr{H}_{-\gamma}} \leqq C\left\{\|f_l - f_k\|_{s,\mathscr{H}_{-\gamma}} + \sum_{j=1}^{m_+} \langle g_{jl} - g_{jk}\rangle_{m-1-r_j+s,\mathscr{H}_{-\gamma}}\right\} \to 0, \quad l,k \to +\infty.$$

Hence there exists u such that

$$\|\Lambda^{-1}(u_l - u)\|_{m+s,\mathscr{H}_{-\gamma}} \to 0, \quad l \to +\infty.$$

Also, u satisfies $\Lambda^{-1}u \in \mathscr{H}_{-\gamma}^{m+s}(\mathbb{R}_+^{n+1})$ and

$$Au = f, \quad x>0,$$
$$B_j u|_{x=0} = g_j \quad (j=1,\ldots,m_+). \qquad \blacksquare$$

Notice that in the above argument we fixed the parameter γ and obtained the existence theorem for $\mathscr{H}_{-\gamma}$-space. But, even though we have some freedom in the choice of γ we are still able to obtain a unique solution u which does not depend on the choice of γ. To see this we establish

Theorem 4(a)′
Let $s \geqq 0$ be an integer. For $\gamma_s \geqq \gamma' \leqq \gamma''$, if

$$f \in \bigcap_{\gamma' \leqq \gamma \leqq \gamma''} \mathscr{H}_{-\gamma}{}^{s}(\mathbb{R}_+{}^{n+1}),$$

$$g_j \in \bigcap_{\gamma' \leqq \gamma \leqq \gamma''} \mathscr{H}_{-\gamma}{}^{m-1-r_j+s}(\mathbb{R}^n) \quad (j = 1, \ldots, m_+),$$

then there exists

$$\Lambda^{-1} u \in \bigcap_{\gamma' \leqq \gamma \leqq \gamma''} \mathscr{H}_{-\gamma}{}^{m+s}(\mathbb{R}_+{}^{n+1})$$

and

$$(\mathrm{P}'') \qquad \begin{cases} Au = f, & (t, x, y) \in \mathbb{R}_+{}^{n+1}, \\ B_j u|_{x=0} = g_j & (j = 1, \ldots, m_+), \quad (t, y) \in \mathbb{R}^n \end{cases}$$

Proof
For an arbitrary $\gamma \geqq \gamma_s$, by Theorem 4(a), we know that a solution u_γ exists and satisfies

$$\Lambda^{-1} u_\gamma \in \mathscr{H}_{-\gamma}{}^{m+s}(\mathbb{R}_+{}^{n+1}).$$

Also, if $\gamma' \leqq \gamma''$ then

$$\mathscr{H}_{-\gamma'}{}^{k}, \mathscr{H}_{-\gamma''}{}^{k} \subset \mathscr{H}_{-\gamma', -\gamma''}{}^{k},$$

therefore

$$u_{\gamma''} - u_{\gamma'} \in \mathscr{H}_{-\gamma', -\gamma''}{}^{m+s}.$$

By Lemma 3.7, if $\gamma'' - \gamma'$ is sufficiently small, then it follows that $u_{\gamma''} - u_{\gamma'} = 0$. Continuing this process, we obtain $u_{\gamma''} = u_{\gamma'}$ for arbitrary γ', γ''. ■

4.3
In the foregoing argument we discussed the boundary problem (P″) instead of the initial boundary value problems (P) or (P′) with the initial value zero. Our conclusion was that a solution of (P′) can be regarded as one of (P″), but a solution of (P″) does not automatically become one of (P′); some auxiliary conditions are needed to establish this. More precisely, if the following condition is satisfied

> if the support of the given data $\{f, g_j\}$ is contained in
> $\{t \geqq 0, (x, y) \in \mathbb{R}^n\}$ then so is the solution u of (P″)

then the existence of a solution for (P″) implies the existence of a solution for (P′). We shall exaime this shortly. First we prove

Lemma 4.4

For $f \in \mathscr{H}_{-\gamma_0}{}^s(\mathbb{R}^1 \times \mathbb{R}_+{}^n)$ (where $s \geqq 0$ is an integer), a necessary and sufficient condition for

$$\operatorname{supp}[f] \subset \overline{\mathbb{R}_+{}^1 \times \mathbb{R}_+{}^n}$$

is that

$$f \in \bigcap_{\gamma \geqq \gamma_0} \mathscr{H}_{-\gamma}{}^s(\mathbb{R}^1 \times \mathbb{R}_+{}^n), \; \|f\|_{s, \mathscr{H}_{-\gamma}} \leqq \|f\|_{s, \mathscr{H}_{-\gamma_0}}.$$

Proof

Let the support of f be contained in $\{t \geqq 0, (x, y) \in \mathbb{R}^n\}$. Then for $\gamma \geqq \gamma_0$ we have

$$\begin{aligned}
\|f\|^2_{s, \mathscr{H}_{-\gamma}(R_1 \times \mathbb{R}_+{}^n)} &= \int_0^\infty e^{-2\gamma t}\, dt \int_{\mathbb{R}_+{}^n} \sum_{|\nu| \leqq s} |D_{t,x,y}{}^\nu f(t, x, y)|^2 \, dx\, dy \\
&\leqq \int_0^\infty e^{-2\gamma_0 t}\, dt \int_{\mathbb{R}_+{}^n} \sum_{|\nu| \leqq s} |D_{t,x,y}{}^\nu f(t, x, y)|^2 \, dx\, dy \\
&= \|f\|^2_{s, \mathscr{H}_{-\gamma_0}(\mathbb{R}^1 \times \mathbb{R}_+{}^n)} .
\end{aligned}$$

This shows the necessity of the condition. Conversely, if we know that the inequality

$$\begin{aligned}
\|f\|^2_{s, \mathscr{H}_{-\gamma}} &= \int_{-\infty}^\infty e^{-2\gamma t}\, dt \int_{\mathbb{R}_+{}^n} \sum_{|\nu| \leqq s} |D_{t,x,y}{}^\nu f(t, x, y)|^2 \, dx\, dy \\
&\leqq \|f\|^2_{s, \mathscr{H}_{-\gamma_0}} = M
\end{aligned}$$

holds, then we assume that a certain part of the support of f is not contained in $\{t \geqq 0, (x, y) \in \mathbb{R}^n\}$. That is, there exists $(t_1, t_2) \subset (-\infty, 0)$ such that

$$\int_{t_1}^{t_2} dt \int_{\mathbb{R}_+{}^n} |f(t, x, y)|^2 \, dx\, dy = c > 0.$$

From this we see that

$$\begin{aligned}
M &\geqq \int_{t_1}^{t_2} e^{-2\gamma t}\, dt \int_{\mathbb{R}_+{}^n} |f(t, x, y)|^2 \, dx\, dy \\
&\geqq e^{-2\gamma t_2} \int_{t_1}^{t_2} dt \int_{\mathbb{R}_+{}^n} |f(t, x, y)|^2 \, dx\, dy \\
&= c e^{-2\gamma t_2} \to +\infty \text{ as } \gamma \to +\infty,
\end{aligned}$$

which is a contradiction. ■

From Theorem 4(a) and Lemma 4.4, we have

Theorem 4(b) *(the existence of a solution of (P))*
Let $s \geqq 0$ be an integer. If

$$f \in \mathscr{H}_{-\gamma}^{s}(\mathbb{R}^1 \times \mathbb{R}_+{}^n), \quad \operatorname*{supp}_t [f] \subset [0, \infty),$$

$$g_j \in \mathscr{H}_{-\gamma}^{m-1-r_j+s}(\mathbb{R}^1 \times \mathbb{R}^{n-1}), \quad \operatorname*{supp}_t [g_j] \subset [0, \infty) \quad (j = 1, \ldots, m_+)$$

is true for γ satisfying $\gamma \geqq \gamma_s$, then there exists

$$\Lambda^{-1} u \in \mathscr{H}_{-\gamma}^{m+s}(\mathbb{R}^1 \times \mathbb{R}_+{}^n), \quad \operatorname*{supp}_t [u] \subset [0, \infty)$$

such that

$$Au = f, \quad (t, x, y) \in \mathbb{R}^1 \times \mathbb{R}_+{}^n,$$

$$B_j u|_{x=0} = g_j \quad (j = 1, \ldots, m_+), \quad (t, y) \in \mathbb{R}^1 \times \mathbb{R}^{n-1}.$$

Let us look at Theorem 4(b) again. The theorem implies that, as far as the problem (P″) is concerned, if the supports of the data lie in $\{t \geqq 0, (x, y) \in \overline{\mathbb{R}_+{}^n}\}$ or $\{t \geqq 0, y \in \mathbb{R}^{n-1}\}$, then that of the solution also lies in $\{t \geqq 0, (x, y) \in \overline{\mathbb{R}_+{}^n}\}$. In other words, the behaviour of the solution at $\{t < 0, (x, y) \in \mathbb{R}_+{}^n\}$ does not depend on that of the data at $\{t > 0, (x, y) \in \mathbb{R}_+{}^n\}$. More precisely, we have

Proposition 4.5 *(the energy inequality of Dirichlet type in $(-\infty, 0)$).*
If $\gamma \geqq \gamma_0, u \in \mathscr{H}_{-\gamma}^{m}((-\infty, 0) \times \mathbb{R}_+{}^n)$ then

$$\gamma \|u\|^2_{m-1, \mathscr{H}_{-\gamma}((-\infty,0)\times \mathbb{R}_+{}^n)} + \sum_{k=0}^{m-1} \langle D_x{}^k u \rangle^2_{m-1-k, \mathscr{H}_{-\gamma}((-\infty,0)\times \mathbb{R}^{n-1})}$$

$$\leqq C\left\{\frac{1}{\gamma}\|Au\|^2_{\mathscr{H}_{-\gamma}((-\infty,0)\times \mathbb{R}_+{}^n)} + \sum_{j=1}^{m_+} \langle B_j u \rangle^2_{m-1-r_j, \mathscr{H}_{-\gamma}((-\infty,0)\times \mathbb{R}^{n-1})}\right\}.$$

Proof
Consider the extension $\tilde{u}$ of u such that

$$\tilde{u} \in \mathscr{H}_{-\gamma}^{m}((-\infty, \infty) \times \mathbb{R}_+{}^n)$$

and $\tilde{u} = u$ at $\{t < 0, (x, y) \in \mathbb{R}_+{}^n\}$. Write $A\tilde{u} = f$, $B_j \tilde{u}|_{x=0} = g_j$. Then decompose $\{f, g_j\}$ as follows:

$$f = f_0 + f_1, \quad \operatorname*{supp}_t [f_1] \subset [0, \infty),$$

$$g_j = g_{j0} + g_{j1}, \quad \operatorname*{supp}_t [g_{j1}] \subset [0, \infty),$$

where

$$\|f_0\|_{\mathscr{H}_{-\gamma}(\mathbb{R}^1 \times \mathbb{R}_+{}^n)} \leqq \|f\|_{\mathscr{H}_{-\gamma}((-\infty,0)\times \mathbb{R}_+{}^n)},$$

$$\langle g_{j0} \rangle_{m-1-r_j, \mathscr{H}_{-\gamma}(\mathbb{R}^1 \times \mathbb{R}^{n-1})} \leqq \langle g_j \rangle_{m-1-r_j, \mathscr{H}_{-\gamma}((-\infty,0)\times \mathbb{R}^{n-1})}.$$

For these components let us consider the solution $u_0 \in \mathscr{H}_{-\gamma}{}^{m-1}$ of (P″) with respect to the data $\{f_0, g_{j0}\}$ and the solution $u_1 \in \mathscr{H}_{-\gamma}{}^{m-1}$ of (P″) with respect to the data $\{f_1, g_{j1}\}$. We now see that $\tilde{u} = u_0 + u_1$. Also we know that $\text{supp}\,[f_1] \subset \{t \leqq 0, (x, y) \in \overline{\mathbb{R}_+{}^n}\}$ and $\text{supp}\,[g_{j1}] \subset \{t \geqq 0, y \in \mathbb{R}^{n-1}\}$. By Theorem 4(b), $\text{supp}\,[u_1] \subset \{t \geqq 0, (x, y) \in \mathbb{R}_+{}^n\}$. Hence we have, $u = \tilde{u} = u_0$ for $\{t < 0, (x, y) \in \overline{\mathbb{R}_+{}^n}\}$. Applying the energy inequality of Dirichlet type to u_0 we see that

$$\gamma \|u\|^2_{m-1,\mathscr{H}_{-\gamma}((-\infty,0)\times \mathbb{R}_+{}^n)} + \sum_{k=0}^{m-1} \langle D_x{}^k u \rangle^2_{m-1-k,\mathscr{H}_{-\gamma}((-\infty,0)\times \mathbb{R}^{n-1})}$$

$$\leqq \gamma \|u_0\|^2_{m-1,\mathscr{H}_{-\gamma}(\mathbb{R}^1\times \mathbb{R}_+{}^n)} + \sum_{k=0}^{m-1} \langle D_x{}^k u_0 \rangle^2_{m-1-k,\mathscr{H}_{-\gamma}(\mathbb{R}^1\times \mathbb{R}^{n-1})}$$

$$\leqq C\left\{\frac{1}{\gamma}\|f_0\|^2_{\mathscr{H}_{-\gamma}(\mathbb{R}^1\times \mathbb{R}_+{}^n)} + \sum_{j=1}^{m_+} \langle g_{j0} \rangle^2_{m-1-r_j,\mathscr{H}_{-\gamma}(\mathbb{R}^1\times \mathbb{R}^{n-1})}\right\}$$

$$\leqq C\left\{\frac{1}{\gamma}\|f\|^2_{\mathscr{H}_{-\gamma}((-\infty,0)\times \mathbb{R}_+{}^n)} + \sum_{j=1}^{m_+} \langle g_j \rangle^2_{m-1-r_j,\mathscr{H}_{-\gamma}((-\infty,0)\times \mathbb{R}^{n-1})}\right\}. \qquad \blacksquare$$

5. The energy inequalities of Dirichlet type (H)

In §3, we observed the energy inequality of Dirichlet type ($\mathscr{H}$). In this section we shall focus our attention on the relation between the energy inequalities ($\mathscr{H}$) and (H), and we shall demonstrate that if ($\mathscr{H}$) is true, then (H) is also true.

To do this, we shall first derive Green's formulas in the domain $\{t_0 < t < t_1, (x, y) \in \mathbb{R}_+{}^n\}$ instead of $\{-\infty < t < +\infty, (x, y) \in \mathbb{R}_+{}^n\}$. But, to begin with, we review the basic notion of a divergence form, in order to establish two versions of the generalised Green's formula.

5.1

Throughout this subsection we shall use the notation

$$x = (x_1, \ldots, x_n), \quad \xi = (\xi_1, \ldots, \xi_n).$$

Lemma 5.1

Let $P(\xi)$, $Q(\xi)$ be polynomials of degree $p, q (q \leqq p)$, respectively. Then there exists a polynomial $G_j(\xi, \bar{\xi})$ of degree $p + q - 1$ with respect to $(\xi, \bar{\xi})$ which satisfies

$$P(\xi)Q(\bar{\xi}) - P(\bar{\xi})Q(\xi) = \sum_{j=1}^{m} (\xi_j - \bar{\xi}_j) G_j(\xi, \bar{\xi}),$$

where if d is the degree of $G_j(\xi,\bar{\xi})$ with respect to ξ and $\bar{d}$ is the degree of $G_j(\xi,\bar{\xi})$ with respect to $\bar{\xi}$, d and $\bar{d}$ have the following property:

$$p > q \Rightarrow d = \bar{d} = p-1,$$
$$p = q \Rightarrow d = p-1, \bar{d} = p \,(\text{or } d = p, \bar{d} = p-1).$$

Proof
Let

$$P(\xi) = \sum_{\alpha=0}^{p} \sum_{i_1,\ldots,i_\alpha=1}^{n} a_{i_1\ldots i_\alpha}\xi_{i_1}\cdots\xi_{i_\alpha},$$

$$Q(\xi) = \sum_{\beta=0}^{q} \sum_{j_1,\ldots\, j_\beta=1}^{n} b_{j_1\ldots j_\beta}\xi_{j_1}\cdots\xi_{j_\beta}.$$

Then we see that

$$P(\xi)Q(\bar{\xi}) - P(\bar{\xi})Q(\xi) = \sum_{\alpha=0}^{p} \sum_{i_1,\ldots,i_\alpha} \sum_{\beta=0}^{q} \sum_{j_1,\ldots,j_\beta} a_{i_1\ldots i_\alpha} b_{j_1\ldots j_\beta}$$
$$\times (\xi_{i_1}\cdots\xi_{i_\alpha}\bar{\xi}_{j_1}\cdots\bar{\xi}_{j_\beta} - \bar{\xi}_{i_1}\cdots\bar{\xi}_{i_\alpha}\xi_{j_1}\cdots\xi_{j_\beta}),$$

therefore in order to establish the lemma, we only need to consider the case

$$P = \xi_{i_1}\cdots\xi_{i_\alpha}, \quad Q = \xi_{j_1}\cdots\xi_{j_\beta} \quad (\alpha \leqq p, \beta \leqq q).$$

Let us assume $\alpha \geq \beta$. We can see that

$$\xi_{i_1}\cdots\xi_{i_\alpha}\bar{\xi}_{j_1}\cdots\bar{\xi}_{j_\beta} - \bar{\xi}_{i_1}\cdots\bar{\xi}_{i_\alpha}\xi_{j_1}\cdots\xi_{j_\beta}$$
$$= (\xi_{i_1}\cdots\xi_{i_\beta}\bar{\xi}_{j_1}\cdots\bar{\xi}_{j_\beta} - \bar{\xi}_{i_1}\cdots\bar{\xi}_{i_\beta}\xi_{j_1}\cdots\xi_{j_\beta})\xi_{i_{\beta+1}}\cdots\xi_{i_\alpha}$$
$$+ (\xi_{i_{\beta+1}}\cdots\xi_{i_\alpha} - \bar{\xi}_{i_{\beta+1}}\cdots\bar{\xi}_{i_\alpha})\bar{\xi}_{i_1}\cdots\bar{\xi}_{i_\beta}\xi_{j_1}\cdots\xi_{j_\beta}.$$

From this we arrive at the following two cases:

$$\text{(a)}\ P = \xi_{i_1}\cdots\xi_{i_\alpha}, \quad Q = \xi_{j_1}\cdots\xi_{j_\alpha},$$
$$\text{(b)}\ P = \xi_{i_1}\cdots\xi_{i_\alpha}, \quad Q = 1.$$

Case (a). Note that

$$P(\xi)Q(\bar{\xi}) - P(\bar{\xi})Q(\xi) = \xi_{i_1}\cdots\xi_{i_\alpha}\bar{\xi}_{j_1}\cdots\bar{\xi}_{j_\alpha} - \bar{\xi}_{i_1}\cdots\bar{\xi}_{i_\alpha}\xi_{j_1}\cdots\xi_{j_\alpha}$$
$$= (\xi_{i_1}\bar{\xi}_{j_1})\ldots(\xi_{i_\alpha}\bar{\xi}_{j_\alpha}) - (\bar{\xi}_{i_1}\xi_{j_1})\ldots(\bar{\xi}_{i_\alpha}\xi_{j_\alpha})$$
$$= (\xi_{i_1}\bar{\xi}_{j_1} - \bar{\xi}_{i_1}\xi_{j_1})(\xi_{i_2}\bar{\xi}_{j_2})\ldots.(\xi_{i_\alpha}\bar{\xi}_{j_\alpha})$$
$$+ (\bar{\xi}_{i_1}\xi_{j_1})(\xi_{i_2}\bar{\xi}_{j_2} - \bar{\xi}_{i_2}\xi_{j_2})(\xi_{i_3}\bar{\xi}_{j_3})\ldots(\xi_{i_\alpha}\bar{\xi}_{j_\alpha})$$
$$+ \ldots + (\bar{\xi}_{i_1}\xi_{j_1})\ldots(\bar{\xi}_{i_{\alpha-1}}\xi_{j_{\alpha-1}})(\xi_{i_\alpha}\bar{\xi}_{j_\alpha} - \bar{\xi}_{i_\alpha}\xi_{j_\alpha})$$

and

$$\xi_i\bar{\xi}_j - \bar{\xi}_i\xi_j = (\xi_i - \bar{\xi}_i)\bar{\xi}_j + \bar{\xi}_i(\bar{\xi}_j - \xi_j).$$

Case (b). In this case we have

$$\begin{aligned} P(\xi) - P(\bar{\xi}) &= \xi_{i_1}\dots\xi_{i_\alpha} - \bar{\xi}_{i_1}\dots\bar{\xi}_{i_\alpha} \\ &= (\xi_{i_1} - \bar{\xi}_{i_1})\xi_{i_2}\dots\xi_{i_\alpha} \\ &\quad + \bar{\xi}_{i_1}(\xi_{i_2} - \bar{\xi}_{i_2})\xi_{i_3}\dots\xi_{i_\alpha} \\ &\quad + \dots + \bar{\xi}_{i_1}\dots\bar{\xi}_{i_{\alpha-1}}(\xi_{i_\alpha} - \bar{\xi}_{i_\alpha}). \end{aligned}$$ ■

We can now give a more general result

Lemma 5.1′

Let $R(\xi, \bar{\xi})$ be a polynomial of degree r with respect to $(\xi, \bar{\xi})$, and of degree p, q with respect to $\xi, \bar{\xi}$. If $R(\xi, \xi) = 0$ then there exists a polynomial $G_j(\xi, \bar{\xi})$ of degree $r - 1$ with respect to $(\xi, \bar{\xi})$ such that

$$R(\xi, \bar{\xi}) = \sum_{j=1}^{n} (\xi_j - \bar{\xi}_j)G_j(\xi, \bar{\xi}).$$

If d is the degree of $G_j(\xi, \bar{\xi})$ with respect to ξ, and $\bar{d}$ is the degree of $G_j(\xi, \bar{\xi})$ with respect to $\bar{\xi}$, then d and $\bar{d}$ have the following property:

$$p + q > r \Rightarrow d = p - 1, \bar{d} = q - 1;$$
$$p + q = r \Rightarrow d = p - 1, \bar{d} = q \quad (\text{or } d = p, \bar{d} = q - 1).$$

Proof

We divide our proof into three steps.

Step 1. We may assume that the coefficients of $R(\xi, \bar{\xi})$ are all reals, say

$$R(\xi, \bar{\xi}) = \sum_{|\nu| \leqq r} \sum_{\mu \in I_\nu} c_{\nu\mu}\xi^{\nu - \mu}\bar{\xi}^{\mu},$$

where

$$I_\nu = \{\mu | c_{\nu\mu} \neq 0\}.$$

Since

$$R(\xi, \xi) = \sum_{|\nu| \leqq r} (\sum_{\mu \in I_\nu} c_{\nu\mu})\xi^{\nu}$$

and $R(\xi, \xi) = 0$, for each ν it follows that

$$\sum_{\mu \in I_\nu} c_{\nu\mu} = 0.$$

Step 2. Consequently, we concentrate only on the case

$$R(\xi, \bar{\xi}) = \sum_{\mu \in I} c_\mu \xi^{\nu - \mu}\bar{\xi}^{\mu}, \quad \sum_{\mu \in I} c_\mu = 0.$$

Let

$$I_+ = \{\mu | c_\mu > 0\} = \{\mu_1{}^+, \ldots, \mu_{N_+}{}^+\},$$
$$I_- = \{\mu | c_\mu < 0\} = \{\mu_1{}^-, \ldots, \mu_{N_-}{}^-\}.$$

We see that

$$\sum_{j=1}^{N+} |c_{\mu_j+}| = \sum_{j=1}^{N-} |c_{\mu_j-}| = a,$$

say. Putting

$$\sum_{j=1}^{k} |c_{\mu_j^\pm}| = a_k{}^\pm$$

we obtain two partitions of the interval $(0, a)$ as

$$0 < a_1{}^+ < \ldots < a_{N_+}{}^+ = a,$$
$$0 < a_1{}^- < \ldots < a_{N_-}{}^- = a.$$

We take the common refinement of these partitions to obtain

$$0 < a_1 < a_2 < \ldots < a_N = a.$$

Setting

$$b_1 = a_1, \quad b_2 = a_2 - a_1, \quad \ldots, \quad b_N = a_N - a_{N-1},$$

we have

$$|c_{\mu_j^\pm}| = \sum_{k=N_{j-1}{}^\pm+1}^{N_j\pm} b_k.$$

We write

$$\tilde{\mu_k}^\pm = \mu_j{}^\pm$$

for $N_{j-1}^\pm + 1 \leqq k \leqq N_j{}^\pm$. As a result we have

$$\sum_{\mu \in I} c_\mu \xi^{\nu-\mu} \bar{\xi}^\mu = \sum_{k=1}^{N} b_k (\xi^{\nu-\mu_k{}^+} \bar{\xi}^{\mu_k{}^+} - \xi^{\nu-\mu_k{}^-} \bar{\xi}^{\mu_k{}^-}).$$

Step 3. Now we wish to prove the lemma for the case

$$R(\xi, \bar{\xi}) = \xi^{\nu-\mu^+} \bar{\xi}^{\mu^+} - \xi^{\nu-\mu^-} \bar{\xi}^{\mu^-}.$$

By writing

$$M = \max\{\mu^+, \mu^-\} = (\max\{\mu_1{}^+, \mu_1{}^-\}, \ldots, \max\{\mu_n{}^+, \mu_n{}^-\}),$$
$$\mu = \min\{\mu^+, \mu^-\} = (\min\{\mu_1{}^+, \mu_1{}^-\}, \ldots, \min\{\mu_n{}^+, \mu_n{}^-\}),$$
$$\alpha = \mu^+ - \mu = M - \mu^-, \quad \beta = \mu^- - \mu = \mathrm{M} - \mu^+,$$

we can express R as

$$R = \xi^{\nu-M} \bar{\xi}^\mu (\xi^\beta \bar{\xi}^\alpha - \xi^\alpha \bar{\xi}^\beta).$$

Using Lemma 5.1, we have

$$\xi^\beta \bar{\xi}^\alpha - \xi^\alpha \bar{\xi}^\beta = \sum_{j=1}^{n} (\xi_j - \bar{\xi}_j) g_j(\xi, \bar{\xi}).$$

Therefore, putting

$$\xi^{\nu - M} \bar{\xi}^\mu g_j(\xi, \bar{\xi}) = G_j(\xi, \bar{\xi})$$

we get

$$R = \sum_{j=1}^{n} (\xi_j - \bar{\xi}_j) G_j(\xi, \bar{\xi}).$$

Hence, if we choose $g_j(\xi, \bar{\xi})$ according to the condition stated in Lemma 5.1, then $G_j(\xi, \bar{\xi})$ will have the required polynomial degree. ∎

Let us consider two fixed polynomials $P(\xi), Q(\xi)$ of degree $m, m-1$, respectively. Given $P(\xi), Q(\xi)$, by Lemma 5.1, there exists a polynomial $G_j(\xi, \bar{\xi})$ of degree $2m-2$ with respect to $(\xi, \bar{\xi})$, and of degree $m-1$ with respect to ξ or $\bar{\xi}$ such that

$$P(\xi)Q(\bar{\xi}) - P(\bar{\xi})Q(\xi) = \sum_{j=1}^{n} (\xi_j - \bar{\xi}_j) G_j(\xi, \bar{\xi}).$$

Of course, there may be a number of sets of polynomials $\{G_j(\xi, \bar{\xi})\}$ having this property. However, as is shown immediately, it is necessary for $\{G_j(\xi, \bar{\xi})\}$ to satisfy the condition: if

$$\xi^{(j)} = (0, \ldots, 0, \xi_j^{(j)}, 0, \ldots, 0), \quad \xi^{(j)\prime} = (\xi_1, \ldots, \xi_{j-1}, 0, \xi_{j+1}, \ldots, \xi_n),$$
$$\xi = \xi^{(j)} + \xi^{(j)\prime},$$

then

$$G_j(\xi^{(j)} + \xi^{(j)\prime}, \bar{\xi}^{(j)} + \xi^{(j)\prime}) = \frac{P(\xi^{(j)} + \xi^{(j)\prime})Q(\bar{\xi}^{(j)} + \xi^{(j)\prime}) - P(\bar{\xi}^{(j)} + \xi^{(j)\prime})Q(\xi^{(j)} + \xi^{(j)\prime})}{\xi_j - \bar{\xi}_j}.$$

In the following lemma we show that at least one polynomial among $\{G_j(\xi, \bar{\xi})\}$ can be arbitrarily given to satisfy only the conditions mentioned above.

Lemma 5.2

Let $P(\xi), Q(\xi)$ be polynomials of degree $m, m-1$, respectively. If there exists a polynomial $G_1(\xi, \bar{\xi})$ of degree $2(m-1)$ with respect to $(\xi, \bar{\xi})$, and of degree $m-1$ with respect to ξ or $\bar{\xi}$, such that G_1 satisfies the relation

$$G_1(\xi_1, \xi'; \bar{\xi}_1, \xi') = \frac{P(\xi_1, \xi')Q(\bar{\xi}_1, \xi') - P(\bar{\xi}_1, \xi')Q(\xi_1, \xi')}{\xi_1 - \bar{\xi}_1},$$

where $\xi' = (\xi_2, \ldots, \xi_n)$, then there exist polynomials $\{G_j(\xi, \bar{\xi})\}_{j=2,\ldots,n}$ such that each of them is of degree $2m-2$ with respect to $(\xi, \bar{\xi})$ and of degree $m-1$ with respect to ξ on $\bar{\xi}$, and satisfies the relation

$$P(\xi)Q(\bar{\xi}) - P(\bar{\xi})Q(\xi) = \sum_{j=1}^{n} (\xi_j - \bar{\xi}_j)G_j'(\xi, \bar{\xi}).$$

Proof

Let us write $\{G_j'(\xi, \bar{\xi})\}$ for the polynomials whose existence is guaranteed by Lemma 5.1. These polynomials satisfy

$$P(\xi)Q(\bar{\xi}) - P(\bar{\xi})Q(\xi) = \sum_{j=1}^{n} (\xi_j - \bar{\xi}_j)G_j'(\xi, \bar{\xi}).$$

Set

$$G_j(\xi, \bar{\xi}) - G_j'(\xi, \bar{\xi}) = g_j(\xi, \bar{\xi}).$$

Then given $g_1(\xi, \bar{\xi})$ our problem is to find $\{g_j(\xi, \bar{\xi})\}_{j=2,\ldots,n}$ which satisfy the condition

$$\sum_{j=1}^{n} (\xi_j - \bar{\xi}_j)g_j(\xi, \bar{\xi}) = 0$$

given that $g_1(\xi, \bar{\xi})$ satisfies the condition

$$g_1(\xi_1, \xi'; \bar{\xi}_1, \xi') = 0.$$

That is, if we write

$$g_1(\xi, \bar{\xi}) = \sum_{\alpha,\beta=0}^{m-1} \varphi_{\alpha\beta}(\xi', \bar{\xi}')\xi_1^{m-1-\alpha}\bar{\xi}_1^{m-1-\beta},$$

then $\varphi_{\alpha\beta}(\xi, \bar{\xi})$ is of degree $\alpha + \beta$ with respect to $(\xi, \bar{\xi})$, of degree α with respect to ξ', and of degree β with respect to $\bar{\xi}'$ and it satisfies the relation

$$\varphi_{\alpha\beta}(\xi', \xi') = 0.$$

Therefore, we have $\varphi_{00} = \varphi_{01} = \varphi_{10} = 0$. By Lemma 5.1′, we see that there exist polynomials $\varphi_{\alpha\beta j}(\xi', \bar{\xi}')$ of degree $\alpha - 1, \beta$ with respect to $\xi', \bar{\xi}'$; and $\psi_{\alpha\beta j}(\xi', \bar{\xi}')$ of degree $\alpha, \beta - 1$ with respect to $\xi', \bar{\xi}'$, such that

$$\varphi_{\alpha\beta}(\xi', \bar{\xi}') = \sum_{j=2}^{n} (\xi_j - \bar{\xi}_j)\varphi_{\alpha\beta j}(\xi', \bar{\xi}') = \sum_{j=2}^{n} (\xi_j - \bar{\xi}_j)\psi_{\alpha\beta j}(\xi', \xi').$$

From this we can take

$$(\xi_1 - \bar{\xi}_1)g_1(\xi, \bar{\xi}) = \sum_{\alpha,\beta=1}^{m-1} \varphi_{\alpha\beta}(\xi', \xi')\xi_1^{m-\alpha}\xi_1^{m-1-\beta}$$

$$- \sum_{\alpha,\beta=1}^{m-1} \varphi_{\alpha\beta}(\xi', \bar{\xi}')\xi_1^{m-1-\alpha}\bar{\xi}_1^{m-\beta}$$

$$= \sum_{j=1}^{n} (\xi_j - \bar{\xi}_j)\left\{ \sum_{\alpha,\beta=1}^{m-1} \varphi_{\alpha\beta j}(\xi', \bar{\xi}')\xi_1{}^{m-\alpha}\bar{\xi}_1{}^{m-1-\beta} - \sum_{\alpha,\beta=1}^{m-1} \psi_{\alpha\beta j}(\xi', \bar{\xi}')\xi_1{}^{m-1-\alpha}\bar{\xi}_1{}^{m-\beta}\right\}.$$

And then we let

$$g_j(\xi, \bar{\xi}) = \sum_{\alpha,\beta=1}^{m-1} \varphi_{\alpha\beta j}(\xi', \bar{\xi}')\xi_1{}^{m-\alpha}\bar{\xi}_1{}^{m-1-\beta} - \sum_{\alpha,\beta=1}^{m-1} \psi_{\alpha\beta j}(\xi', \bar{\xi}')\xi_1{}^{m-1-\alpha}\bar{\xi}_1{}^{m-\beta}$$

$$(j = 2, \ldots, n). \quad \blacksquare$$

Now, suppose that we have a polynomial R whose coefficients are smooth functions of $x = (x_1, \ldots x_n)$, and which is written as

$$R(x;\xi,\bar{\xi}) = \sum_{\nu,\mu} a_{\nu\mu}(x)\xi^{\nu}\bar{\xi}^{\mu}.$$

Let $u(x), v(x)$ be smooth functions. For these functions u, v we can define a map by the rule

$$R(x;\mathrm{D}_x, \overline{\mathrm{D}_x})u(x)\cdot\overline{v(x)} = \sum_{\nu,\mu} a_{\nu\mu}(x)\mathrm{D}_x{}^{\nu}u(x)\cdot\overline{\mathrm{D}_x{}^{\mu}v(x)}.$$

Note that we then have a bijection such that

$$R(x;\xi,\bar{\xi}) \to R(x;\mathrm{D}_x, \overline{\mathrm{D}_x}).$$

In fact, letting $u = \mathrm{e}^{\mathrm{i}x.\xi}, \quad v = \mathrm{e}^{\mathrm{i}x.\eta}$ we have

$$R(x;\mathrm{D}_x, \overline{\mathrm{D}_x})u(x)\cdot\overline{v(x)} = R(x;\xi,\bar{\eta})\mathrm{e}^{\mathrm{i}x.\xi}\cdot\overline{\mathrm{e}^{\mathrm{i}x.\eta}},$$

therefore

$$R(x;\xi,\eta) \equiv 0 \Leftrightarrow R(x;\mathrm{D}_x, \overline{\mathrm{D}_x}) = 0.$$

Lemma 5.3 *(the divergence form)*

If

$$P(x,\xi)Q(x,\bar{\xi}) - P(x,\bar{\xi})Q(x,\xi) = \sum_{j=1}^{n} (\xi_j - \bar{\xi}_j)G_j(x;\xi,\bar{\xi})$$

then

$$P(x,\mathrm{D}_x)u\cdot\overline{\bar{Q}(x,\mathrm{D}_x)v} - Q(x,\mathrm{D}_x)u\cdot\overline{\bar{P}(x;\mathrm{D}_x)v}$$

$$= \sum_{j=1}^{n} \mathrm{D}_{x_j}\{G_j(x;\mathrm{D}_x, \overline{\mathrm{D}_x})u\cdot\bar{v}\} - \sum_{j=1}^{n} (\mathrm{D}_{x_j}G_j)(x;\mathrm{D}_x, \overline{\mathrm{D}_x})u\cdot\bar{v},$$

where $\bar{P}(x,\xi)$ is a polynomial with respect to ξ obtained by replacing the coefficients of $P(x,\xi)$ with their complex conjugates.

Proof
Let

$$P(x,\xi)=\sum_{\nu} a_\nu(x)\xi^\nu,\quad Q(x,\xi)=\sum_{\mu} b_\mu(x)\xi^\mu,$$

and let

$$\begin{aligned}R(x;\xi,\bar\xi)&=P(x,\xi)Q(x,\bar\xi)-P(x,\bar\xi)Q(x,\xi)\\&=\sum a_\nu(x)b_\mu(x)(\xi^\nu\bar\xi^\mu-\bar\xi^\nu\xi^\mu).\end{aligned}$$

Then we have

$$\begin{aligned}R(x;\mathrm{D}_x,\overline{\mathrm{D}_x})u\cdot\bar v&=\sum a_\nu(x)b_\mu(x)(\mathrm{D}_x{}^\nu u\overline{\mathrm{D}_x{}^\mu v}-\mathrm{D}_x{}^\mu u\overline{\mathrm{D}_x{}^\nu v})\\&=P(x,\mathrm{D}_x)u\overline{\bar Q(x,\mathrm{D}_x)v}-Q(x,\mathrm{D}_x)u\overline{\bar P(x,\mathrm{D}_x)v}.\end{aligned}$$

On the other hand, we can write

$$\begin{aligned}R(x;\xi,\bar\xi)&=\sum_j(\xi_j-\bar\xi_j)G_j(x;\xi,\bar\xi)\\&=\sum_j(\xi_j-\bar\xi_j)\sum_{\nu,\mu}c_{j\nu\mu}(x)\xi^\nu\bar\xi^\mu,\end{aligned}$$

which gives

$$\begin{aligned}R(x;\mathrm{D}_x,\overline{\mathrm{D}_x})u\cdot\bar v&=\sum_{j,\nu,\mu}c_{j\nu\mu}(x)\{\mathrm{D}_{x_j}\mathrm{D}_x{}^\nu u\cdot\overline{\mathrm{D}_x{}^\mu v}-\mathrm{D}_x{}^\nu u\cdot\overline{\mathrm{D}_{x_j}\mathrm{D}_x{}^\mu v}\}\\&=\sum_{j,\nu,\mu}c_{j\nu\mu}(x)\mathrm{D}_{x_j}\{\mathrm{D}_x{}^\nu u\cdot\overline{\mathrm{D}_x{}^\mu v}\}\\&=\sum_{j=1}^n\left\{\mathrm{D}_{x_j}\left(\sum_{\nu,\mu}c_{j\nu\mu}(x)\mathrm{D}_x{}^\nu u\cdot\overline{\mathrm{D}_x{}^\mu v}\right)\right.\\&\qquad\left.-\sum_{\nu,\mu}(\mathrm{D}_{x_j}c_{j\nu\mu})(x)\mathrm{D}_x{}^\nu u\cdot\overline{\mathrm{D}_x{}^\mu v}\right\}\\&=\sum_{j=1}^n[\mathrm{D}_{x_j}\{G_j(x;\mathrm{D}_x,\overline{\mathrm{D}_x})u\cdot\bar v\}\\&\qquad-(\mathrm{D}_{x_j}G_j)(x;\mathrm{D}_x,\overline{\mathrm{D}_x})u\cdot\bar v].\end{aligned}$$

■

By performing integration on the divergence form which has just been obtained, we can establish Green's formula. However, if we give appropriate and concrete definitions of P, Q and their domains of integration, then we get a concrete form of Green's formula. We shall do this in the following subsection to establish two versions of concrete Green's formulas.

5.2

First, we seek a generalised Green's formula in $I\times\Omega$ which is fundamental with respect to a strongly hyperbolic operator A. For the principal symbol

A_0 we have

$$A_0(t,x,y;\tau,\xi,\eta) = \prod_{j=1}^{m} (\tau - \tau_j(t,x,y;\xi,\eta)),$$

$$|\tau_j(t,x,y;\xi,\eta) - \tau_j(t,x,y;\xi,\eta)| \geqq c(|\xi| + |\eta|), \quad i \neq j, \quad (c > 0).$$

Letting

$$A_0'(t,x,y;\tau,\xi,\eta) = \frac{\partial}{\partial\tau} A_0(t,x,y;\tau,\xi,\eta)$$

$$= \sum_{k=1}^{m} \prod_{j\neq k} (\tau - \tau_j(t,x,y;\xi,\eta))$$

if we put A_0, A_0' for P, Q in Lemma 5.1, we see that

$$\begin{aligned} &A_0(t,x,y;\tau,\xi,\eta)A_0'(t,x,y;\bar\tau,\bar\xi,\bar\eta) \\ &\quad - A_0(t,x,y;\bar\tau,\bar\xi,\bar\eta)A_0'(t,x,y;\tau,\xi,\eta) \\ &= (\tau - \bar\tau)G_t(t,x,y;\tau,\xi,\eta;\bar\tau,\bar\xi,\bar\eta) \\ &\quad + (\xi - \bar\xi)G_x(t,x,y;\tau,\xi,\eta;\bar\tau,\bar\xi,\bar\eta) \\ &\quad + \sum_{j=1}^{n-1} (\eta_j - \bar\eta_j)G_{y_j}(t,x,y;\tau,\xi,\eta;\bar\tau,\bar\xi,\bar\eta). \end{aligned}$$

To simplify things, we ignore (t,x,y) and write

$$G_t(\tau,\xi,\eta;\bar\tau,\bar\xi,\bar\eta) = \sum_{i,j=0}^{m-1} g_{ij}(\xi,\eta;\bar\xi,\bar\eta)\tau^{m-1-i}\bar\tau^{m-1-j}$$

$$= (\bar\tau^{m-1},\ldots,1) \quad (\xi,\eta;\bar\xi,\bar\eta) \begin{bmatrix} \tau^{m-1} \\ \vdots \\ 1 \end{bmatrix}.$$

Then, g_{ij} is a polynomial of degree i with respect to (ξ,η), and of degree j with respect to $(\bar\xi,\bar\eta)$ and

$$G_t(\tau,\xi,\eta;\bar\tau,\xi,\eta) = \frac{A_0(\tau,\xi,\eta)A_0'(\bar\tau,\xi,\eta) - A_0(\bar\tau,\xi,\eta)A_0'(\tau,\xi,\eta)}{\tau - \bar\tau}.$$

Also, if $(\xi,\eta)\in\mathbb{R}^n\backslash\{0\}$, the real roots of $A_0 = 0$ with respect to τ, and the real roots of $A_0' = 0$ with respect to τ are mutually separated from each other. Therefore, as we saw in Lemma 3.5, Bezout's form of A_0 and A_0' with respect to τ becomes positive-definite, i.e. $\mathscr{G}(\xi,\eta;\xi,\eta)$ is a positive-definite matrix. Keeping this fact in mind, and applying Lemma 5.3, we obtain the divergence form of the corresponding operator. Then, we make the form weighted by multiplying by $e^{-2\gamma t}$ and perform integration over $I \times \Omega = (S,T) \times \mathbb{R}_+{}^n$. As the result we obtain

Lemma 5.4 *(the generalised Green's formula for $\{A_0, A'_0\}$ in $I \times \Omega$)*
For $u, v \in \mathscr{H}_{-\gamma}{}^m(I \times \Omega)$, the relation

$$(A_0 u, A_0{}'v)_{\mathscr{H}_{-\gamma}(I\times\Omega)} - (A_0{}'u, A_0 v)_{\mathscr{H}_{-\gamma}(I\times\Omega)}$$
$$= -\mathrm{i}(G_t u, v)_{\mathscr{H}_{-\gamma}(\Omega), t=T} + \mathrm{i}(G_t u, v)_{\mathscr{H}_{-\gamma}(\Omega), t=S} + \mathrm{i}(G_x u, v)_{\mathscr{H}_{-\gamma}(I\times\partial\Omega)}$$
$$- 2\mathrm{i}\gamma(G_t u, v)_{\mathscr{H}_{-\gamma}(I\times\Omega)} - (\mathrm{D}Gu, v)_{\mathscr{H}_{-\gamma}(I\times\Omega)}$$

holds, where

$$\mathrm{D}G = \mathrm{D}_t G_t + \mathrm{D}_x G_x + \sum_{j=1}^{n-1} \mathrm{D}_{y_j} G_{y_j}$$

and

$$(Gu, v)_{\mathscr{H}_{-\gamma}(I\times\Omega)} = \int_{I\times\Omega} \mathrm{e}^{-2\gamma t} G(\mathrm{D}, \bar{\mathrm{D}}) u\bar{v}\, \mathrm{d}t\, \mathrm{d}x\, \mathrm{d}y.$$

We now state a result about the operator G_t which appeared in Lemma 5.4.

Lemma 5.5
Let $u \in \mathscr{H}_{-\gamma}{}^m(I \times \Omega)$. Then the following inequality holds

$$(G_t u, u)_{\mathscr{H}_{-\gamma}(\Omega), t=T} \geqq c \sum_{j=0}^{m-1} \|\mathrm{D}_t{}^j u(T)\|^2_{m-1-j, \mathscr{H}_{-\gamma}(\Omega)}$$
$$- C\Bigg\{ \sum_{j=0}^{m-2} \|\mathrm{D}_t{}^j u(T)\|^2_{m-2-j, \mathscr{H}_{-\gamma}(\Omega)}$$
$$+ \sum_{j=0}^{m-1} \|\mathrm{D}_x{}^j u\|^2_{m-1-j, \mathscr{H}_{-\gamma}(I\times\partial\Omega)}$$
$$+ \sum_{j=0}^{m-1} \|\mathrm{D}_t{}^j u(S)\|^2_{m-1-j, \mathscr{H}_{-}(\Omega)} \Bigg\},$$

where the positive constants c, C do not depend on T, S.

Proof
The norm of $\mathrm{D}_x{}^k \mathrm{D}_t{}^j u$ at $\{t = T, x = 0, y \in \mathbb{R}^{n+1}\}$ can be estimated in the following two ways: If we regard the norm as a boundary value of a function of $\{t = T, x > 0, y \in \mathbb{R}^{n-1}\}$ on $\{t = T, x = 0, y \in \mathbb{R}^{n-1}\}$ then we get an estimation

$$\sum_{j=0}^{m-2} \sum_{k=0}^{m-2-j} \|\mathrm{D}_x{}^k \mathrm{D}_t{}^j u(T)\|_{m-1-j-k-1/2, \mathscr{H}_{-\gamma}(\partial\Omega)}$$
$$\leqq C \sum_{j=0}^{m-2} \|\mathrm{D}_t{}^j u(T)\|_{m-1-j, \mathscr{H}_{-\gamma}(\Omega)}.$$

On the other hand, as a boundary value of a function of $\{x=0, S<t<T, y\in\mathbb{R}^{n-1}\}$ on $\{x=0, t=T, y\in\mathbb{R}^{n-1}\}$ we get another estimation

$$\sum_{k=0}^{m-2}\sum_{j=0}^{m-2-k}\|D_t{}^jD_x{}^ku(T)\|_{m-1-j-k-1/2,\mathscr{H}_{-\gamma}(\partial\Omega)}$$

$$\leqq C\left\{\sum_{k=0}^{m-2}\|D_x{}^ku\|_{m-1-k,\mathscr{H}_{-\gamma}(I\times\partial\Omega)}\right.$$
$$\left.+\sum_{k=0}^{m-2}\sum_{j=0}^{m-2-k}\|D_t{}^jD_x{}^ku(S)\|_{m-1-j-k-1/2,\mathscr{H}_{-\gamma}(\partial\Omega)}\right\}$$

$$\leqq C'\left\{\sum_{k=0}^{m-2}\|D_x{}^ku\|_{m-1-k,\mathscr{H}_{-\gamma}(I\times\partial\Omega)}\right.$$
$$\left.+\sum_{j=0}^{m-2}\|D_t{}^ju(S)\|_{m-1-j,\mathscr{H}_{-\gamma}(\Omega)}\right\}.$$

Now, for a fixed $j(\leqq m-2)$ we consider $v_j\in\mathscr{H}_{-\gamma}{}^{m-1-j}(\Omega)$ such that

$$D_x{}^kv_j|_{x=0}=D_x{}^kD_t{}^ju(T)|_{x=0}\quad(k=0,1,\ldots,m-2-j)$$

and

$$\|v_j\|_{m-1-j,\mathscr{H}_{-\gamma}(\Omega)}\leqq C\sum_{k=0}^{m-2-j}\|D_x{}^kD_t{}^ju(T)\|_{m-1-j-k-1/2,\mathscr{H}_{-\gamma}(\partial\Omega)}.$$

For such v_j we have

$$\sum_{j=0}^{m-2}\|v_j\|_{m-1-j,\mathscr{H}_{-\gamma}(\Omega)}\leqq C\left\{\sum_{k=0}^{m-2}\|D_x{}^ku\|_{m-1-k,\mathscr{H}_{-\gamma}(I\times\partial\Omega)}\right.$$
$$\left.+\sum_{j=0}^{m-2}\|D_t{}^ju(S)\|_{m-1-j,\mathscr{H}_{-\gamma}(\Omega)}\right\}.$$

Now we let $v_{m-1}=0$ and

$$w_j=D_t{}^ju(T)-v_j\in\mathscr{H}_{-\gamma}{}^{m-1-j}(\Omega)\quad(j=0,\ldots,m-1).$$

Writing $\tilde{w}_j$ for the extension of w_j to 0 in $\{x<0, t=T, y\in\mathbb{R}^{n-1}\}$ we find that $\tilde{w}_j\in\mathscr{H}_{-\gamma}{}^{m-1-j}(\mathbb{R}^n)$.

Consider

$$(G_tu,u)_{\mathscr{H}_{-\gamma}(\Omega),t=T}=\left(\mathscr{G}\begin{bmatrix}D_t{}^{m-1}u(T)\\ \vdots\\ u(T)\end{bmatrix},\begin{bmatrix}D_t{}^{m-1}u(T)\\ \vdots\\ u(T)\end{bmatrix}\right)_{\mathscr{H}_{-\gamma}(\Omega)}.$$

Setting

$$\begin{bmatrix}D_t{}^{m-1}u(T)\\ \vdots\\ u(T)\end{bmatrix}=\begin{bmatrix}w_{m-1}\\ \vdots\\ w_0\end{bmatrix}+\begin{bmatrix}v_{m-1}\\ \vdots\\ v_0\end{bmatrix}\doteq w+v,$$

we have

$$(G_t u, u)_{\mathscr{H}_{-\gamma}(\Omega), t=T} = (\mathscr{G} w, w)_{\mathscr{H}_{-\gamma}(\Omega)} + (\mathscr{G} w, v)_{\mathscr{H}_{-\gamma}(\Omega)} + (\mathscr{G} v, w)_{\mathscr{H}_{-\gamma}(\Omega)} + (\mathscr{G} v, v)_{\mathscr{H}_{-\gamma}(\Omega)},$$

therefore

$$|(G_t u, u)_{\mathscr{H}_{-\gamma}(\Omega), t=T} - (\mathscr{G} w, w)_{\mathscr{H}_{-\gamma}(\Omega)}| \leqq C\left\{\sum_{j=0}^{m-1} \|w_j\|_{m-1-j, \mathscr{H}_{-\gamma}(\Omega)} \cdot \sum_{j=0}^{m-1} \|v_j\|_{m-1-j, \mathscr{H}_{-\gamma}(\Omega)} + \sum_{j=0}^{m-1} \|v_j\|^2_{m-1-j, \mathscr{H}_{-\gamma}}\right\}.$$

At the same time we have an estimation

$$\begin{aligned}(\mathscr{G} w, w)_{\mathscr{H}_{-\gamma}(\Omega)} &= (\mathscr{G} \tilde{w}, \tilde{w})_{\mathscr{H}_{-\gamma}(\mathbb{R}^n)} \\ &\geqq c \sum_{j=0}^{m-1} \|\tilde{w}_j\|^2_{m-1-j, \mathscr{H}_{-\gamma}(\mathbb{R}^n)} - C \sum_{j=0}^{m-2} \|\tilde{w}_j\|^2_{m-2-j, \mathscr{H}_{-\gamma}(\mathbb{R}^n)} \\ &= c \sum_{j=0}^{m-1} \|w_j\|^2_{m-1-j, \mathscr{H}_{-\gamma}(\Omega)} - C \sum_{j=0}^{m-2} \|w_j\|^2_{m-2-j, \mathscr{H}_{-\gamma}(\partial\Omega)}.\end{aligned}$$

From this we obtain

$$\begin{aligned}(G_t u, u)_{\mathscr{H}_{-\gamma}(\Omega), t=T} &\geqq c \sum_{j=0}^{m-1} \|w_j\|^2_{m-1-j, \mathscr{H}_{-\gamma}(\Omega)} - C\left\{\sum_{j=0}^{m-2} \|w_j\|^2_{m-2-j, \mathscr{H}_{-\gamma}(\Omega)} + \sum_{j=0}^{m-1} \|v_j\|^2_{m-1-j, \mathscr{H}_{-\gamma}(\Omega)}\right\} \\ &\geqq c' \sum_{j=0}^{m-1} \|D_t{}^j u(T)\|^2_{m-1-j, \mathscr{H}_{-\gamma}(\Omega)} - C'\left\{\sum_{j=0}^{m-2} \|D_t{}^j u(T)\|^2_{m-2-j, \mathscr{H}_{-\gamma}(\Omega)} + \sum_{j=0}^{m-1} \|v_j\|^2_{m-1-j, \mathscr{H}_{-\gamma}(\Omega)}\right\}\end{aligned}$$

$$\geqq c' \sum_{j=0}^{m-1} \|D_t^j u(T)\|^2_{m-1-j,\mathscr{H}_{-\gamma}(\Omega)}$$

$$- C'' \left\{ \sum_{j=0}^{m-2} \|D_t^j u(T)\|^2_{m-2-j,\mathscr{H}_{-\gamma}(\Omega)} \right.$$

$$+ \sum_{j=0}^{m-2} \|D_x^j u\|^2_{m-1-j,\mathscr{H}_{-\gamma}(I\times\partial\Omega)}$$

$$\left. + \sum_{j=0}^{m-1} \|D_t^j u(S)\|^2_{m-1-j,\mathscr{H}_{-\gamma}(\Omega)} \right\}. \qquad \blacksquare$$

If we set $u = v$ in Lemma 5.4, then, from Lemma 5.4, we obtain a fundamental inequality concerning the strongly hyperbolic operator A as follows.

Lemma 5.6
Let γ_0 be sufficiently large. Then, for $\gamma \geqq \gamma_0$, $u \in \mathscr{H}_{-\gamma}{}^m(I \times \Omega), (I = (S, T), \Omega = \mathbb{R}_+{}^n)$ the estimation

$$\gamma \|u\|^2_{m-1,\mathscr{H}_{-\gamma}(I\times\Omega)} + \sum_{j=0}^{m-1} \|D_t^j u(T)\|^2_{m-1-j,\mathscr{H}_{-\gamma}(\Omega)}$$

$$\leqq C \left\{ \frac{1}{\gamma} \|Au\|^2_{\mathscr{H}_{-\gamma}(I\times\Omega)} + \sum_{j=0}^{m-1} \|D_x^j u\|^2_{m-1-j,\mathscr{H}_{-\gamma}(I\times\partial\Omega)} \right.$$

$$\left. + \sum_{j=0}^{m-1} \|D_t^j u(S)\|^2_{m-1-j,\mathscr{H}_{-\gamma}(\Omega)} \right\}$$

holds, where the positive constant C is independent of S, T as well as u, γ.

Note. We can set $S = -\infty$ or $T = +\infty$ in this lemma. If $S = -\infty$, then the norm of the third term on the right hand side of the inequality can be made zero, and if $T = +\infty$, then the second term on the left hand side can be made zero.

Notice that the norm of the second term on the right hand side of the inequality, which is defined on the boundary, should be treated under the given boundary condition. Recall that we have already done this in Proposition 4.5 where we considered the case $I = (-\infty, T)$.

Proposition 5.7 *(the energy inequality of Dirichlet type in $(-\infty, T)$)*
For $\gamma \geqq \gamma_0$, $u \in \mathscr{H}_{-\gamma}{}^m(I \times \Omega)$, $I = (-\infty, T)$ the estimation

$$\gamma \|u\|^2_{m-1,\mathscr{H}_{-\gamma}(I\times\Omega)} + \sum_{j=0}^{m-1} \|D_t^j u(T)\|^2_{m-1-j,\mathscr{H}_{-\gamma}(\Omega)}$$

$$+\sum_{j=0}^{m-1} \|D_x{}^j u\|^2_{m-1-j,\mathscr{H}_{-\gamma}(I\times\partial\Omega)}$$

$$\leqq C\left\{\frac{1}{\gamma}\|Au\|^2_{\mathscr{H}_{-\gamma}(I\times\Omega)}+\sum_{j=1}^{m_+}\|B_j u\|^2_{m-1-r_j,\mathscr{H}_{-\gamma}(I\times\partial\Omega)}\right\}$$

holds.

Up to now our argument has centred on $\{A, B_j\}$ in relation to problem (P). Parallel arguments exist for $\{A^*, B_j{}^*\}$ and problem (P*). This implies that instead of Lemma 5.7 we have

Proposition 5.7′

For $\gamma \geqq \gamma_0$, $v \in \mathscr{H}_\gamma{}^m(I\times\Omega)$, $I=(S, +\infty)$ the estimation

$$\gamma\|v\|^2_{m-1,\mathscr{H}_\gamma(I\times\Omega)}+\sum_{j=0}^{m-1}\|D_x{}^j v\|^2_{m-1-j,\mathscr{H}_\gamma(I\times\partial\Omega)}$$

$$+\sum_{j=0}^{m-1}\|D_t{}^j v(S)\|^2_{m-1-j,\mathscr{H}_\gamma(\Omega)}$$

$$\leqq C\left\{\frac{1}{\gamma}\|A^*v\|^2_{\mathscr{H}_\gamma(I\times\Omega)}+\sum_{j=m_++1}^{m}\|B_j{}'^*v\|^2_{m-1-r_j',\mathscr{H}_\gamma(I\times\partial\Omega)}\right\}$$

holds.

5.3

Next, we wish to derive another generalised Green's formula. To do this, we first set $P=A_0, Q=\tau^{\nu_0}\eta^\nu(\nu_0+|\nu|=m-1)$ and use Lemma 5.3. Note that for $0\leqq j\leqq m-1$ we can decompose Q as $Q_j Q_j{}'$ where

$$Q_j=\tau^{\nu_0'}\eta^{\nu'} \quad (\nu_0{}'+|\nu'|=r_j{}'),$$
$$Q_j{}'=\tau^{\nu_0''}\eta^{\nu''} \quad (\nu_0{}''+|\nu''|=r_j).$$

Then, setting

$$G_x(t,x,y;\tau,\xi,\eta;\bar\tau,\bar\xi,\bar\eta)$$
$$=\sum_{j=1}^{m} Q_j(\tau,\eta)B_{j0}(t,y;\tau,\xi,\eta)Q_j{}'(\bar\tau,\bar\eta)B_{j0}{}'(t,x,y;\bar\tau,\bar\xi,\bar\eta),$$

we have

$$G_x(t,x,y;\tau,\xi,\eta;\tau,\bar\xi,\eta)=Q(\tau,\eta)\,\frac{A_0(t,x,y;\tau,\xi,\eta)-A_0(t,x,y;\tau,\bar\xi,\eta)}{\xi-\bar\xi}.$$

Now, by Lemma 5.2, we see that there exist G_t, G_{y_j} such that

$$\begin{aligned}
&A_0(t,x,y;\tau,\xi,\eta)Q(\bar{\tau},\bar{\eta}) - A_0(t,x,y;\bar{\tau},\bar{\xi},\bar{\eta})Q(\tau,\eta)\\
&\quad = (\tau-\bar{\tau})G_t(t,x,y;\tau,\xi,\eta;\bar{\tau},\bar{\xi},\bar{\eta})\\
&\qquad + (\xi-\bar{\xi})G_x(t,x,y;\tau,\xi,\eta;\bar{\tau},\bar{\xi},\bar{\eta})\\
&\qquad + \sum_{j=1}^{n-1}(\eta_j-\bar{\eta}_j)G_{y_j}(t,x,y;\tau,\xi,\eta;\bar{\tau},\bar{\xi},\bar{\eta}).
\end{aligned}$$

From this we can establish

Lemma 5.8 *(the generalised Green's formula for $\{A_0, Q\}$ in $I \times \Omega$)*
Let $\gamma \geqq \gamma_0$, $u \in \mathscr{H}_{-\gamma}{}^m(I \times \Omega)$, $v \in \mathscr{H}_\gamma{}^m(I \times \Omega)$, $I = (S, T)$. Then the relation

$$\begin{aligned}
&(A_0u, Qv)_{H(I\times\Omega)} - (Qu, A_0v)_{H(I\times\Omega)}\\
&\quad = -\mathrm{i}(G_tu, v)_{H(\Omega),t=T} + \mathrm{i}(G_tu, v)_{H(\Omega),t=S}\\
&\qquad + \mathrm{i}\sum_{j=1}^{m}(B_{j0}Q_ju, \bar{B}_{j0}{}'Q_j{}'v)_{H(I\times\partial\Omega)} - (\mathrm{D}Gu, v)_{H(I\times\Omega)}
\end{aligned}$$

holds.

In this version of the generalised Green's formula, if we adjust the lower-order terms and regard the equation as a relation between $\{A, B_j\}$ and $\{A^*, B_j{}^*\}$, then we obtain

Corollary to Lemma 5.8
Let $\gamma \geqq \gamma_0$, $u \in \mathscr{H}_{-\gamma}{}^m(I \times \Omega)$, $v \in \mathscr{H}_\gamma{}^m(I \times \Omega)$, $I = (S, T)$. Then

$$\begin{aligned}
(Au, Qv)_{H(I\times\Omega)} - (Qu, A^*v)_{H(I\times\Omega)} &\cong -\mathrm{i}(G_tu, v)_{H(\Omega),t=T} + \mathrm{i}(G_tu, v)_{H(\Omega),t=S}\\
&\quad + \mathrm{i}\sum_{j=1}^{m}(Q_jB_ju, Q_j{}'B_j{}'^*v)_{H(I\times\partial\Omega)}
\end{aligned}$$

where the difference involved in $\cong$ is bounded by a constant times

$$\begin{aligned}
&\frac{1}{\gamma}\left\{\gamma\|u\|^2_{m-1,\mathscr{H}_{-\gamma}(I\times\Omega)} + \sum_{j=0}^{m-1}\|\mathrm{D}_x{}^ju\|^2_{m-1-j,\mathscr{H}_{-\gamma}(I\times\partial\Omega)}\right\}^{1/2}\\
&\quad\times\left\{\gamma\|v\|^2_{m-1,\mathscr{H}_\gamma(I\times\Omega)} + \sum_{j=0}^{m-1}\|\mathrm{D}_x{}^jv\|^2_{m-1-j,\mathscr{H}_\gamma(I\times\partial\Omega)}\right\}^{1/2}.
\end{aligned}$$

Proof
Obviously, the error in the part of the integration over $I \times \Omega$ is bounded by a constant times

$$\|u\|_{m-1,\mathscr{H}_{-\gamma}(I\times\Omega)}\cdot\|v\|_{m-1,\mathscr{H}_\gamma(I\times\Omega)}$$

For the error in the part of the integration over $I \times \partial\Omega$ first note that

$$\begin{aligned} I_j &= (B_{j0}Q_ju, \bar{B}_{j0}'Q_j'v)_{H(I\times\partial\Omega)} - (Q_jB_ju, Q_j'B_j'^*v)_{H(I\times\partial\Omega)} \\ &= ((B_{j0}Q_j - Q_jB_j)u, \bar{B}_{j0}'Q_j'v)_{H(I\times\partial\Omega)} \\ &\quad + (Q_jB_ju, (\bar{B}_{j0}'Q_j' - Q_j'B_j'^*)v)_{H(I\times\partial\Omega)}. \end{aligned}$$

From this we see that

$$\sum_{j=1}^{m} |I_j| \leqq \frac{C}{\gamma} \sum_{j=0}^{m-1} \|D_x{}^j u\|_{m-1-j,\mathscr{H}_{-\gamma}(I\times\partial\Omega)} \cdot \sum_{j=0}^{m-1} \|D_x{}^j v\|_{m-1-j,\mathscr{H}_{\gamma}(I\times\partial\Omega)}. \quad \blacksquare$$

We now consider an alternative expression for the norm over the boundary $I \times \partial\Omega$.

Lemma 5.9

For a fixed integer $k(\geqq 0)$, the estimation

$$c_1\|u\|_{\mathscr{H}_\gamma(I\times\partial\Omega)} \leqq \sum_{\nu_0+|\nu|=k} \sup_{v\in B_k} |(u, D_t{}^{\nu_0}D_y{}^{\nu}v)_{H(I\times\partial\Omega)}| \leqq c_2\|u\|_{\mathscr{H}_{-\gamma}(H\times\partial\Omega)}$$

$$B_k = \{v; \|v\|_{k,\mathscr{H}_\gamma(I\times\partial\Omega)} \leqq 1\}$$

holds, where c_1, c_2 are positive constants and independent of I as well as u, γ.

Proof

For convenience, we set

$$\|u\|'_{\mathscr{H}_{-\gamma}} = \sum_{\nu_0+|\nu|=k} \sup_{v\in B_k} |(u, D_t{}^{\nu_0}D_y{}^{\nu}v)_H|.$$

Then obviously

$$\|u\|'_{\mathscr{H}_{-\gamma}} \leqq C\|u\|_{\mathscr{H}_{-\gamma}}.$$

Therefore we have to demonstrate that

$$\|u\|'_{\mathscr{H}_{-\gamma}} \leq C\|u\|_{\mathscr{H}_{-\gamma}}.$$

To do this consider that if for $\varphi \in \mathscr{D}(I \times \partial\Omega)$, we take

$$v_\varphi(t, y) = \mathscr{L}^{-1}_{(\tau,\eta)\to(t,y)}[\{\tau + i(|\eta|^2 + \gamma^2)^{1/2}\}^{-k}(\mathscr{L}\varphi)(\tau, \eta)], \operatorname{Im}\tau = \gamma, \operatorname{Im}\eta = 0,$$

then we see that

$$(D_t + i\Lambda_\gamma')^k v_\varphi = \varphi,$$

where

$$\Lambda_\gamma' v = \mathscr{L}^{-1}_{(\tau,\eta)\to(t,y)}[(|\eta|^2 + \gamma^2)^{1/2}(\mathscr{L}v)(\tau, \eta)], \quad \operatorname{Im}\tau = \gamma, \quad \operatorname{Im}\eta = 0.$$

Hence we find

$$\|\Lambda'^k v_\varphi\|_{\mathscr{H}_\gamma(\mathbb{R}^n)} \leqq \|\varphi\|_{\mathscr{H}_\gamma(\mathbb{R}^n)}.$$

That is,

$$\|v_\varphi\|_{k,\mathscr{H}_\gamma(I\times\partial\Omega)} \leqq \|v_\varphi\|_{k,\mathscr{H}_\gamma(\mathbb{R}^n)} \leqq \|\varphi\|_{\mathscr{H}_\gamma(\mathbb{R}^n)} = \|\varphi\|_{\mathscr{H}_\gamma(I\times\partial\Omega)}.$$

Therefore, we see that

$$\begin{aligned}\|u\|_{\mathscr{H}_{-\gamma}(I\times\partial\Omega)} &= \sup_{\varphi\in\mathscr{D}(I\times\partial\Omega)\cap B_0} |(u,\varphi)_H| \\ &= \sup_{\varphi\in\mathscr{D}(I\times\partial\Omega)\cap B_0} |(u,(\mathrm{D}_t+\mathrm{i}\Lambda_y{}')^k v_\varphi)_H| \\ &\leqq C' \sum_{\nu_0=0}^{k} \sup_{v\in B_k} |(u,\mathrm{D}_t{}^{\nu_0}\Lambda_y{}'^{k-\nu_0}v)_H| \\ &\leqq C' \sum_{\nu_0+|\nu|=k} \sup_{v\in B_k} |(u,\mathrm{D}_t{}^{\nu_0}\mathrm{D}_y{}^{\nu}v)_H| \\ &= C'\|u\|'_{\mathscr{H}_{-\gamma}}. \qquad \square\end{aligned}$$

Now, using Lemma 5.8 (the generalised Green's formula for $\{A_0, Q\}$) we can derive the energy inequality of Dirichlet type in $(S, +\infty)$ for the original problem from that of its adjoint problem, (see Proposition 5.7′). We can establish

Proposition 5.10 *(the energy inequality of Dirichlet type in $(S, +\infty)$)*
For $\gamma \geqq \gamma_0$, $u\in\mathscr{H}_{-\gamma}{}^m(I\times\Omega)$, $I=(S,+\infty)$ the estimation

$$\begin{aligned}\gamma\|u\|^2_{m-1,\mathscr{H}_{-\gamma}(I\times\Omega)} &+ \sum_{j=0}^{m-1}\|\mathrm{D}_x{}^j u\|^2_{m-1-j,\mathscr{H}_{-\gamma}(I\times\partial\Omega)} \\ &\leqq C\left\{\frac{1}{\gamma}\|Au\|^2_{\mathscr{H}_{-\gamma}(I\times\Omega)} + \sum_{j=1}^{m_+}\|B_j u\|^2_{m-1-r_j,\mathscr{H}_{-\gamma}(I\times\partial\Omega)}\right. \\ &\qquad \left. + \sum_{j=0}^{m-1}\|\mathrm{D}_t{}^j u(S)\|^2_{m-1-j,\mathscr{H}_{-\gamma}(\Omega)}\right\}\end{aligned}$$

holds.

Proof
First, we wish to estimate

$$\sum_{j=0}^{m-1}\|\mathrm{D}_x{}^j u\|^2_{m-1-j,\mathscr{H}_{-\gamma}(I\times\partial\Omega)}.$$

To this end, noticing that

$$\sum_{j=0}^{m-1}\|\mathrm{D}_x{}^j u\|^2_{m-1-j,\mathscr{H}_{-\gamma}(I\quad\Omega)} \leqq C\sum_{j=1}^{m}\|B_j u\|^2_{m-1-r_j,\mathscr{H}_{-\gamma}(I\times\partial\Omega)},$$

we have only to estimate

$$\sum_{j=m_++1}^{m}\|B_j u\|^2_{m-1-r_j,\mathscr{H}_{-\gamma}(I\times\partial\Omega)}.$$

To do this we fix l at $m_+ + 1 \leqq l \leqq m$, and consider the adjoint problem

$$A^*v = 0,$$
$$B_l'^*v|_{x=0} = \varphi,$$
$$B_j'^*v|_{x=0} = 0, \quad (m_+ + 1 \leqq j \leqq m, \quad j \neq l),$$

for $\varphi \in \mathscr{D}(\mathbb{R}^1 \times \partial\Omega)$. Let $v \in \mathscr{H}_\gamma^m(\mathbb{R}^1 \times \Omega)$ be a solution of the problem. Taking v as satisfying the condition of the corollary to Lemma 5.8, we have

$$(Au, Qv)_{H(I\times\Omega)} = \mathrm{i}(G_t u, v)_{H(\Omega), t=S} + \mathrm{i}\sum_{j=1}^{m_+} (Q_j B_j u, Q_j' B_j'^* v)_{H(I\times\partial\Omega)}$$
$$+ \mathrm{i}(Q_l B_l u, Q_l'\varphi)_{H(I\times\partial\Omega)}.$$

Therefore we find

$$|(Q_l B_l u, Q_l'\varphi)_{H(I\times\partial\Omega)}|$$
$$\leqq C\Bigg\{\frac{1}{\gamma}\|Au\|^2_{\mathscr{H}_{-\gamma}(I\times\Omega)} + \sum_{j=1}^{m_+}\|B_j u\|^2_{m-1-r_j,\mathscr{H}_{-\gamma}(I\times\partial\Omega)}$$
$$+ \sum_{j=0}^{m-1}\|\mathrm{D}_t^{\,j}u(S)\|^2_{m-1-j,\mathscr{H}_{-\gamma}(\Omega)}$$
$$+ \frac{1}{\gamma^2}\Bigg(\gamma\|u\|^2_{m-1,\mathscr{H}_{-\gamma}(I\times\Omega)} + \sum_{j=0}^{m-1}\|\mathrm{D}_x^{\,j}u\|^2_{m-1-j,\mathscr{H}_{-\gamma}(I\times\partial\Omega)}\Bigg)\Bigg\}^{1/2}$$
$$\times\Bigg\{\gamma\|v\|^2_{m-1,\mathscr{H}_\gamma(I\times\Omega)} + \sum_{j=0}^{m-1}\|\mathrm{D}_x^{\,j}v\|^2_{m-1-j,\mathscr{H}_\gamma(I\times\partial\Omega)}$$
$$+ \sum_{j=0}^{m-1}\|\mathrm{D}_t^{\,j}v(S)\|^2_{m-1-j,\mathscr{H}_\gamma(\Omega)}\Bigg\}^{1/2}.$$

Since v is a solution of the adjoint problem, by Proposition 5.7′, we see that

$$\gamma\|v\|^2_{m-1,\mathscr{H}_\gamma(I\times\Omega)} + \sum_{j=0}^{m-1}\|\mathrm{D}_x^{\,j}v\|^2_{m-1-j,\mathscr{H}_\gamma(I\times\partial\Omega)} + \sum_{j=0}^{m-1}\|\mathrm{D}_t^{\,j}v(\mathrm{S})\|^2_{m-1-j,\mathscr{H}_\gamma(\Omega)}$$
$$\leqq C\|\varphi\|^2_{m-1-r_l',\mathscr{H}_\gamma(I\times\partial\Omega)} = C\|\varphi\|^2_{r_l,\mathscr{H}_\gamma(I\times\partial\Omega)},$$

therefore

$$|Q_l B_l u, Q_l'\varphi)_{H(I\times\partial\Omega)}|$$
$$\leqq C\Bigg\{\frac{1}{\gamma}\|Au\|^2_{\mathscr{H}_{-\gamma}(I\times\Omega)} + \sum_{j=1}^{m_+}\|B_j u\|^2_{m-1-r_j,\mathscr{H}_{-\gamma}(I\times\partial\Omega)}$$
$$+ \sum_{j=0}^{m-1}\|\mathrm{D}_t^{\,j}u(S)\|^2_{m-1-j,\mathscr{H}_{-\gamma}(\Omega)}$$

$$+\frac{1}{\gamma^2}\left(\gamma\|u\|^2_{m-1,\mathscr{H}_{-\gamma}(I\times\Omega)}\right.$$

$$\left.\left.+\sum_{j=0}^{m-1}\|D_x{}^ju\|^2_{m-1-j,\mathscr{H}_{-\gamma}(I\times\partial\Omega)}\right)\right\}^{1/2}\|\varphi\|_{r_l,\mathscr{H}_\gamma(I\times\partial\Omega)}.$$

Given the freedom in the choice of Q_l' and in the choice of φ, from Lemma 5.9 we have

$$\|Q_lB_lu\|^2_{\mathscr{H}_{-\gamma}(I\times\partial\Omega)}$$

$$\leqq C\left\{\frac{1}{\gamma}\|Au\|^2_{\mathscr{H}_{-\gamma}(I\times\Omega)}+\sum_{j=0}^{m_+}\|B_ju\|^2_{m-1-r_j,\mathscr{H}_{-\gamma}(I\times\partial\Omega)}\right.$$

$$+\sum_{j=0}^{m-1}\|D_t{}^ju(S)\|^2_{m-1-j,\mathscr{H}_{-\gamma}(\Omega)}$$

$$\left.+\frac{1}{\gamma^2}\left(\gamma\|u\|^2_{m-1,\mathscr{H}_{-\gamma}(I\times\Omega)}+\sum_{j=0}^{m-1}\|D_x{}^ju\|^2_{m-1-j,\mathscr{H}_{-\gamma}(I\times\partial\Omega)}\right)\right\}.$$

Also, considering the freedom in the choice of Q_l we see that the left hand side can be replaced by

$$\|B_lu\|^2_{m-1-r_l,\mathscr{H}_{-\gamma}(I\times\partial\Omega)}.$$

Since l is arbitrary in $m_+ + 1 \leqq l \leqq m$ it can also be replaced by

$$\sum_{l=m_++1}^{m}\|B_lu\|^2_{m-1-r_l,\mathscr{H}_{-\gamma}(I\times\partial\Omega)}.$$

This implies that

$$\sum_{j=0}^{m-1}\|D_x{}^ju\|^2_{m-1-j,\mathscr{H}_{-\gamma}(I\times\partial\Omega)}$$

$$\leqq C\left\{\frac{1}{\gamma}\|Au\|^2_{\mathscr{H}_{-\gamma}(I\times\Omega)}+\sum_{j=1}^{m_+}\|B_ju\|^2_{m-1-r_j,\mathscr{H}_{-\gamma}(I\times\partial\Omega)}\right.$$

$$+\sum_{j=0}^{m-1}\|D_t{}^ju(S)\|^2_{m-1-j,\mathscr{H}_{-\gamma}(\Omega)}$$

$$\left.+\frac{1}{\gamma^2}\left(\gamma\|u\|^2_{m-1,\mathscr{H}_{-\gamma}(I\times\Omega)}+\sum_{j=0}^{m-1}\|D_x{}^ju\|^2_{m-1-j,\mathscr{H}_{-\gamma}(I\times\partial\Omega)}\right)\right\}.$$

We see by Lemma 5.6 that we can replace the left hand side of the above inequality with

$$\gamma\|u\|^2_{m-1,\mathscr{H}_{-\gamma}(I\times\Omega)}+\sum_{j=0}^{m-1}\|D_x{}^ju\|^2_{m-1-j,\mathscr{H}_{-\gamma}(I\times\partial\Omega)}.$$

Then, we take a new γ_0 to make $\gamma \geqq \gamma_0$. We now see that the last term of the right hand side of the above inequality can be absorbed in the new left hand side term. ■

5.4

[For brevity, we shall write E.I.D. for 'energy inequality of Dirichlet type'.]

Let us look back at the argument which led us to Lemma 5.7 (the E.I.D. in $(-\infty, T)$). Starting from the E.I.D. in $(-\infty, +\infty)$ (Theorem 3), via the existence of a solution in $(T, +\infty)$ (Theorem 4(b)), we obtained the E.I.D. in $(-\infty, T)$ in an 'incomplete' form (Proposition 4.5). Then, using Lemma 5.6 we arrived at the E.I.D. in the 'complete' form (Proposition 5.7). Now instead of taking the E.I.D. in $(-\infty, +\infty)$ as our starting point, if we take the E.I.D. in $(S, +\infty)$ (Proposition 5.10) and repeat the same argument as before, then eventually we can obtain the E.I.D. at (S, T). In other words, using the theorem for the existence of a solution in $(T, +\infty)$ (Theorem 4(b)), we first obtain an 'incomplete' form of the E.I.D. in (S, T), and by the aid of Lemma 5.6 we can obtain the 'complete form' of the E.I.D. in (S, T). Summing up we have

Theorem 5(a) *(the energy inequality of Dirichlet type (H))*
If the uniform Lopatinski condition for $\{A, B_j\}$ is valid, then the energy inequality of Dirichlet type $(\mathrm{H})_\gamma$ holds; i.e. there exist positive numbers γ_0, C such that for any $\gamma \geqq \gamma_0$, $I = (S, T)$, $u \in \mathscr{H}_{-\gamma}{}^m(I \times \Omega)$, the estimation

$$\gamma \|u\|^2_{m-1, \mathscr{H}_{-\gamma}(I \times \Omega)} + \sum_{j=0}^{m-1} \|D_x{}^j u\|^2_{m-1-j, \mathscr{H}_{-\gamma}(I \times \partial\Omega)}$$

$$+ \sum_{j=0}^{m-1} \|D_t{}^j u(T)\|^2_{m-1-j, \mathscr{H}_{-\gamma}(\Omega)}$$

$$\leqq C\left\{\frac{1}{\gamma}\|Au\|^2_{\mathscr{H}_{-\gamma}(I \times \Omega)} + \sum_{j=1}^{m_+} \|B_j u\|^2_{m-1-r_j, \mathscr{H}_{-\gamma}(I \times \partial\Omega)}\right.$$

$$\left. + \sum_{j=0}^{m-1} \|D_t{}^j u(S)\|^2_{m-1-j, \mathscr{H}_{-\gamma}(\Omega)}\right\}$$

holds.

As we observed in §3, we can derive the E.I.D. for higher orders from the E.I.D. $(\mathscr{H})_\gamma$ (see Lemma 3.7). Similarly, we can derive the E.I.D. for higher orders from the E.I.D. $(\mathrm{H})_\gamma$.

Theorem 5(a)′ *(the energy inequality of Dirichlet type (H) for higher orders)* If the uniform Lopatinski condition holds for $\{A, B_j\}$ then for any integer $s \geqq 0$ there exist positive numbers γ_s, C_s such that for any $\gamma \geqq \gamma_s$, $I = (S, T)$, $u \in \mathscr{H}_{-\gamma}{}^{m+s}(I \times \Omega)$ the estimation

$$\gamma \|u\|^2_{m-1+s, \mathscr{H}_{-\gamma}(I\times\Omega)} + \sum_{j=0}^{m-1+s} \|D_x{}^j u\|^2_{m-1-j+s, \mathscr{H}_{-\gamma}(I\times\partial\Omega)}$$
$$+ \sum_{j=0}^{m-1+s} \|D_t{}^j u(T)\|^2_{m-1-j, \mathscr{H}_{-\gamma}(\Omega)}$$
$$\leqq C_s \left\{ \frac{1}{\gamma} \|Au\|^2_{s, \mathscr{H}_{-\gamma}(I\times\Omega)} + \sum_{j=1}^{m_+} \|B_j u\|^2_{m-1-r_j+s, \mathscr{H}_{-\gamma}(I\times\partial\Omega)} \right.$$
$$\left. + \sum_{j=0}^{m-1} \|D_t{}^j v(S)\|^2_{m-1-j+s, \mathscr{H}_{-\gamma}(\Omega)} \right\}$$

holds.

Finally, we shall give the more precise form of the theorem for the existence of a solution for the initial value problem (P) than the one given in Theorem 4(b). If we approximate the data with smooth functions, and get successive approximations of the solution by using the E.I.D. we have

Theorem 5(b)

Under the uniform Lopatinski condition for $\{A, B_j\}$ and for an arbitrary integer $s \geqq 0$ there exists a sufficiently large γ such that given the following data

$$\{f \in \mathscr{H}_{-\gamma}{}^s((S, T) \times \mathbb{R}_+{}^n), g_j \in \mathscr{H}_{-\gamma}{}^{m-1-r_j+s}((S, T) \times \mathbb{R}^{n-1}) \quad (j = 1, \ldots, m_+),$$
$$h_j \in \mathscr{H}_{-\gamma}{}^{m-1-j+s}(\mathbb{R}_+{}^n) \quad (j = 0, \ldots, m-1)\}$$

satisfying the necessary compatibility condition of appropriate order (see 2.2, Chapter 2), there exists a solution $u \in \mathscr{H}_{-\gamma}{}^{m-1+s}((S, T) \times \mathbb{R}_+{}^n)$ satisfying

$$\text{(P)} \qquad \begin{cases} Au = f, & (t, x, y) \in (S, T) \times \mathbb{R}_+{}^n, \\ B_j u|_{x=0} = g_j \quad (j = 1, \ldots, m_+), & (t, y) \in (S, T) \times \mathbb{R}^{n-1}, \\ D_t{}^j u|_{t=S} = h_j \quad (j = 0, \ldots, m-1) & (x, y) \in \mathbb{R}_+{}^n. \end{cases}$$

6. The domain of dependence†

In §§3–5, assuming that the uniform Lopatinski condition holds for the operator $\{A, B_j\}$ we observed that the initial value problem (P) was solvable

† Translator's note: See Courant & Hilbert, *Methods of Mathematical physics*, p. 209 for the exact definition of 'domain of dependence'.

within a Sobolev space. In this section, we shall see that under the same conditions, we can derive the local uniqueness of the solution for (P). More precisely, we can demonstrate that (P) has a finite speed of propagation. If this property is demonstrated in a way analogous to the one in Chapter 2, we decompose the data into functions with compact supports, and then, after obtaining solutions for these functions in Sobolev space, we synthesise the solutions in order to demonstrate the $\mathscr{E}$-well-posedness of (P). To demonstrate the local uniqueness of a solution we must consider domains which are surrounded by curved surfaces instead of planes. Therefore we wish to deal in this section with the initial value problem in the *general domain*. Here, by 'general' domain we mean the domain which can be locally transformed into a known such as those we treated in the foregoing argument.

6.1

Consider a surface $\{\varphi(t,x,y)=0\}$. Let $\Gamma_{t,x,y}$ be the component of

$$\{(\tau,\xi,\eta)\in\mathbb{R}^{n+1}\,|\,A_0(t,x,y;\tau,\xi,\eta)\neq 0\}$$

containing the point $(1,0,0)$. The surface is said to be *spacial* iff for every point (t,x,y) on the surface,

$$(\varphi_t,\varphi_x,\varphi_y)=\left(\frac{\partial\varphi}{\partial t},\frac{\partial\varphi}{\partial x},\frac{\partial\varphi}{\partial y}\right)\in\Gamma_{t,x,y}.$$

Let the curved surfaces $\{\varphi(t,x,y)=0\}$ and $\{\psi(t,x,y)=0\}$ both be spacial. Then we assume that the domain of $\mathbb{R}^1\times\mathbb{R}_+{}^n$ surrounded by these surfaces,

$$D=\{(t,x,y)\in\mathbb{R}^1\times\mathbb{R}_+{}^n\,|\,\varphi(t,x,y)>0,\psi(t,x,y)<0\},$$

is bounded, and let the boundaries, S, T, δ of D be given by

$$S=\{x>0,\varphi(t,x,y)=0,\psi(t,x,y)<0\},$$
$$T=\{x>0,\varphi(t,x,y)>0,\psi(t,x,y)=0\},$$
$$\delta=\{x=0,\varphi(t,x,y)>0,\psi(t,x,y)<0\}.$$

(See Fig. 10.) Then the initial boundary value problem in the domain D can

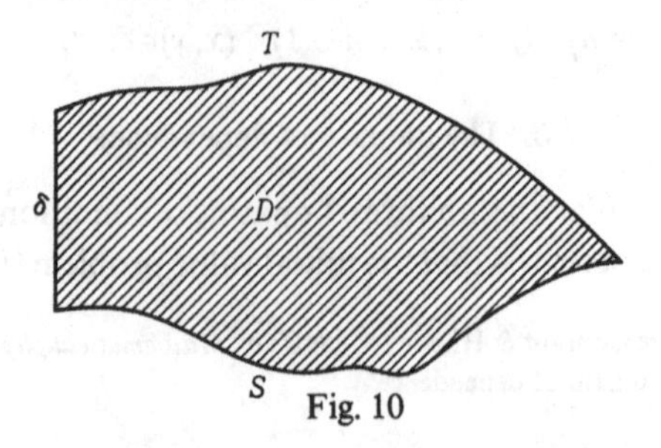

Fig. 10

be given as

$$(\tilde{\mathrm{P}}) \qquad \begin{cases} Au = f & (D), \\ B_j u = g_j \quad (j = 1, \ldots, m_+) & (\delta), \\ \mathrm{D}_t{}^j u = h_j \quad (j = 0, \ldots, m-1) & (S). \end{cases}$$

We now assume that δ in $(\tilde{\mathrm{P}})$ is *not* spacial. From this we can see that $(\varphi_t, \varphi_y) \neq (0,0)$ on $S \cap \bar{\delta}$. For if this is not the case then $(0, \varphi_x, 0) \in \Gamma_{t,x,y}$ for a point $(t, 0, y)$ where $(\varphi_t, \varphi_y) = (0,0)$ holds. Hence, since δ is non-characteristic (see (ii) of Hypothesis (A)) it follows that δ is spacial; a contradiction. For the time being, we assume that $\{\varphi(t,x,y) = 0\}$ is spacial, and at the same time satisfies

$$(*) \qquad \begin{aligned} &\text{(i) } (\varphi_t, \varphi_x, \varphi_y) \in \Gamma_{t,x,y}, \\ &\text{(ii) } (\varphi_t, \varphi_y) \neq (0,0), \\ &\text{(iii) } \varphi(t,x,y) = t \text{ outside a sufficiently large ball in } \mathbb{R}^{n+1} \end{aligned}$$

for all $(t,x,y) \in \mathbb{R}^{n+1}$. We now establish

Proposition 6.1
For the data

$$f \in H^s(D), \quad g_j \in H^{m-1-r_j+s}(\delta), \quad h_j \in H^{m-1-j+s}(S)$$

there exists a solution $u \in H^{m-1+s}(D)$ of (P) in D.

Proof
Following (ii) and (iii) of $(*)$ we can choose an appropriate smooth function $\varphi_j(t,x,y)$ such that

$$t' = \varphi(t,x,y), \quad y_j' = \varphi_j(t,x,y) \quad (j = 1, \ldots, n-1), \quad x' = x$$

defines a smooth change of variables, which is the identity outside a ball. Then

$$\begin{aligned} &S \text{ is transformed into } S' \subset \{t' = 0, x' > 0, y' \in \mathbb{R}^{n-1}\}, \\ &\delta \text{ into } \delta' \subset \{x' = 0, t' > 0, y' \in \mathbb{R}^{n-1}\} \end{aligned}$$

and

$$D \text{ into } D' \subset \{t' > 0, x' > 0, y' \in \mathbb{R}^{n-1}\}.$$

Suppose $A \to A'$, $B_j \to B_j'$ by the same transformation. Then we have

$$A_0'(\tau', \xi', \eta') = A_0\left(\varphi_t \tau' + \sum \varphi_{jt} \eta_j', \varphi_x \tau' + \sum \varphi_{jx} \eta_j' + \xi', \varphi_y \tau' + \sum \varphi_{jy} \eta_j'\right)$$

and a similar equality holds for B_j'. Note that A', B_j' are polynomials with respect to (τ', ξ', η') with smooth coefficients depending on (t', x', y') which are constant outside a ball, and satisfy Hypothesis (A′) by (i) of

(∗). Also, we define

$$A_{0+}'(\tau',\xi',\eta') = \prod(\xi' - \xi_j'^{+}(\tau',\eta'))$$

in the same way as

$$A_{0+}(\tau,\xi,\eta) = \prod(\xi - \xi_j^{+}(\tau,\eta)).$$

From the expression of A_0' in terms of A_0 we find that

$$\xi_j'^{+}(\tau',\eta') = -(\varphi_x\tau' + \sum \varphi_{jx}\eta_j') \\ + \xi_j^{+}(\varphi_t\tau' + \sum \varphi_{jt}\eta_j', \varphi_y\tau' + \sum \varphi_{jy}\eta_j'),$$

so that

$$A_{0+}'(\tau',\xi',\eta') = A_{0+}(\varphi_t\tau' + \sum \varphi_{jt}\eta_j', \varphi_x\tau' + \sum \varphi_{jx}\eta_j' + \xi', \varphi_y\tau' + \sum \varphi_{jy}\eta_j').$$

Hence, putting

$$R_0'(\tau,\eta) = \det\left(\frac{1}{2\pi i}\oint\frac{B_{j0}'(\tau',\xi',\eta')\xi'^{k-1}}{A_{0+}'(\tau',\xi',\eta')}d\xi'\right)_{j,k=1,\ldots,m_+}$$

we see that

$$R_0'(\tau',\eta') = R_0(\varphi_t\tau' + \sum \varphi_{jt}\eta_j', \varphi_y\tau' + \sum \varphi_{jy}\eta_j').$$

Recall the property of a hyperbolic function given in Chapter 2. Using the results from there we find that

$$R_0'(\tau',\eta') \neq 0 \quad \operatorname{Im}\tau' \leqq 0, \quad \eta' \in \mathbb{R}^{n-1}, \quad (\tau',\eta') \neq 0,$$

therefore $\{A', B_j'\}$ also satisfies the uniform Lopatinski condition.

We now extend the data in an appropriate way, and use Theorem 5(a)′ to establish the existence of the solution u' of the problem

$$\begin{aligned} &A'u' = f' && t' > 0, \quad x' > 0, \quad y' \in \mathbb{R}^{n-1}, \\ &B_j'u'|_{x'=0} = g_j' \quad (j = 1,\ldots,m_+), && t' > 0, \quad y' \in \mathbb{R}^{n-1}, \\ &D_t^{\,j}u'|_{t'=0} = h_j' \quad (j = 0,\ldots,m-1), && x' > 0, \quad y' \in \mathbb{R}^{n-1}. \end{aligned}$$

If we restrict the domain of u' to D' and change the variables back to the original variables we can obtain the desired result. ■

For the conjugate problem the situation is exactly the same. That is, the initial boundary value problem in D becomes

$$(\tilde{\mathrm{P}}^*)\quad \begin{cases} A^*v = f & (D), \\ B_j'^*v = g_j \quad (j = m_+ + 1,\ldots,m) & (\delta), \\ D_t^{\,j}v = h_j \quad (j = 0,\ldots,m-1) & (T) \end{cases}$$

and the same form of the existence theorem can be found as we had in Proposition 6.1. From the argument in §5, we have

$$A(\tau,\xi,\eta) - \overline{A^*(\bar\tau,\bar\xi,\bar\eta)} - (D_xG_x)(\tau,\xi,\eta;\bar\tau,\bar\xi,\bar\eta) = (\tau - \bar\tau)G_t(\tau,\xi,\eta;\bar\tau,\bar\xi,\bar\eta) \\ + (\xi - \bar\xi)G_x(\tau,\xi,\eta;\bar\tau,\bar\xi,\bar\eta) + \sum(\eta_j - \bar\eta_j)G_{y_j}(\tau,\xi,\eta;\bar\tau,\bar\xi,\bar\eta),$$

where

$$G_x(\tau,\xi,\eta;\bar{\tau},\bar{\xi},\bar{\eta}) = \sum_{j=1}^{m} B_j(\tau,\xi,\eta)\overline{B_j'^*(\bar{\tau},\bar{\xi},\bar{\eta})}.$$

This implies

Lemma 6.2 *(Green's formula in D)*

$$(Au,v)_{L^2(D)} - (u,A^*v)_{L^2(D)} = -\mathrm{i}(\tilde{G}_\psi u,v)_{L^2(T)} + \mathrm{i}(\tilde{G}_\varphi u,v)_{L^2(S)} + \mathrm{i}\sum_{j=1}^{m}(B_j u, B_j'^* v)_{L^2(\delta)}$$

where

$$\tilde{G}_\psi = (\psi_t{}^2 + \psi_x{}^2 + \textstyle\sum \psi_{y_j}{}^2)^{-1/2} G_\psi, \quad G_\psi = \psi_t G_t + \psi_x G_x + \textstyle\sum \psi_{y_j} G_{y_j},$$
$$\tilde{G}_\varphi = (\varphi_t{}^2 + \varphi_x{}^2 + \textstyle\sum \varphi_{y_j}{}^2)^{-1/2} G_\varphi, \quad G_\varphi = \varphi_t G_t + \varphi_x G_x + \textstyle\sum \varphi_{y_j} G_{y_j}.$$

We can also derive the uniqueness of the solution for the original problem from the existence of a solution for the adjoint problem as follows:

Lemma 6.3

For $u \in H^m(D)$ if

$$\begin{aligned} &Au = 0 && && (D),\\ &B_j u = 0 && (j = 1,\ldots,m_+) && (\delta),\\ &\mathrm{D}_t{}^j u = 0 && (j = 0,\ldots,m-1) && (S) \end{aligned}$$

then $u \equiv 0$.

Proof

From the theorem for the existence of a solution for (P*) in D, we can see that there exists a solution $v \in H^m$, such that

$$\begin{aligned} &A^* v = u && && (D),\\ &B_j'^* v = 0 && (j = m_+ + 1,\ldots,m) && (\delta),\\ &\mathrm{D}_t{}^j v = 0 && (j = 0,\ldots,m-1) && (T). \end{aligned}$$

For u, and so also for v, using Green's formula in Lemma 6.2, we find $\|u\|_{L^2(D)} = 0$. ■

Note that up to now, we have considered spacial surfaces satisfying the conditions (*) (see the paragraph just before Proposition 6.1), but if the domain of interest is far from $x = 0$, then we may simply observe a pure initial problem. Therefore, if we decompose the domain under consideration into small components we can remove the condition (*).

6.2

Next we look at the domain of dependence. Since $\bigcap_{t,x,y}\Gamma_{t,x,y}$ is a convex cone containing $(1,0,0)$ as one of its interior points, we can find an open convex cone satisfying the condition

$$(1,0,0)\in\Gamma_0\subset\bigcap_{t,x,y}\Gamma_{t,x,y},\quad \Gamma_0=\{(\tau,\xi,\eta)\in\mathbb{R}^{n+1}\,|\,\tau>g(\xi,\eta)\}$$

such that $g(\xi,\eta)$ is infinitely differentiable at $(\xi,\eta)\neq(0,0)$ and satisfies the condition

$$\operatorname{rank}\begin{bmatrix} g_{\xi\xi}(\xi,\eta) & g_{\xi\eta}(\xi,\eta) \\ g_{\eta\xi}(\xi,\eta) & g_{\eta\eta}(\xi,\eta)\end{bmatrix}=n-1,\quad (\xi,\eta)\neq(0,0).$$

For the adjoint cone Γ_0' of Γ_0, we can write

$$H=\partial\Gamma_0'\cap\{t=1\}=\{x=-g_\xi(\xi,\eta),\,y=-g_\eta(\xi,\eta)\}.$$

Therefore, using an infinitely differentiable function $F(x,y)$ with the property $(F_x,F_y)\neq(0,0)$, we can write

$$H=\{(x,y)\in\mathbb{R}^n\,|\,F(x,y)=0\},$$
$$\dot{H}=\{(x,y)\in\mathbb{R}^n\,|\,F(x,y)>0\}.$$

Hence we have

$$\dot{\Gamma}_0'=\left\{(t,x,y)\in\mathbb{R}^{n+1}\,\middle|\,t>0,\,F\left(\frac{x}{t},\frac{y}{t}\right)>0\right\}.$$

Letting

$$(1,a,b)\in\dot{\Gamma}_0'$$

we find that the differential of $F(x/t,y/t)$ in this direction is not zero. Therefore, by the change of variables

$$t'=t,\quad x'=x-at,\quad y'=y-bt$$

we consider the equation

$$F\left(\frac{x'+at'}{t'},\frac{y'+bt'}{t'}\right)=0,$$

which has a solution for $t'>0$, such that

$$t'=f(x',y')>0,$$

where $f(x',y')$ is a homogeneous polynomial of degree 1, and is infinitely differentiable at $(x',y')\neq(0,0)$. Thus we can write

$$\Gamma_0'=\{(t,x,y)\in\mathbb{R}^{n+1}\,|\,t\geqq f(x-at,y-bt)\}.$$

Let us now assume that $\{\varphi(t,x,y)=0\}$ is spacial and that

$$D_0=((t_0,x_0,y_0)-\dot{\Gamma}_0')\cap\{(t,x,y)\in\mathbb{R}^1\times\mathbb{R}_+{}^n\,|\,\varphi(t,x,y)>0\}$$

is bounded for

$$(t_0, x_0, y_0) \in \{\varphi(t, x, y) > 0\}.$$

Under this assumption we have

Lemma 6.4

For an arbitrary compact set K contained in D_0, there exists an open set D_1 such that $K \subset D_1 \subset D_0$ and

$$D_1 = \{(t, x, y) \in \mathbb{R}^1 \times \mathbb{R}_+{}^n \mid \varphi(t, x, y) > 0, \psi_1(t, x, y) < 0\},$$

where $\{\psi_1(t, x, y) = 0\}$ is spacial.

Proof

Since $\operatorname{dis}(K, \partial D_0) > 0$ there exists $(t_1, x_1, y_1) \in D_0$ such that

$$K \subset \{(t_1, x_1, y_1) - \dot{\Gamma}_0{}'\} \cap \{x > 0 \mid \varphi(t, x, y) > 0\}.$$

(See Fig. 11). Also we can write

$$\Gamma_0{}' = \{(t, x, y) \in \mathbb{R}^{n+1} \mid t \geqq f(x - at, y - bt)\}.$$

Because the normal of $\partial\Gamma_0{}'\backslash\{0\}$ is contained in $\bar{\Gamma}_0$, if we set

$$\Gamma_\varepsilon{}' = \{(t, x, y) \in \mathbb{R}^{n+1} \mid t + \varepsilon t \geqq f(x - at, y - bt)\} \supset \Gamma_0{}' \quad (\varepsilon > 0),$$

we see that the normal of $\partial\Gamma_\varepsilon{}'\backslash\{0\}$ is contained in Γ_0. At the same time, if we take a sufficiently small $\varepsilon > 0$, then we find

$$K \subset \{(t_1, x_1, y_1) - \dot{\Gamma}_\varepsilon{}'\} \cap \{x > 0, \varphi(t, x, y) > 0\} \subset D_0.$$

Now, by multiplication by Friedrichs' mollifier† $f(x', y')$ becomes

$$f_\sigma(x', y') = \rho_\sigma(x', y') * f(x', y').$$

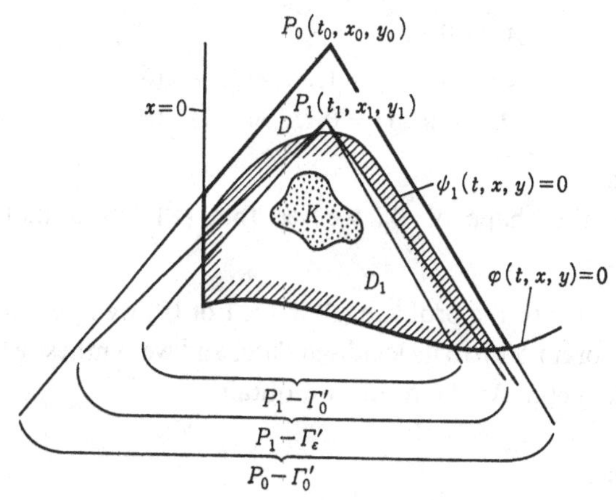

Fig. 11

† Translator's note: See Mizohata, *Theory of partial differential equations.*

Now, the above $f(x',y')$ is differentiable at $(x',y') \neq (0,0)$ and $\partial f/\partial x', \partial f/\partial y'$ are bounded, therefore

$$\frac{\partial f_\sigma}{\partial x'} = \rho_\sigma * \frac{\partial f}{\partial x'}, \quad \frac{\partial f_\sigma}{\partial y'} = \rho_\sigma * \frac{\partial f}{\partial y'}.$$

Setting

$$\Gamma_{\varepsilon,\sigma}' = \{(t,x,y) \in \mathbb{R}^{n+1} \,|\, t + \varepsilon t \geqq f_\sigma(x - at, y - bt)\},$$

we see that for $\sigma > 0$ sufficiently small,

$$K \subset \{(t_1, x_1, y_1) - \dot{\Gamma}_{\varepsilon,\sigma}'\} \cap \{x > 0 \,|\, \varphi(t,x,y) > 0\} \subset D_0$$

and that the normal of $\partial \Gamma_{\varepsilon,\sigma}' \backslash \{0\}$ is contained in Γ_0. Therefore, if we set

$$\psi_1(t,x,y) = f_\sigma(x_1 - x - a(t_1 - t), y_1 - y - b(t_1 - t)) - (1 + \varepsilon)(t_1 - t),$$

we obtain the desired result. ■

Now, note that for D_1 (in Lemma 6.4) the uniqueness of solutions of $(\tilde{\mathrm{P}})$ can be established (use Lemma 6.3), therefore the uniqueness of the solution of $(\tilde{\mathrm{P}})$ for D itself follows, that is, D_0 is a dependence domain of the point (t_0, x_0, y_0) for $(\tilde{\mathrm{P}})$.

Summing up we have

Theorem 6(a)

Let

$$\delta_0 = \partial D_0 \cap \{x = 0\},$$
$$S_0 = \partial D_0 \cap \{\varphi(t,x,y) = 0\}.$$

If $u \in H^m(D_0)$ satisfies

$$\begin{aligned} &Au = 0 && && (D_0), \\ &B_j u = 0 && (j = 1, \ldots, m_+) && (\delta_0), \\ &\mathrm{D}_t{}^j u = 0 && (j = 0, \ldots, m-1) && (S_0) \end{aligned}$$

then $u \equiv 0$ in D_0.

Note: From the shape of D_0 we see that $(\tilde{\mathrm{P}})$ has a finite speed of propagation.

We now return to the problem (P) in §5. For (P) we apply the existence theorem (Theorem 5(b)) to its localised data, and we synthesise the solution using Theorem 6(a). As the result, we obtain

Theorem 6(b)

If $\{A, B_j\}$ satisfies Hypothesis (A) and the uniform Lopatinski condition, then the problem (P) is $\mathscr{E}$-well-posed.

6.3

In this last subsection, we briefly discuss the initial value problem in a more general domain. We write

$$x=(x_1,\ldots,x_n)\in\mathbb{R}^n,$$
$$\xi=(\xi_1,\ldots,\xi_n)\in\mathbb{R}^n.$$

Now let $A(t,x;\tau,\xi)$ be a polynomial of degree m which is strongly hyperbolic in the direction of the t-axis. This implies that $A_0(t,x;1,0)\neq 0$ and the roots of $A_0(t,x;\tau,\xi)=0$ with respect of ξ are all distinct real numbers for $\xi\in\mathbb{R}^n\backslash\{0\}$. We write $\Gamma_{t,x}$ for the connected component of

$$\{(\tau,\xi)\in\mathbb{R}^{n+1}\,|\,A_0(t,x;\tau,\xi)\neq 0\}$$

containing $(1,0)$. If we can express a bounded domain D of $\mathbb{R}^{n+1}$ as

$$D=\{(t,x)\in\mathbb{R}^{n+1}\,|\,\varphi(t,x)>0,\psi(t,x)<0,\chi(t,x)>0\}$$

then the boundary ∂D of D can be divided into three parts as follows:

$$S=\partial D\cap\{\varphi(t,x)=0\},$$
$$T=\partial D\cap\{\psi(t,x)=0\},$$
$$\delta=\partial D\cap\{\chi(t,x)=0\},$$

where we assume that S,T are spacial, and that δ is neither spacial nor characteristic with respect to A_0 (see Fig. 12). This means that $(\varphi_t,\varphi_x)\in\Gamma_{t,x}$

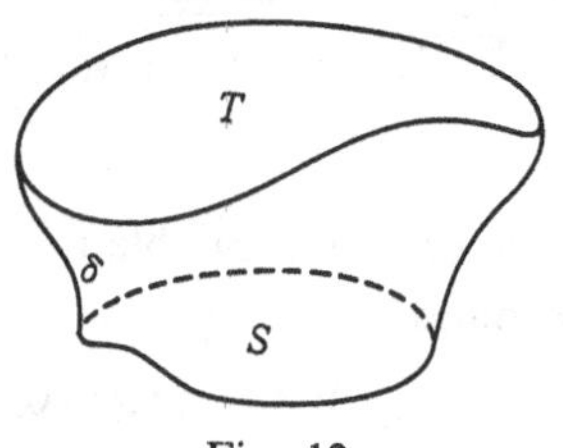

Fig. 12

for each $(t,x)\in S$, $(\psi_t,\psi_x)\in\Gamma_{t,x}$ for each $(t,x)\in T$, and $(\chi_t,\chi_x)\in\Gamma_{t,x}{}^1$ for each $(t,x)\in\delta$ where $\Gamma_{t,x}{}^1$ is a connected component of $\{A_0(t,x;\tau,\xi)\neq 0\}$ and is different from $\Gamma_{t,x}$ and $-\Gamma_{t,x}$.

Suppose we are given a Cauchy initial condition in S and a boundary condition in δ. Under such conditions we consider the initial value problem. That is, for the data $\{f,g_j(j=1,\ldots,m_+),h_j(j=0,\ldots,m-1)\}$ we seek a solution u satisfying

$$(\tilde{\mathrm{P}})\qquad\begin{cases}A(t,x;\mathrm{D}_t,\mathrm{D}_x)u=f, & (t,x)\in D,\\ B_j(t,x;\mathrm{D}_t,\mathrm{D}_x)u=g^j\,(j=1,\ldots,m_+), & (t,x)\in\delta,\\ \mathrm{D}_t{}^ju=h_j\,(j=0,\ldots,m-1), & (t,x)\in S,\end{cases}$$

where m $B_j(t,x;\tau,\xi)$ are defined as follows. For m_+, we arrange the roots of $A_0(t,x;\tau,\xi)=0$ with respect to τ in ascending order as follows:

$$\tau_1(t,x;\xi)<\tau_2(t,x;\xi)<\ldots<\tau_m(t,x;\xi),\quad \xi\in\mathbb{R}^n\backslash\{0\}.$$

Then we choose m_+ to satisfy

$$\Gamma_{t,x}{}^1=\{(\tau,\xi)|\tau_{m_+}(t,x;\xi)<\tau<\tau_{m_++1}(t,x\ \xi)\}.$$

$B_j(t,x;\tau,\xi)$ is a polynomial of degree r_j with respect to (τ,ξ) and it satisfies

$$\mathrm{B}_{j0}(t,x;\chi_t,\chi_x)\neq 0$$

for each $(t,x)\in\delta$ assuming that $0\leqq r_j\leqq m-1$ and $r_i\neq r_j (i\neq j)$.

Since $A_0(t,x;\tau,\xi)$ is hyperbolic in the direction of $\Gamma_{t,x}$ for $\operatorname{Im}\mu<0, (\nu_0,\nu)\in\mathbb{R}^{n+1}, (\nu_0,\nu)\perp\{(\chi_t,\chi_x),(\phi_t,\phi_x)\}$ the roots of

$$A_0(t,x;\lambda\chi_t+\mu\varphi_t+\nu_0,\lambda\chi_x+\mu\varphi_x+\nu)=0$$

with respect to λ are not reals; $\{\lambda=\lambda_j{}^+(t,x;\mu,\nu_0,\nu)\}_{j=1,\ldots,m_+}$ are positive imaginaries, and $\{\lambda=\lambda_j{}^-(t,x;\mu,\nu_0,\nu)\}_{j=1,\ldots,m_-}$ are negative imaginaries, where $m_++m_-=m$. We are now able to define the Lopatinski determinant as

$$R_0(t,x;\mu,\nu_0,\nu)$$
$$=\det\left(\frac{1}{2\pi i}\oint\frac{B_{j0}(t,x;\lambda\chi_t+\mu\varphi_t+\nu_0,\lambda\chi_x+\mu\varphi_x+\nu)\lambda^{k-1}}{\prod_{i=1}^{m+}(\lambda-\lambda_i(t,x;\mu,\nu_0,\nu))}\mathrm{d}\lambda\right)_{j,k=1,\ldots,m}.$$

This determinant is continuously extendable to $\operatorname{Im}\mu\leqq 0$, so the uniform Lopatinski condition becomes

$$R_0(t,x;\mu,\nu_0,\nu)\neq 0,\quad (t,x)\in\delta,\quad \operatorname{Im}\mu\leqq 0,\quad (\nu_0,\nu)\in\mathbb{R}^{n+1},$$
$$(\nu_0,\nu)\perp\{(\chi_t,\chi_x),(\varphi_t,\varphi_x)\},\quad (\mu,\nu_0,\nu)\neq(0,0,0).$$

Notice that, as in the case of $(\tilde{\mathrm{P}})$, if this is true, then by a local change of variables, the existence and uniqueness of solutions can be established. Therefore, $(\tilde{\tilde{\mathrm{P}}})$ becomes $\mathscr{E}$-well-posed and the energy inequality of Dirichlet type is valid.

Summing up we conclude

Theorem 6(c)
If the uniform Lopatinski condition is satisfied by $(\tilde{\mathrm{P}})$, then $(\tilde{\tilde{\mathrm{P}}})$ is $\mathscr{E}$-well-posed, and if $u\in H^m(D)$ the estimation

$$\|u\|^2_{m-1,H(D)}+\sum_{j=0}^{m-1}\|\mathrm{D}_n{}^ju\|^2_{m-1-j,H(\delta)}+\sum_{j=0}^{m-1}\|\mathrm{D}_t{}^ju\|^2_{m-1-j,H(T)}$$

$$\leqq C\left\{ \|Au\|^2_{H(D)} + \sum_{j=1}^{m_+} \|B_j u\|^2_{m-1-r_j, H(\delta)} + \sum_{j=0}^{m-1} \|\mathrm{D}_t{}^j u\|^2_{m-1-j, H(S)} \right\}$$

holds, where

$$\mathrm{D}_n = (\chi_t{}^2 + |\chi_x|^2)^{-1/2}\left(\chi_t \mathrm{D}_t + \sum_{j=1}^{n} \chi_{x_j} \mathrm{D}_{x_j} \right).$$

BIBLIOGRAPHY

The following list contains books and articles which the author has used as references. References with an asterisk are cited in the text.

General (methodology, overall views, etc.)

*Courant, R. & Hilbert, D. *Methods of mathematical physics* vols. 1, 2. Interscience, New York (1953, 1962). (German edition Springer, Berlin 1931.)

Hadamard, J. *Lectures on Cauchy's problem.* Yale University Press (1923).

Hörmander, L. *Linear partial differential operators.* Springer, Berlin (1963).

*Mizohata, S. *The theory of partial differential equations.* Cambridge University Press (1973). (Translated from Japanese.)

Petrovski, I. G. *Lectures on partial differential equations.* Interscience, New York, (1954). (Russian edition Ukrainian Academy, Moscow 1953.)

*Schwartz, L. *Mathematics for the physical sciences.* Addison–Wesley, Reading, Mass. (1966). (French edition Hermann, Paris 1961).

Equations with constant coefficients (in relation to Chapter 2, especially about Fourier–Laplace transforms)

Atiyah, M. F., Bott, R. & Gårding, L. Lacunas for hyperbolic differential operators. *Acta Math.* **24** (1970).

Duff, G. F. D. On wave fronts and boundary waves. *Communs pure appl. Math.* **17** (1964).

*Gårding, L. Linear hyperbolic partial differential equations with constant coefficients. *Acta Math.* **85** (1950).

Hersh, R. Mixed problems in several variables. *J. Math. Mech.* **12** (1963).

Hersh, R. Boundary conditions for equations of evolution. *Archs ration. Mech. Analysis* **16** (1964).

Hörmander, L. On the regularity of the general boundary problems. *Acta Math.* **99** (1958).

Kaneko, A. *Partial differential equations with constant coefficients* (in Japanese) in *Iwanami Kōza–A Mathematical Lecture Series.* Iwanami Shoten, Tokyo (1976).

Kasahara, K. On weak well posedness of mixed problems for hyperbolic systems. *Publ. Res. Inst. Math. Sci. Kyoto Univ.* **6** (1971).

Leray, J. *Hyperbolic equations.* Princeton Lecture Note (1954).

Lopatinski, Ya. B. On a method of reducing boundary problems for a system of differential equations of elliptic type to regular integral equations (in Russian) *Ukrainian Math.* **5**, no. 2 (1953), 123–51.

Petrowsky, I. G. Über das Cauchysche Problem für ein System linearer partieller Differentialgleichungen im Gebiete der nichtanalytischen Functionen. *Bull. Univ. Etate Moskow* (1938).

Petrowsky, I. G. On the diffusion of waves and the lacunas for hyperbolic equations. *Mat. Sbornik* **59** (1945).

Sakamoto, R. $\mathscr{E}$-well posedness for hyperbolic mixed problems with constant coefficients. *J. Math. Kyoto Univ.* **14** (1974).

Shirota, T. On the propagation speed of hyperbolic mixed boundary conditions. *J. Fac. Sci. Hokkaido Univ.* **22** (1972).

Equations with variable coefficients (in relation to Chapter 1 §4 (2nd order) and Chapter 3 (higher orders); about the method of energy estimations)

Agemi, R. & Shirota, T. On necessary and sufficient conditions for L^2-well posedness of mixed problems for hyperbolic equations I, II. *J. Fac. Sci. Hokkaido Univ.* **21** (1970) and **22** (1972).

Agmon, S. Problèmes mixtes pour les équations hyperboliques d'ordre supérieur. Coll. Int. C. N. R. S. (1962).

Balaban, T. On the mixed problem for a hyperbolic equation. *Mem. Am. Math. Soc.* **112** (1970).

Duff, G. F. D. Mixed problems for hyperbolic equations of general order. *Can. J. Math.* **11** (1959).

Friedrichs, K. Symmetric hyperbolic system of linear differential equations. *Communs pure appl. Math.* **7** (1954).

Gårding, L. Dirichlet's problem foı linear partial differential equations. *Math. Scand.* **1** (1953).

Gårding, L. Solution directe de probléme de Cauchy pour les équations hyperboliques. *Proc. Coll. Int. C. N. R. S.* **71** (1956).

Ikawa, M. Mixed problems for second-order hyperbolic equations (in Japanese), *Sūgaku* **22** (1970).

Ikawa, M. Mixed problem for the wave equation with an oblique derivative boundary condition. *Osaka J. Math.* **7** (1970).

Kajitani, K. First order hyperbolic mixed problems. *J. Math. Kyoto Univ.* **11** (1971).

Kreiss, H. O. Initial boundary value problems for hyperbolic systems. *Communs pure appl. Math.* **23** (1970).

Krzyzanski, M. & Schauder, J. Quasilineare Differentialgleichungen zweiter Ordnung vom hyperbolischen Typus. Gemischte Randwertaufgaben. *Studia Math.* **6** (1936).

Miyatake, S. Mixed problems for hyperbolic equations of second order with first order complex boundary operators. *Japanese J.* new ser. **1** (1975).

Mizohata, S. Systemes hyperboliques. *J. Math. Soc. Japan* **11** (1959).

Mizohata, S. *Quelques problèmes au bord, du type mixte, pour des equations hyperboliques.* Collège de France (1966–1967).

Sakamoto, R. Mixed problems for hyperbolic equations I, II. *J. Math. Kyoto Univ.* **10** (1970).

Schechter, M. General boundary value problems for elliptic partial differential equations. *Communs pure appl. Math.* **12** (1959).

NOTATION

Spaces and objects

$\tilde{a}$	standard extension of the symbol a
$A(x, \xi, \eta)$	$= (a_{ij}(x, \xi, \eta))_{i, j = 1, \dots, N}$
$A_0(\tau, \xi, \eta)$	principal part of $A(\tau, \xi, \eta)$
$\mathscr{B}$	Banach space of functions whose derivatives of any order are bounded
$\mathscr{B}^p(K)$	Banach space of functions having continuous and bounded partial derivatives up to order p
C_+	simple closed curve defined on the complex ξ-plane
$C(\Gamma)$	convex hull of Γ
$\mathscr{D}$	vector space of the family of complex-valued functions with compact supports on $\mathbb{R}^n$, which are infinitely differentiable
$\mathscr{D}'$	the dual space of $\mathscr{D}$, i.e. the space of distributions.
D^ν	$= \mathrm{D}_{x_1}^{\nu_1} \dots \mathrm{D}_{x_n}^{\nu_n}$, $\mathrm{D}_{x_j} = \frac{1}{\mathrm{i}} \frac{\partial}{\partial x_j}$ $\quad (\mathrm{i} = \sqrt{-1})$.
δ_x	Dirac distribution
Δ_f	$= \{\eta \in \mathbb{R}^n \mid \mathrm{e}^{x \cdot \eta} f \in \mathscr{S}'_x, f \in \mathscr{D}'\}$ (also written as Ω)
$\mathring{\Delta}$	interior of Δ
$\partial\Omega$	boundary of the domain Ω
$\mathscr{E}(X)$	Fréchet space of infinite differentiable functions defined over X.
$\{f, g, h_j\}$	data
$G_k(t, x, y)$	Poisson function
Γ	$(\subset \mathbb{R}^n)$ cone
Γ'	$= \{x \in \mathbb{R}^n \mid x \cdot \eta \geq 0, \forall \eta \in \Gamma\}$ (i.e. the conjugate cone of Γ)
$\dot{\Gamma}$	$= \{(\tau, \eta) \in \mathbb{R}^n \mid (\tau, \xi, \eta) \in \Gamma\}$ (i.e. the orthogonal projection of Γ

	on to the (τ,η)-space)
$H^m(\Omega)$	Sobolev space defined over Ω
$\mathcal{H}^m(\Omega)$	weighted Sobolev space defined over Ω
$\mathcal{H}^k_{\gamma}$, $\mathcal{H}^k_{-\gamma}$	$=\mathcal{H}^k_{\gamma(1,0,0)}$, $\mathcal{H}^k_{-\gamma(1,0,0)}$, i.e. Sobolev spaces weighted in the direction of the t-axis
$M(\subset \mathbb{R}^m)$	algebraic or quasi-algebraic set
M_s	section of M of s.
(p,q)	signature of a quadratic form
$R(\tau,\eta)$	Lopatinski's determinant
$\mathbb{R}^n_x$	$=\mathbb{R}^n$ where x indicates the independent variable of functions belonging to $\mathcal{S}$
$\mathbb{R}^n_+$	$=\{(x,y)\mid x>0,\ y=(y_1,\ldots,y_{n-1})\in\mathbb{R}^{n-1}\}$ (i.e. a half-space)
$\mathcal{S}$	Fréchet space of all infinitely differentiable complex-valued functions defined on $\mathbb{R}^n$ whose derivatives are bounded on $\mathbb{R}^n$ modulo multiplications of polynomials
$\mathcal{S}'$	the dual space of $\mathcal{S}$ whose elements are called tempered distributions
$\mathcal{S}_x$	$=\mathcal{S}$ where x indicates the independent variable of functions belonging to $\mathcal{S}$
$\mathcal{S}'_\Omega$	$=\{f\mid\Delta_f\supset\Omega$, a domain of $\mathbb{R}^n\}$
supp $[f]$	support of the distribution f

Operations and operators

$a^{(*)}$	(formal) adjoint operator of a with respect to the inner product in weighted L^2
$A(D_t,D_x,D_y)$	differential operator
$A^*(D_t,D_x,D_y)$	formal adjoint of the differential operator A with respect to the inner product in L^2
$\alpha(x,D'_x,\eta)$	singular differential operator for the symbol $\alpha(x,\xi,\eta)$
$*$	convolution
$\underset{(t,y)}{*}$	convolution in (t,y)-space
$\mathcal{F}$	Fourier transform in (t,x,y)-space
$\mathcal{F}_0$	Fourier transform in (t,y)-space
$\varphi(x,D'_x,\eta)$	localising operator
G	Green's operator
$G_K f$	Poisson operator
$\mathcal{L}$	Fourier–Laplace transform in (t,x,y)-space
$\mathcal{L}_0$	Fourier–Laplace transform in $(t,\ y)$-space.

Λ^s	singular integral operator defined on a symbol
l.i.m.	limit in $L^2(\mathbb{R})$-space
T	continuous linear map
T $'$	transpose map of T(a continuous linear map)
$\otimes$	tensor product

INDEX